Effects of Low Temperatures on Biological Membranes

Based on a meeting "Effects of Low Temperatures on Biological Membranes" held 25 September 1980 at the Royal Free Hospital, London, UK

Effects of Low Temperatures on Biological Membranes

Edited by

G. J. MORRIS

*Natural Environment Research Council
Culture Centre of Algae and Protozoa, 36 Storey's Way
Cambridge, UK*

A. CLARKE

*Natural Environment Research Council
British Antarctic Survey, High Cross, Madingley Road
Cambridge, UK*

1981

ACADEMIC PRESS
A Subsidiary of Harcourt Brace Jovanovich, Publishers
London New York Toronto Sydney San Francisco

ACADEMIC PRESS INC. (LONDON) LTD.
24/28 Oval Road,
London NW1

United States Edition published by
ACADEMIC PRESS INC.
111 Fifth Avenue
New York, New York 10003

British Library Cataloguing in Publication Data
Effects of low temperature on biological membranes.
  1. Cryobiology – Congresses   2. Cold adaptation
  – Congresses   3. Cold – Physiological effects –
Congresses.
I. Morris, G. J.   II. Clarke, A.
574.88   QH324.9.7

ISBN 0-12-507650-9

LCCCN 81-67921

Printed in Great Britain

LIST OF CONTRIBUTORS

Bullock, G.R. *Ciba-Geigy (ADT) Co., Pharmaceuticals Division, Wimblehurst Road, Horsham, West Sussex, RH12 4AB, U.K.*

Chapman, D. *Department of Biochemistry and Chemistry, Royal Free Hospital School of Medicine, 8 Hunter Street, London, WC1N 1BP, U.K.*

Clarke, A. *British Antarctic Survey, (Natural Environment Research Council), High Cross, Madingley Road, Cambridge, CB3 OET, U.K.*

Clarke, K. *Institute of Terrestrial Ecology, (Natural Environment Research Council), Culture Centre of Algae and Protozoa, 36 Storey's Way, Cambridge, CB3 ODT, U.K.*

Cossins, A. *Department of Zoology, University of Liverpool, Liverpool, L69 3BX, U.K.*

Coulson, G.E. *Institute of Terrestrial Ecology, (Natural Environment Research Council), Culture Centre of Algae and Protozoa, 36 Storey's Way, Cambridge, CB3 ODT, U.K.*

Dowgert, M.F. *Department of Agronomy, Cornell University, Bradfield and Emerson Halls, Ithaca, N.Y. 14853, U.S.A.*

Ellory, J.C. *Department of Physiology, University of Cambridge, Downing Street, Cambridge, CB2 3EG, U.K.*

Franks, F. *Department of Botany, University of Cambridge, Downing Street, Cambridge, CB2 3EA, U.K.*

Fujikawa, S. *Institute of Low Temperature Science, Hokkaido University, Sapporo, Japan 060.*

Goodall, M.A. *Ciba-Geigy (ADT) Co., Pharmeuticals Division, Wimblehurst Road, Horsham, West Sussex, RH12 4AB, U.K.*

Grout, B.W.W. *Department of Biology, North East London Polytechnic, Romford Road, London E15 4LZ, U.K.*

Heber, U. *Institute of Botany and Pharmazeutical Biology, University of Würzburg, D-8700 Würzburg, F.D.R.*

Herbert, R.A. *Department of Biological Sciences, The University, Dundee, DD1 4HN, Scotland, U.K.*

Klosson, R.J. *Institute of Botany, University of Düsseldorf, D-4000 Düsseldorf 1, F.D.R.*

Krause, G.H. *Institute of Botany, University of Düsseldorf, D-4000 Düsseldorf 1, F.D.R.*

McGrath, J.J. *Bioengineering Transport Processes Laboratory,*
Mechanical Engineering Department, Michigan State University,
East Lansing, Michigan 48824, U.S.A.

McMurdo, A. *School of Plant Biology, University College of*
North Wales, Bangor LL57 2UW, Wales, U.K.

Morris, G.J. *Institute of Terrestrial Ecology, (Natural*
Environment Research Council), Culture Centre of Algae
and Protozoa, 36 Storey's Way, Cambridge, CB3 ODT, U.K.

Pringle, M.J. *Department of Cell Physiology, Boston Biomedical*
Research Institute, 20 Staniford Street, Boston, Massachu-
setts, 02114, U.S.A.

Robards, A.W. *Department of Biology, University of York,*
Heslington, York YO1 5DD, U.K.

Santarius, K.A. *Institute of Botany, University of Düsseldorf,*
D-4000 Düsseldorf 1, F.D.R.

Schmitt, J.M. *Institute of Botany and Pharmazeutical Biology,*
University of Würzburg, D-8700 Würzburg, F.D.R.

Sibbons, P.D. *Ciba-Geigy (ADT) Co., Pharmaceuticals Division,*
Wimblehurst Road, Horsham, West Sussex, RH12 4AB, U.K.

Simon, E. *Department of Botany, The Queen's University of*
Belfast, Belfast BT7 1NN, Northern Ireland, U.K.

Steponkus, P. *Department of Agronomy, Cornell University,*
Bradfield and Emerson Halls, Ithaca, N.Y. 14853, U.S.A.

Watson, P. *Department of Physiology, The Royal Veterinary*
College, Royal College Street, London NW1 OTU, U.K.

Willis, J.S. *Department of Physiology and Biophysics, 524*
Burrill Hall, University of Illinois at Urbana-Champaign,
Urbana, Illinois 61801, U.S.A.

Wilson, J. *School of Plant Biology, University College of*
North Wales, Bangor LL57 2UW, Wales, U.K.

Wolfe, J. *Department of Agronomy, Cornell University,*
Bradfield and Emerson Halls, Ithaca, N.Y. 14853, U.S.A.

PREFACE

In the course of evolution, organisms have adapted to a wide
range of thermal environments. Marine organisms are abundant
in polar seas at -1.9°C and some insect pupae successfully
overwinter at temperatures below -60°C. At the upper end of
the range, thermophilic bacteria grow in thermal springs with
temperatures as high as 101°C. No single organism, however,
can withstand the whole range of these temperatures and
environmental temperature is one of the major ecological
factors controlling the distribution of plants and animals.
The precise way in which organisms achieve adaptation to
temperature is of obvious scientific interest, but there are
also practical and economic aspects. For example, it has
been calculated that a 2°C increment in the frost hardiness
of wheat, allowing cultivation in areas which are presently
marginal because of cold stress, would result in a potential
increase in world wheat production of between 25 and 40 percent (Weiser, 1978).

In this volume we have tried to collate work from varied
sources to focus on the cellular response to low temperatures.
In particular we have concentrated on the relationship between
low temperature and membranes, for membranes must adapt to
temperature and there is increasing evidence that membranes
are the site of the primary effects of cold injury. Following
an introductory section on the biophysics of water and membrane structure in relation to temperature, a series of
chapters examine the way a variety of organisms adapt their
physiology to low temperatures, highlighting the importance
of the physical state of the cellular membranes.

The following sections deal with chilling and freezing
injury. Although these are very different they may both be
seen as a failure to adapt to low temperatures. It is now
generally accepted that the damage caused during both chilling
and freezing injury is located in the cellular membranes,
but the specific mechanisms are still poorly understood.
Although much of the work discussed in these chapters is
primarily ecological, rarely dealing with temperatures below

-10°C, the basic concepts dealt with are nevertheless of relevance to fields such as cryopreservation and cryosurgery. The book concludes with an analysis of the mechanical properties of membranes to account for freezing damage in plant protoplasts and a thermodynamic investigation of damage to erythrocytes at slow rates of cooling.

We feel that this publication is the first attempt at an integrated approach to low temperature damage, utilising evidence from a wide range of investigations. It is timely because the understanding of the membrane and its response to alterations in temperature has become increasingly sophisticated and there is now a wealth of new biochemical, biophysical and thermodynamic information. Secondly, the Lyons-Raison hypothesis of chilling injury (1973), so long a useful working model, is now being increasingly questioned, and thirdly, for the first time a general model of freezing injury at low rates of cooling is emerging.

Some of the contributions in this book were originally presented at a symposium of the Society for Low Temperature Biology on September 25th, 1980. We would like to thank the Society for organising this symposium. We would also particularly like to thank Glyn Coulson for assistance in editing and her painstaking preparation of the camera-ready manuscripts.

October 1981

G.J. Morris
A. Clarke

CONTENTS

FREEZING INJURY

NOMENCLATURE

A number of terms used widely in the literature have assumed
different meaning for various authors. In this book we have
standardised on the following terminology.

Acclimation:
: The process of adjusting an organism to a new temperature in the laboratory.

Acclimatisation:
: The seasonal adjustment of an organism's physiology to changes in environmental temperature.

Adaptation:
: The evolutionary adjustment of an organism's physiology to its environment. This can obviously include adjustment to a season-ally varying temperature requiring acclimatisation.

Chilling injury:
: Damage to an organism caused by a reduction in temperature, but where there is no water/ice phase change in the system. This is also referred to as cold-shock or thermal injury.

Ectotherm/Endotherm:
: These terms, often incorrectly used synonymously with poikilo-therm/homoiotherm, describe the major sources of body heat used by an organism. In ectotherms heat is supplied by the environ-ment, in endotherms a large pro-portion of heat is generated metabolically.

Freezing injury:
: Damage to an organism during a reduction in temperature when there is a water/ice phase change in the system.

Homoiotherm/Poikilotherm: These terms describe the degree
                              to which an organism can regulate
                              its internal temperature indep-
                              endent of that of the environment.
                              A homoiotherm can maintain a
                              constant internal temperature
                              (usually by the production of
                              metabolic heat), whereas a poik-
                              ilotherm although able to reg-
                              ulate its body temperature within
                              limits (often by behavioural
                              means) has a body temperature
                              which tends to follow that of the
                              environment.

Normothermic: This term is often used by workers
                              studying hibernation to indicate
                              the normal body temperature of an
                              aroused individual.

LIPIDS

The literature concerning lipid biochemistry and temperature
acclimation contains a variety of nomenclature for describing
lipids. In this book we have attempted to standardise on one
of the generally accepted terminologies. The terms used and
the data given in the following tables are based on the revised
(1979) edition of Information to Contributors to Biochimica
et Biophysica Acta, Fasman (1975), Gurr and James (1971),
Snyder (1972), Szoka and Papahadjopoulos (1980) and Tanford
(1980). Some common synonyms are given, but we have attempted
to avoid the use of these.

Fatty Acid Structure

Several methods have been devised to describe the structure
of fatty acids. The convention adopted in this book is that
used widely by marine lipid biochemists, where a fatty acid
structure is designated:

$$C:n$$

where C is the number of carbon atoms in the chain, including
       that present in the carboxyl group.
       n is the number of double bonds in the chain.

Unless otherwise designated, it is assumed that these double
bonds are all *cis*, and when n >1, are methylene-interrupted:

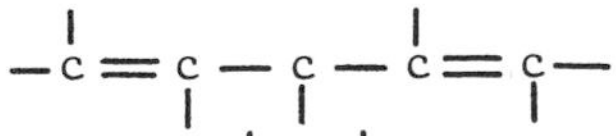

$$-\overset{|}{c}=\overset{}{c}-\overset{}{c}-\overset{|}{c}=\overset{}{c}-$$ methylene-interrupted *cis* double bonds

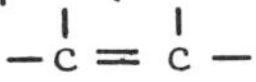

$$-\overset{|}{c}=\overset{|}{c}-$$ *trans* double bond

where the precise structure of the fatty acid is known from either structural analysis or high precision capillary gas chromatography, then the following shorthand convention is used:

18:1ω9   oleic acid, single *cis* double bond 9 carbon atoms from the methyl terminal (*i.e.* between carbon atoms 9 and 10)

18:3ω3   α-linolenic acid; 3 methylene-interrupted *cis* double bonds, with the first bond 3 carbon atoms from the methyl terminal (*i.e.* between carbon atoms 3 and 4)

Other abbreviations are:

iso 16:0   a single methyl branch in the iso
$$CH_3-CH-$$
$$\qquad\quad |$$
$$\qquad\quad CH_3 \qquad position$$

anteiso 16:0   a single methyl branch in the anteiso
$$CH_3-CH_2-CH-$$
$$\qquad\qquad |$$
$$\qquad\qquad CH_3 \qquad position$$

18:1 *trans*   double bond has *trans* structure

16:0 cyc C_{16} acid containing a cyclopropane ring

2-OH.16:0 C_{16} acid containing an hydroxyl group on carbon atom 2.

C_{14} represent a 14 carbon fatty acid,

whereas

c-14   represents carbon atom number 14.

the structures and melting points of selected fatty acids are given in Table 1.

It should be noted that although the above terminology is described in terms of fatty acids, the same conventions also apply to other fatty acyl chains (for example, fatty aldehydes released from alk-1-enyl linked fatty acyl chains, and the alkyl-linked hydrocarbons of glyceryl ethers).

Table 1

Structures and melting points of selected fatty acids

Saturated fatty acids

			mol. wt as R.COOH	M. Pt.
12:0	Dodecanoic	lauric	200.3	44.2
14:0	Tetradecanoic	myristic	228.4	53.9
15:0	Pentadecanoic		242.2	52.3
16:0	Hexadecanoic	palmitic	256.4	63.1
17:0	Heptadecanoic	margaric	270.4	61.3
18:0	Octadecanoic	stearic	284.5	69.6
19:0	Nonadecanoic		298.5	68.6
20:0	Eicosanoic	arachidic	312.5	76.5
22:0	Docosanoic	behenic	340.6	81.5
24:0	Tetracosanoic	lignoceric	368.6	86.0

Monounsaturated fatty acids

				mol. wt as R.COOH	M. Pt.
16:1ω7		9-Hexadecenoic	palmitoleic	254.4	-0.5 to 0.5
18:1ω9		9-Octadecenoic	oleic	282.5	13.4(α) 16.3(β)
18:1ω9	trans	trans-9-Octadecenoic	elaidic	282.5	44.5
18:1ω7		11-Octadecenoic	vaccenic	282.5	44.0
20:1ω11		9-Eicosenoic	gadoleic	310.5	24.5
22:1ω11		11-Docosenoic	cetoleic	338.6	32.5
22:1ω9		13-Docosenoic	erucic	338.6	34.7
24:1ω9		15-Tetracosenoic	nervonic	366.6	42.5

Table 1 (Continued).

			mol. wt as R.COOH	M. Pt.
Polyunsaturated fatty acids				
16:2ω6	7,10-Hexadecadienoic		252.4	
18:2ω6	9,12-Octadecadienoic	α-linoleic	280.5	-5.0 to -5.2
18:2ω7	6,11-Octadecadienoic	cilienic	280.5	
16:3ω3	7,10,13-Hexadecatrienoic		250.4	
18:3ω3	9,12,15-Octadecatrienoic	α-linolenic	278.4	-10 to -11.3
18:3ω6	6,9,12-Octadecatrienoic	γ-linolenic	278.4	
16:4ω1	6,9-12,15-Hexadecatetraenoic		248.4	
18:4ω3	6,9,12,15-Octadecatetraenoic		276.4	
18:4ω3	9,11,13,15-Octadecatetraenoic	parinaric	276.4	85-86(α)
20:4ω6	5,8,11,14-Eicosatetraenoic	arachidonic	304.5	-49.5
20:4ω3	8,11,14,17-Eichosatetraenoic		304.5	
18:5ω3	3,6,9,12,15-Octadecapentaenoic		274.4	
20:5ω3	5,8,11,14,17-Eicosapentaenoic		302.5	-54.4 to -53.8
22:5ω6	4,7,10,13,16-Docosapentaenoic		330.5	
22:5ω3	7,10,13,16,19-Docosapentaenoic	clupanodonic	330.5	
22:6ω3	4,7,10,13,16,19-Docosahexaenoic		328.5	-44.5 to -44.1
Branched chain fatty acids				
iso 16:0	14-Methylpentadecanoic	isopalmitic	256.4	62.4
anteiso 16:0	13-Methylpentadecanoic		256.4	

Phospholipid structure

Phospholipids are a complex group of lipids in which both the
fatty acyl chains and the head group may be attached to the
glycerol backbone by a variety of linkages. A further complication is that phospholipids are usually isolated as a
mixture by chromatographic techniques and this mixture will
contain varying amounts of other polar lipids such as glycolipids or sulpholipids.

The three linkages by which fatty acyl chains are attached
to the phosphoglycerol backbone are:

C - COOR   acyl (fatty acid)

C - O.CH_2.CH_2.R   alkyl (saturated ether)

C - O.CH=CH.R   alk-1-enyl (vinyl ether or
                               aldehyde)

where R represents a fatty acyl or hydrocarbon chain.

Although these linkages theoretically allow for a large
number of different phospholipid types, in practice only 3
major structures are generally found. These are, using choline
phosphoglycerides as an example:

CH_2.O.CH_2CH_2R   1-alkyl-2-acyl-*sn*-glycero-
       |
RCOO.CH   3-phosphoryl choline
       |
CH_2.PO_4.choline

CH_2.O.CH=CH.R   1-alk-1-enyl-2-acyl-*sn*-
       |
RCOO.CH   glycero-3-phosphoryl
       |
CH_2.PO_4.choline   choline (choline plasmalogen)

CH_2OCOR   1-2-diacyl-*sn*-glycero-3-
       |
RCOO.CH   phosphoryl choline
       |
CH_2.PO_4.choline   (phosphatidylcholine)

The fatty acid composition of phosphatidylcholine may be
defined, as in dipalmitoylphosphatidylcholine.

Where the precise structure is not known, usually because
isolation has been performed only with a single chromatographic step, the mixture of phospholipids is referred to as
choline phosphoglyceride (choline PG), Lecithin is a trivial
name for choline phosphoglyceride.

The head groups of the major biological phospholipids are
given in Table 2.

Table 2

Structures of phospholipids and glycolipids

Assuming a   1-2-diacyl-*sn*-glycero-3-phosphoryl structure:

$$CH_2.O.COR$$
$$RCOO.CH$$
$$CH_2.PO_4.headgroup$$

headgroup	phospholipid
H	Phosphatidic acid
$CH_2.CH_2.NH_2$	Phosphatidylethanolamine
$CH_2.CH_2.C(NH_3)_3$	Phosphatidylcholine
$CH_2.CH(COOH).NH_3$	Phosphatidylserine
$CH_2.CH_2.NH(CH_3)$	Phosphatidyl(N-methyl)ethanolamine
$CH_2.CHOH.CH_2OH$	Phosphatidylglycerol
$CH_2.CHOH.CH_2O.PO_3.CH_2$ $RCOO.CH$ $RCOO.CH_2$	Diphosphatidylglycerol (cardiolipin)
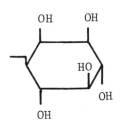	Phosphatidyl(myo)inositol

Phosphonolipids are a small class of phospholipids present in trace amounts in marine invertebrates, but important in many microorganisms. In these, the head group is attached by a C-P bond giving rise to a series of compounds analogous to the more usual animal phospholipids, for example:

phospholipid   Phosphonolipid

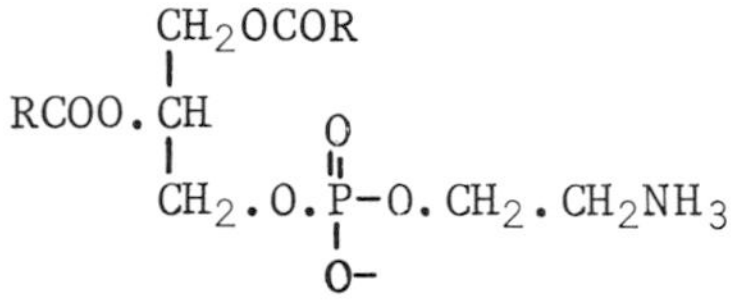

$$CH_2OCOR$$
$$|$$
$$RCOO.CH$$
$$|$$
$$CH_2.O.\overset{O}{\overset{||}{P}}-O.CH_2.CH_2NH_3$$
$$|$$
$$O-$$

$$CH_2.OCOR$$
$$|$$
$$RCOO.CH$$
$$|$$
$$CH_2.O\overset{O}{\overset{||}{P}}-CH_2.CH_2.NH_3{}^+$$
$$|$$
$$O-$$

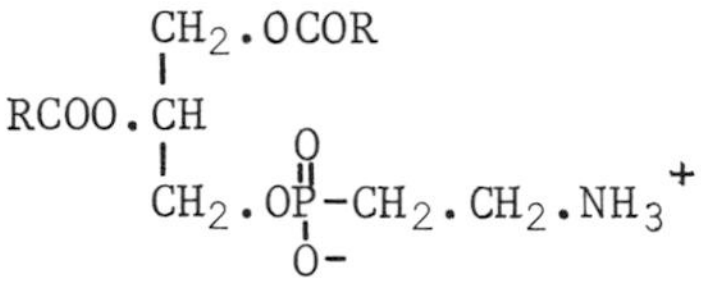

Phosphatidylcholine   2-aminoethylphosphonolipid

Phosphonolipid analogues are also known of sphingosyl phosphatides, for example:

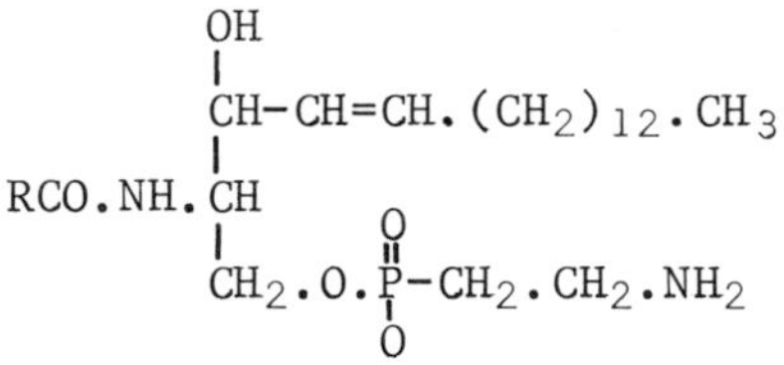

$$OH$$
$$|$$
$$CH-CH=CH.(CH_2)_{12}.CH_3$$
$$|$$
$$RCO.NH.CH$$
$$|$$
$$CH_2.O.\overset{O}{\overset{||}{P}}-CH_2.CH_2.NH_2$$
$$|$$
$$O$$

ceramide-2-aminoethylphosphonolipid
(ceramide ciliatine)

Also important in biological membranes are polar lipids with structures based on N-acyl-sphingosine (ceramide) and related compounds, and glycolipids (Table 3).

Some microorganisms contain very unusual polar lipids, for example sulpholipids and the di-alkyl-linked phytanic acid containing phospholipids of halophilic bacteria (Langworthy, 1977), but these are rarely, if ever, encountered in other organisms.

Some thermodynamic properties of selected phospholipids are given in Table 4.

Table 3

Some selected sphingolipids and glycolipids important in biomembranes

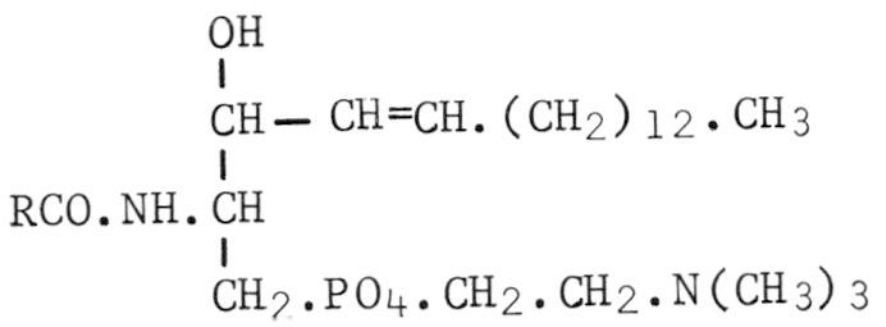

$$\text{OH}$$
$$|$$
$$\text{CH} - \text{CH}=\text{CH}.(\text{CH}_2)_{12}.\text{CH}_3$$
$$|$$
$$\text{RCO}.\text{NH}.\text{CH}$$
$$|$$
$$\text{CH}_2.\text{PO}_4.\text{CH}_2.\text{CH}_2.\text{N}(\text{CH}_3)_3$$

Sphingomyelin
(compare with ceramide
-2-aminoethylphospho-
nolipid)

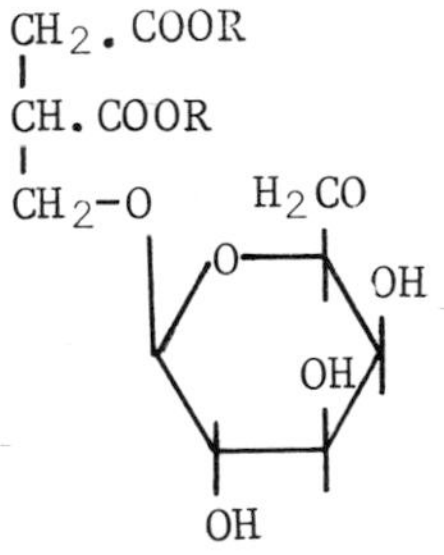

$$\text{CH}_2.\text{COOR}$$
$$|$$
$$\text{CH}.\text{COOR}$$
$$|$$
$$\text{CH}_2\text{-O} \qquad \text{H}_2\text{CO}$$

Monogalactosyldiacyl-
glycerol

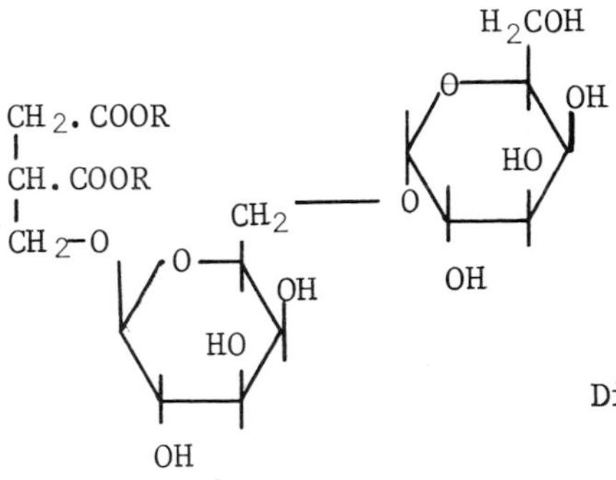

$$\text{CH}_2.\text{COOR}$$
$$|$$
$$\text{CH}.\text{COOR}$$
$$|$$
$$\text{CH}_2\text{-O}$$

Digalactosyldiglyceride

Table 4

Thermodynamic properties of selected phospholipids[1]

Lipid		Transition temperature $T_c(^{\circ}C)$	Free energy of transition ΔH kcal mol^{-1}
	egg lecithin	-15 to -7	
di C_{12}	phosphatidylcholine	-2	–
di C_{14}	phosphatidylcholine	23	5.4
di C_{16}	phosphatidylcholine	41	8.7
di C_{18}	phosphatidylcholine	55	10.6
di C_{22}	phosphatidylcholine	75	14.9
di $C_{18:1}$	phosphatidylcholine	-22	7.6
di C_{14}	phosphatidylethanolamine	50	5.8
di C_{16}	phosphatidylethanolamine	60	8.5
di C_{14}	phosphatidylserine	38	–
di C_{16}	phosphatidylserine	51-55[2]/72[3]	3[2]/9[3]

[1] *Measured by differential scanning calorimetry on hydrated multilayers.*

[2] *At high pH. Exact pH depends on electrolyte (Tanford, 1980).*

[3] *At low pH.*

Sterols and Tetrahymenol

Structures are given below:

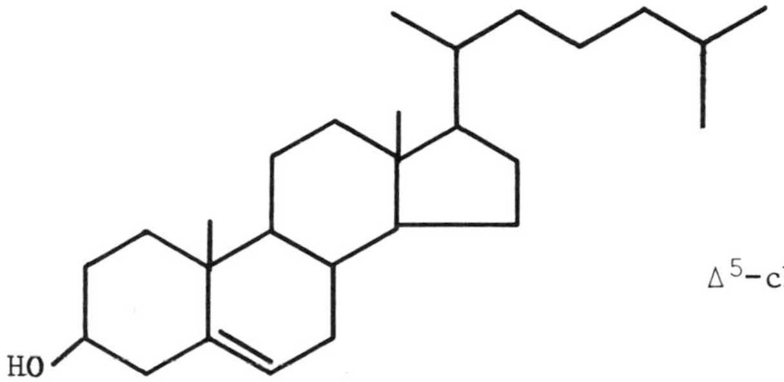

cholesterol
Δ^5-cholesta-1-en-3β-ol

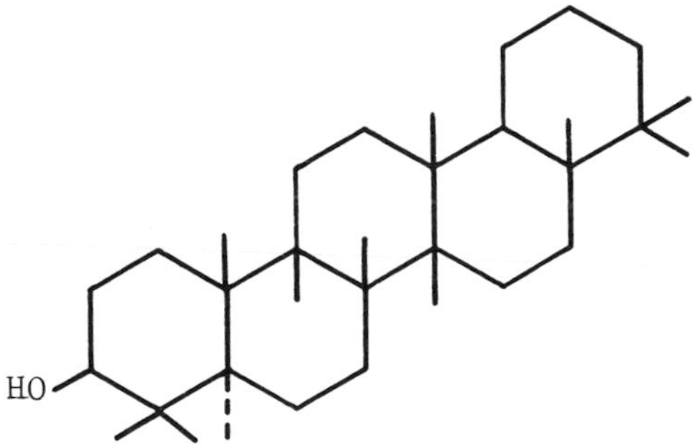

Tetrahymenol

Spectroscopic Probes

The structures of some commonly used fluorescent or electron
spin resonant probes are given below

1,6-diphenyl-1,3,5-hexatrience (DPH)

$CH_3(CH_2)_5$ — C — $(CH_2)_{10}COOH$ 7-(doxyl)-stearic acid

(Doxyl)-stearic acids may be synthesised with the spin
label 2,2-dimethyl-N-oxyloxazolidine moiety attched at different parts of the C_{18} chain.   This enables information about
the behaviour of different parts of the hydrophobic interior
of lipid bilayer to be obtained by ESR.

BASIC PRINCIPLES

BIOPHYSICS AND BIOCHEMISTRY OF LOW TEMPERATURES AND FREEZING

F. Franks

Department of Botany, University of Cambridge, England.

INTRODUCTION

At the outset, and before discussing the biophysical and
biochemical consequences of low temperatures it is necessary
to dispel the widely held belief that low temperature and
freezing are synonymous. In the first place, low temperature
is a relative term; to the physicist it conjures up quantum
fluids, superconductivity and the milliKelvin temperature
range. In the life sciences the term is usually identified
with subzero temperatures, with a lower limit of about -70°C
below which no life processes can persist. This is a somewhat arbitrary definition of low temperature, because many
organisms, *e.g.* tropical plants and thermophilic bacteria,
can show all the signs of chill injury well above 0°C. The
total physiological temperature range can usefully be divided
into those temperatures at which the aqueous substrate is
liquid, including the undercooled state, and those temperatures at which the aqueous phase will be partly frozen.
For *in vivo* systems capable of cold survival this dividing
temperature is frequently in the neighbourhood of -20°C, but
it depends on many factors, such as the species, the environment, the degree of cold acclimation and the state of development of the particular organism.

One generalization of universal validity however, is that
because of the intimate involvement of water in life
processes at every level, all such processes are sensitively
attuned to the physical properties of water (Franks, 1977).
Since these properties show some remarkable changes as the
temperature is decreased, especially to below 0°C (Angell,
1981), it is to be expected that such changes will be

reflected in the biochemical and biophysical relationships
responsible for the maintenance of viability. At the
simplest level, such relationships are often expressed in
terms of the thermodynamic water activity (a_w), but this
quantity itself, although convenient to use, is only a
manifestation of the delicate hydrogen bonding patterns that
exist between water molecules and also between water and
other polar molecules. This type of weak association is
extremely sensitive to changes in temperature and such
sensitivity is then reflected in the ability (or inability)
of a complex organism to cope with changes in temperature.

TEMPERATURE DEPENDENCE OF BIOLOGICAL EQUILIBRIA

It is now necessary to distinguish between the effects produced by low temperatures *per se* and those that are specifically due to freezing. The preceding paragraphs suggest that
most, if not all, physical properties of water and aqueous
systems are sensitive to temperature changes. Let us examine
a few of the properties most relevant to biological equilibria.

Ionic Dissociation

One such fundamental property is the ionic dissociation of
water, expressed through an equilibrium constant K_w which
forms the basis of the pH scale. A thermodynamic analysis
of the dissociation equilibrium of water indicates that K_w
decreases (pK_w increases) with decreasing temperature.
Although direct measurements only extend down to -20 C,
various interpolation equations have been proposed which
relate K_w to temperature with a reasonable degree of accuracy.
Using the best experimental data, the following equation is
believed to provide a reliable representation (Hepler and
Woolley, 1973):

$$\ln K_w = -(34865/T) + 939.8563 + 0.22645T - 161.94 \ln T \quad (1)$$

Substituting $T = 238°K$ ($-35°C$), a temperature which is not
uncommon in the natural environment of many organisms, yields
a pK of approximately 17. Since the strengths of acids and
bases are defined in terms of the dissociation of water, the
degree of ionization of acidic and basic residues of proteins
must also be markedly affected by temperature. It is not
certain how much such changes alter the native stability of
biopolymers, but the phenomenon of denaturation is well
established and suggests that a perturbation in the delicate
electrostatic balance between ionic residues may well
decrease the stability of native states. We shall presently

return to this subject.

Dielectric permittivity

Another property of water which is of importance in modulating interactions between charged residues is the dielectric
permittivity (ε). Water has a relatively high permittivity
which makes it a good solvents for electrolytes. It should
be noted in passing that usually a high permittivity arises
from a large molecular dipole moment (D), but this is not so
in the case of water which has a molecular dipole moment of
only 1.84D. By comparison, ammonia with a dipole moment of
1.48D has ε = 17.8, and acetone with a dipole moment of
2.85D has ε = 21.5. The high value of ε for water (80 at
20°C) arises from the unique molecular association pattern
that exists in the liquid phase, and which becomes more
pronounced with a decrease in temperature. Recent measurements on undercooled water indicate that -35°C, ε is
approximately 100 (Hasted and Sahidi, 1976), a 25 percent
increase over the value at the physiological temperature.
This means that at low temperatures water is even better
able to reduce the attraction between ions of unlike charge.
A direct consequence of this change in ε is revealed in the
temperature dependence of the constant A in the Debye-Hückel
limiting law

$$\ln \gamma_{\pm} = -A_{z+z-} \, I^{\frac{1}{2}} \qquad (2)$$

where $\gamma_{\pm}$ is the mean ionic activity coefficient, I is the
ionic strength and A = 44290.76 $\{\rho/(\varepsilon T)^3\}^{\frac{1}{2}}$, ρ being the
density in kg m^{-3}. Equation (2) is a measure of the electro-
static contribution to the deviation from ideal solution
behaviour which arises from interion attractions and repulsions. A change in the temperature from +20 to -20°C is
accompanied by a 6 percent decrease in the contribution to
the electrostatic free energy equation (2) (Clarke and Glew,
1980). This may not seem a great deal, but considering that
the stability of higher (3°,4°) levels of protein structure is exceedingly delicate, a 6 percent reduction in the
electrostatic interactions may be quite significant.

Protein Stability at Low Temperatures

The phenomenon of thermal denaturation of proteins is well
documented. Thus, most proteins as well as nucleic acids
and many polysaccharides, are able to exist in their
biologically active states only up to certain temperatures,
characteristic of a given protein and its environment (pH,

ionic strength, specific ions and so on). Although the
phenomenon of heat denaturation is well documented little is
known about the behaviour of proteins at low temperatures.
The main reason for this is that the solvent medium often
freezes in the temperature range of interest (Brandts *et al.*,
1970). However, there are well-described examples of cold
denaturation, where a protein spontaneously unfolds or where
a multisubunit structure dissociates into biologically inactive species which may or may not reassemble when the
system is restored to its physiological temperature (Dixon
et al., 1981). All this is not too surprising, in view of
the complex energy balance which is responsible for the conformational stability of native proteins. Thus, the free
energy of stabilization which is usually of the order of no
more than -50 to -100 kJ mol^{-1} is the algebraic sum of
several contributions, some of which are very large (Pain,
1979). These include entropic terms, such as the conformational entropy of the peptide chain, and a term arising
from the hydrophobic hydration of apolar amino acid residues,
but there are also electrostatic contributions, peptide
hydrogen bonding and short range van der Waals attractions
and repulsions. All these separate interactions vary in
different ways with temperature. It is to be expected that
all proteins should be stable only within a limited temperature range and that cold denaturation should be as universal as heat denaturation.

Viscosity

Turning now to the dynamic properties of water, here too a
decrease in the temperature gives rise to profound effects.
The viscosity of water rises much more steeply with decreasing temperature than is predicted by the simple Arrhenius
relationship. In other words, the energy of activation
itself is strongly temperature dependent (Osipov *et al.*, 1977).
The bulk viscosity of the liquid is related to the rotational
and translational diffusion by the equation

$$\tau = <r^2> / 12D = kT/6\pi\eta\underline{a} \qquad (3)$$

where D is the self-diffusion coefficient and $<r^2>$ is the
mean square displacement of a molecule during time   ; η is
the bulk viscosity, $\underline{a}$ the molecular radius and T the absolute temperature. Figure 1 shows the temperature dependence
of D, η and τ, relative to their values at 25°C. It is
apparent that below 0°C (T^{-1}= 3.66.10^{-3}) the assumption of
an Arrhenius relationship leads to ever larger errors, which

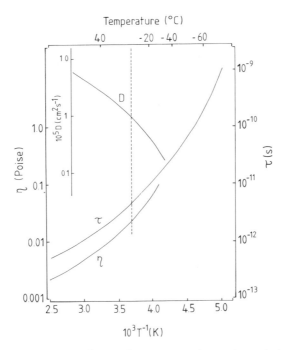

Figure 1. *Temperature dependence of viscosity (η), self-diffusion coefficient (D) and rotational diffusion time (τ) of water. The η and D curves refer to atmospheric pressure and the τ curve to a pressure of 250 MPa.*

reach an order of magnitude at -35°C. The high pressure data for τ are of particular interest, because they extend down to -86°C. At ordinary pressure the undercooling limit for water is -38.5°C at which temperature ice is nucleated spontaneously (Hobbs, 1974). The nucleation temperature decreases with increasing pressure to a minimum value of -90°C at a pressure of 2 Kbar (Kanno *et al.*, 1975). Even allowing for the fact that high pressures tend to reduce the anomalies in the properties of water, the effects shown in Fig. 1 are striking. Equation (3) indicates that at -85°C water is five times as viscous as glycerol is at room temperature (Lang and Lüdemann, 1977). While no claim can be made that such temperatures (and pressures) have any biological significance, it is not improbable that even just a doubling of the viscosity (a 50% reduction in D) might have profound effects on the functioning of a living organism. It must also be borne in mind that, along with changes in the transport properties of water, go similar changes in the transport behaviour of dissolved solutes.

For water flow in highly heterogeneous systems one must also consider the influence of temperature on any barriers present, such as membranes. Thus, the fluidities of the membrane components themselves are subject to a temperature dependence and this affects the water permeability of the membrane structure, even when no temperature induced phase changes occur within the membrane (Leibo, 1980).

Kinetics of Biochemical Reactions

Finally, we must consider the effect of temperature on the kinetics of biochemical reactions, especially in systems of coupled reactions. Many biochemical processes have large activation energies and are therefore very sensitive to changes in temperature. For instance, a reaction with an energy of activation of 20 kJ mol^{-1} and a rate constant of unity at 25°C will be subject to a 75 percent reduction in rate at -20°C. The temperature effect is even more striking if we consider a sequence of two reactions

$$A \xrightarrow{k_1} I \xrightarrow{k_2} B$$

where I may be an intermediate in an enzyme catalysed reaction. Assume that at 25°C $k_1 = k_2$ and that the energies of activation of the formation and decomposition of the intermediate are 50 and 70 kJ mol^{-1} respectively. At -20°C, $k_1 = 3k_2$, and at -40°C $k_1 = 6k_2$, and the intermediate species will thus accumulate. Even the simplest enzyme catalysed reaction consists of several steps, each of them likely to have a different activation energy, so that the overall effect of a temperature reduction is difficult to estimate. For instance the overall lysozyme catalysed degradation of cell wall hexasaccharide is reduced by a factor of 200 for a temperature drop from 40° to -20°C (Douzou, 1977).

FREEZING

Let us now turn from low temperature effects in general to those due to freezing in particular. In the terminology of thermodynamics freezing is referred to as a first order phase change, the melting point being the only temperature at which the solid and the liquid can coexist in stable equilibrium at a given pressure. Figure 2 shows the vapour pressure of ice and liquid water over the temperature range -40 to +10°C. The phase with the lowest vapour pressure is the stable one at a particular temperature. Liquid water is metastable with respect to ice below 0°C and will therefore freeze spontaneously under the right conditions. In the

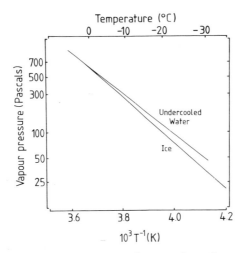

Figure 2. *Vapour pressure of ice and undercooled water.*

foregoing paragraphs emphasis has been placed on the properties of water in the undercooled state, the reason being
that the body fluids of many living organisms readily
undercool, as their temperature drops to below 0°C. Indeed,
the cold resistance mechanisms of some organisms depend on
just this ability to undercool, rather than to freeze (Burke
et al., 1976).

Nucleation Events

The reason why water does not freeze spontaneously at the
equilibrium freezing point is that any crystallization
process must be preceded by a nucleation event, in which a
critical group of water molecules adopts a configuration
that can be "recognized" by other water molecules as an ice
embryo onto which they can condense with a reduction in their
chemical potential. Such critical nuclei can either form
spontaneously through random density and energy fluctuations
within the bulk of the liquid, or their formation can be
aided by a solid particle which acts as a seed. Whatever
the method of nucleation, once a nucleus of a given critical
size and lifetime exists, then crystallization will take
place rapidly, at least at fairly high subzero temperatures.
Unless water is frozen very slowly, that is, with a minimum
of undercooling, the ice crystals that form are metastable
and subsequently recrystallize to a more stable form, a
process which may itself be injurious to living organisms.

Increase in Solute Concentration

Biochemically one of the most important direct effects of
freezing is the concomitant concentration of all soluble
species. This can have far reaching consequences in systems
containing molecules sensitive to high concentrations of
electrolytes. It can also lead to the eventual precipitation
of certain substances, as the total liquid volume is decreased through the removal of water as ice. The residual liquid
solidifies at the eutectic point with a solid phase separation of ice and solutes. In biological systems eutectic
separation is not often observed because of the multicomponent and complex nature of the aqueous phase.

One other aspect that needs to be examined in this context
is the osmotic equilibrium between a cell and its surroundings
(Mazur, 1970). Freezing almost invariably begins in the
extracellular spaces which will then experience the concentration changes referred to above. This will perturb the
osmotic equilibrium and the cell will become partially dehydrated. As the temperature is further decreased and more
water freezes (outside the cell) both the osmotic stress
produced and the increase in intracellular solute concentration, may lead to irreversible injury. It should be noted,
however, that it is not necessarily the ice that causes the
damage, except in an indirect manner. The symptoms of freeze
damage closely resemble those caused by drought and high
salt concentrations generally (Farrant, 1977). The symptoms
are, however, of a completely different nature from those
caused by a lowering of the temperature, without freezing.
In other words, freeze-thaw injury is not necessarily closely
related to chill injury, although some of the subsidiary
effects may be identical since the freezing of water takes
place below the optimum physiological temperature range of
most organisms.

Freezing in the Laboratory and the Environment

In discussion of cold injury, resistance and survival a
distinction must be drawn between low temperatures as they
occur in the natural environment of an organism, and low
temperatures used as a laboratory technique. In the field,
the air and ground temperatures are subject to climatic
fluctuations and rates of change of temperature are low,
with the possibility of cycling about the freezing point.
The gradual decrease in mean temperature during the autumn
is also accompanied by a change in the photoperiod, and it
is this combination of temperature and photoperiod changes

that is implicated in the phenomenon of winter hardening of
plants. In the laboratory, on the other hand, the experi-
menter controls the final temperature and the rates of both
cooling and warming although in practice the maximum warming
rate is limited by the low thermal conductivity of ice. He
also has at his disposal a range of additives, cryoprotectants,
to reduce the degree of freeze-thaw injury. Although the
same basic laws of physics determine cold injury and survival
in the field and in the laboratory, the practical problems
differ considerably. Thus, in the natural environment only
those species will survive that have built-in mechanism for
hardening. In the laboratory one is usually concerned with
using low temperature as a means of storing cells, tissues,
organs, or organisms that do not possess such natural
resistance, in a temporary state of suspended animation.
The storage temperature is then usually -196°C. Under these
conditions all biochemical reactions, except for some
electron transfer processes, are at a standstill. This is
of course not the case at the temperatures to which organisms
are exposed in their natural environment, where the extreme
is in the neighbourhood of -70°C. Under these conditions
metabolic activity still persists, although at low levels.
Indeed, there are few places on Earth where cold completely
prevents the development of life (Lozina-Lozinskii, 1974).
Many species of microorganisms, plants and animals can
survive under extremely severe climatic conditions.

Avoidance of Freezing Injury

It is instructive to study the physical and chemical principles involved in freezing injury and its avoidance. There
are as yet only a few cases where cold resistance can be
traced directly to its physical and chemical origin, but
there are many species where the development of resistance
is accompanied by physico-chemical events which are also
known to influence the undercooling and freezing behaviour
of water. Mention has already been made of the phenomenon
of nucleation which must precede the onset of freezing. In
practice, the nucleation of ice from the liquid phase is
almost invariably triggered by an extraneous particle which
acts as a focus for condensation of water molecules to form
a stable nucleus. In order to be effective as such a
heteronucleator the solid particle must be of greater than
about 10 nm diameter. In other words, molecularly dispersed
solutes cannot facilitate the nucleation of ice (Fletcher,
1970). Little is known about the actual crystal structures
that make certain substances efficient ice nucleators. At

one time it was considered that a match in the lattice
dimensions of the solid and of ice would be essential. It
was this type of reasoning which led to the adoption of
silver iodide as a substance likely to promote the nucleation
of water and ice from supersaturated vapour in the upper
atmosphere. However, more recent work has established that
when lattice defects are introduced into silver iodide, this
treatment enhances rather than diminishes its nucleating
potential (Vonnegut and Chessin, 1971). It has also been
established that biogenic ice nuclei, such as are provided
by certain microorganisms (Maki *et al.*, 1974), are effective
at higher subzero temperatures than is silver iodide. A
strikingly efficient ice nucleation is found in *Lobelia
telekii* which grows on the slopes of Mount Kenya under rapidly
cycling conditions of temperature (Krog *et al.*, 1979). Thus,
the temperature during the day is above freezing, but during
the night drops to $-10^{\circ}C$. The plant therefore has to exist
almost simultaneously under conditions of growth and
resistance. The fluid in its inflorescence contains a
substance, probably of carbohydrate origin, which is able to
nucleate ice at $-0.5^{\circ}C$. The latent heat released during
freezing helps to maintain the temperature of the plant at
close to $0^{\circ}C$, despite the drop in the air temperature during
the night. Here, then we have an example of resistance
through nucleation and freezing.

At the opposite end of the spectrum are plants that
survive by so-called deep undercooling, or freeze avoidance
(Burke *et al.*, 1976). The mechanism of this type of adaptation is unknown, but there appears to be a lower temperature
limit of $-45^{\circ}C$. It can probably be assumed that the tissues
of most living organisms contain supermolecular structures,
be they in the form of organelles or fibrillar aggregates,
that will become active heteronuclei at some particular
subzero temperature (Franks and Bray, 1980). In other words,
it is unlikely that living cells can undercool down to the
homogeneous nucleation temperature, which is $-38.5^{\circ}C$ for
pure water and lower still for aqueous solutions. Because
this temperature is in the neighbourhood of the lowest
observed deep undercooling of plants, it has been suggested
that cold acclimation involves the inactivation of any
structures that might provide heteronuclei for the freezing
of tissue water. This is, however, unlikely, if only
because hardening as well as dehardening can occur over very
short periods of time.

One way of preventing nuclei from promoting the growth of
ice could be to render them inactive by chemical poisoning,
just as catalysts can be poisoned by the irreversible

adsorption of certain molecules at the active site. The
inhibition of enzymes by this type of mechanism is quite
common, so that it might also form the mode of action of the
so-called antifreeze proteins and glycoproteins which have
been isolated from the blood sera of Arctic and Antarctic
fish. These chemically rather simple molecules have been
found to inhibit freezing of the blood of the protected
species at temperatures down to -1.91°C. The effect cannot
be of an osmotic origin (depression of freezing point)
because of the high molecular weight of the proteins. This
has been confirmed by the observation that, unlike the
freezing point, the melting point of the blood, once frozen,
is not depressed. The antifreeze effect must therefore
originate from a direct effect exerted by the peptides on
nucleation and/or crystal growth (DeVries, 1980).

Bound Water

There is yet one more principle that is employed by nature
as a means of survival, namely the exploitation of so-called
"bound", or unfreezable water. When a solution of a polar
compound is cooled, freezing can be induced by seeding the
solution with an ice crystal at the appropriate temperature.
However, in many cases further cooling does not result in a
eutectic separation of the two solids. Instead, the solution
becomes supersaturated and is able to undercool to an
apparently unlimited extent (Kuntz, 1971). Eventually it
undergoes a glass transition, and displays all the mechanical
properties of a glass. It seems therefore, that the prox-
imity of chemical groups which are capable of forming
hydrogen bonds can perturb water to the extent that the
water molecules are prevented from diffusing to the ice
surface and participating in the crystallization process.
The corollary is that the organic component cannot be
crystallized from aqueous solution, because the energy
barrier for dehydration is too high. It must be emphasized
that this phenomenon is not an equilibrium property, and it
is pointless to calculate water binding equilibrium constants.
Rather, it is a kinetic effect, which becomes particularly
marked in systems where the water exists in narrow capill-
aries or pores, and where diffusion is therefore subject to
obstacles (Packer, 1977). All biological macromolecules in
their native states are associated with a certain amount of
unfreezable water which is an absolute requirement for the
maintainance of native structures, even during freezing
(Kuntz and Kauzmann, 1974). This water does not contribute
to the osmotic properties, *i.e.* it does not act as solvent.

It does, however, act as a structural cement in the protection
of complex and vulnerable macromolecular assemblies.

Even more remarkable, several species of microorganisms and
plants are able, under conditions of water stress, to synthesize of otherwise accumulate molecules which bind relatively
large quantities of water. The cells of such organisms can
tolerate concentration increases of more than two orders of
magnitude in the water binding substances, and so resist
dehydration, whether caused by heat, cold or salinity (Gould
and Measures, 1977).

Undercooling, Nucleation and Crystal Growth

Although of no immediate relevance to freezing in the natural
environment, mention should be made of the complex inter-
relationship between undercooling, nucleation and ice crystal
growth (Franks, 1981). The formation of clusters of water
molecules able to act as nuclei for the growth of macroscopic
crystals is a statistical phenomenon which arises from random
density and energy fluctuations within the liquid. In order
to become an active nucleus, such a cluster must be of a
certain minimum size (*i.e.* contain a certain number of
molecules) and must also have a sufficient lifetime to allow
further water molecules to diffuse to the cluster-melt interface and initiate the growth of the crystal. The lifetime of
such a cluster is inversely proportional to the temperature
and directly proportional to the size of the so-called
critical nucleus. Thus, near the normal melting point of ice,
the number of molecules that are required for the formation
of a cluster capable of initiating ice crystallization is
very large, so that probability of the existance of such a
cluster in a given volume of the liquid is very low. In
addition, in the neighbourhood of 0°C the lifetime of such a
cluster is short. However, the lower the temperature, the
smaller is the number of water molecules needed to produce a
critical nucleus; the probability of nucleation thus increases with decreasing temperature. In fact, the rate of
nucleation increases by many order of magnitude over a narrow
range of temperature. It is therefore difficult, if not
impossible, to prevent nucleation during cooling. Calculations based on the theory of nucleation suggest that at -40°C
the number of water molecules required to produce a critical
nucleus is of the order of 400. There must be a very high
probability that at any given moment, somewhere in the bulk
of the liquid a random fluctuation gives rise to a molecular
arrangement involving 400 molecules which could be recognized
by neighbouring water molecules as a structure sufficiently

close to that of ice onto which they can condense. One
further factor enhances this probability: water is unique
among common liquids, in that the amplitude of its random
fluctuations in density increases with decreasing temperature.
In other words, the lower the kinetic energy of the molecules,
the more pronounced are the fluctuations of density and
energy about the mean value. There are a few other known
substances whose structures resemble that of ice and which
also exhibit this strange effect *e.g.* silicon dioxide. This
actually arises from an abnormal temperature dependence of
the compressibility of the liquid in the undercooled state
(Angell, 1981).

The above arguments can also explain the observation that
the homogeneous nucleation temperature is a function of the
volume of the liquid; this effect becomes noticeable for
droplets with diameters of less than about 10 µm. It has
been suggested that organisms that survive winter by deep
undercooling have the ability to redistribute their tissue
water into small domains. They are thus able to reduce the
temperature of spontaneous nucleation. If the water domains
are also segregated from each other, the nucleation and
freezing of one such domain would not automatically cause
freezing of other domains within the tissue. However, to
avoid nucleation altogether, such domains would need to be
smaller than the size of the critical nucleus at the particular temperature.

While the rate of nucleation increases rapidly with decreasing temperature, the opposite is the case for crystal
growth. The propagation of the ice front is a function of
the diffusion of water molecules to the liquid/solid interface and this, in turn, is governed by the activation energy
(Fig. 1). The rate of crystallization is therefore governed,
among other factors, by the rate of cooling. Since nucleation and crystal growth have very different origins, their
temperature dependences are governed by quite different
variables and the calculation of crystal sizes and concentrations as function of temperature and cooling rate is
extremely complex (Franks, 1981). This is particularly so
when the aqueous phase is not pure water but a multicomponent
solution, because concentration gradients are then built up
ahead of the advancing ice front, leading to a local lowering
of the freezing point; also the nucleation behaviour of water
is markedly affected by the presence of solutes.

Phase Change Events

Emphasis has so far been placed on the freezing of water.

However, freezing, as previously defined, need not be limited
to water. Indeed, any component that is capable of undergoing a disorder – order transition (of which the liquid –
solid transition is a special case) at a unique temperature
will give rise to some of the symptoms we associate with
freezing. Processes which answer this description include
liquid – liquid crystalline transitions observed in lipid
bilayer systems, and self-assembly phenomena involving the
aggregation of peptides to form multisubunit structures.

   The "freezing" of lipid bilayer systems has been studied
in great detail and is described elsewhere (Pringle and
Chapman, this volume). This phenomenon has also been implicated in chill and/or freeze injury (Lyons and Raison, 1970).
While this is not the place for a review of the evidence for
and against such correlations, two important points must be
made. A real biological membrane, as distinct from a phospholipid vesicle of homogeneous chemical composition, is a
multicomponent phase; it consists of lipids of varying chain
lengths, degrees of unsaturation and head groups. It may
also contain cholesterol, cholesterol esters and proteins,
all of which will interfere with the cooperative nature of
the liquid-crystal transition. The effect is not different
from that observed when an aqueous solution, rather than
pure water, is subjected to freezing: an aqueous solution
freezes over a range of temperature, rather than at a unique
temperature.

   The other point concerns the correlation between the onset
of chill injury and this type of lipid phase transition. It
is often argued that breaks in an Arrhenius plot of some
dynamic property against the inverse temperature provide
evidence for such correlations (Raison and McMurchie, 1974).
However, such treatments are very weakly based, both
experimentally and statistically. The temperature range that
can be covered by experiment hardly exceeds $35^{\circ}C$ for *in vivo*
systems or even *in vitro* enzymes. This corresponds to
$\Delta (T^{-1}) = 4 \times 10^{-4}$, equivalent to about 10% of the T^{-1} value
corresponding to $35^{\circ}C$ ($T^{-1} = 3.2 \times 10^{-3} K^{-1}$). In order to
establish a statistical significance of a discontinuous
change in the Arrhenius slope over such a narrow range of
T^{-1}, measurements of the other variable (enzyme activity,
diffusion, motion of a label in the membrane, *etc.*) must be
made to within ±0.1% (Krug *et al.*, 1976), a requirement that
has so far only been met in very few studies of the kinetics
of well characterized organic reactions, under ideal conditions and where the accessible temperature range is in any
case considerably larger than it is for biochemical reactions.
Whilst it cannot be denied, therefore, that a freezing

process in a membrane may have far reaching consequences for
the viability of an organism, the occurrence of such processes, especially over narrow ranges of temperature, has yet to
be established for *in vivo* systems. The theoretical and
statistical basis underlying the Arrhenius treatment of
kinetic data is by no means universally appreciated (Wolfe
and Bagnall, 1979).

Combined Effects of Low Temperatures and Freezing

Finally, let us examine briefly the combined effects of low
temperature and freezing; they are common not only in our
terrestrial environment but also in the technology of frozen
storage of labile materials; they also impinge on processes
such as freeze drying and freeze substitution. The general
effect of a lowering of the temperature is to retard all
dynamic processes, *e.g.* diffusion and reaction rates. Such
retardation is quite independent of the occurrence of a
freezing event as such, but it is greatly affected by the
increase in concentration that accompanies freezing. Thus,
the activity of an enzyme (as measured by the rate of substrate turnover) decreases with decreasing temperature above
the freezing point, but may increase below the freezing
point. This means that the effect of enhanced concentration
more than balances the rate decrease due to the low temperature (Kiovsky and Pincock, 1966). In practice this leads
to a break in the Arrhenius plot, but such a break is of
course an artefact, because the conditions under which the
reaction is studied change continuously between the freezing
point and the final eutectic temperature at which the whole
system solidifies. In systems which supersaturate (*i.e.*
do not exhibit eutectic separation) the final concentration
of solute after freezing will be the same, whatever was the
initial concentration, but the solid to liquid phase volume
ratio will depend on the initial concentration.
    The effects of freezing on the kinetics of a simple
organic reaction, the acid catalysed mutarotation of glucose
are shown in Fig. 3. The initial portion of the curve
reflects the normal effect of temperature on the rate constant k. The dilute solution freezes at 0°, whereas the
more concentrated solution freezes at $-1.5^\circ C$. The rate
enhancement observed reflects the concentration increase
due to freezing. Eventually k passes through a maximum at
some temperature, below which no more water freezes; the
k(T) relationship should then once again exhibit an Arrhenius
behaviour.
    In biochemical processes the observed kinetics can be

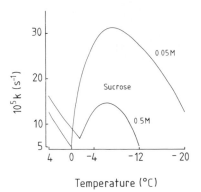

Temperature (°C)

Figure 3. *Effect of temperature on the mutarotation of* α −
*D − glucose in frozen solutions and in the presence of HCl.
Data from Piovsky and Pincock (1966).*

further complicated by the occurrence of a spontaneous
dissociation or aggregation of macromolecular species. These
can be brought about both by a low temperature and/or by an
increasing concentration. They can also lead to irreversible
effects (cold denaturation) such as become apparent in the
cold induced insolubilization of certain food proteins or
the inactivation of enzymes.

Indeed, there are reactions the rates of which are enhanced
by orders of magnitude as a result of low water contents
induced by freezing and subsequent freeze concentration.
They include the oxidation of β−carotene and vitamin A, the
insolubilization of myosin and the oxidation of oxymyoglobin.
Such deleterious effects can be further enhanced by freeze
drying, because during this process the temperature is
usually allowed to rise, so that any benefits gained from a
retardation in the deleterious reactions due to the low
temperature are nullified (Fennema, 1975).

SUMMARY

Low temperature and freezing thus give rise to very different
consequences which, when combined, produce a complex set of
superimposed processes. These include changes in the positions of chemical and biochemical equilibria, changes in
the rates of motion of molecules, changes in concentrations
of the components in a mixture resulting from phase separations, and the possibility of irreversible aggregation or
dissociation events involving macromolecules. Additional
complications arise from the possibility of undercooling of

the aqueous phase and from the phenomenon of bound, or
unfreezable water. Many of the above effects, singly or in
combination, can be identified not only in natural cold and
frost resistance mechanisms but also in cold and frost
injury, whether caused by climatic fluctuations or by man.

ACKNOWLEDGEMENTS

I wish to thank Dr Helen Skaer for her critical, yet
sympathetic reading of the manuscript.

BIOMEMBRANE STRUCTURE AND EFFECTS OF TEMPERATURE

M.J. Pringle and D. Chapman

Department of Biochemistry and Chemistry, Royal Free Hospital School of Medicine, London, England.

INTRODUCTION

The cell membrane is no longer regarded as a simple barrier separating the cytoplasm from extracellular constituents, but is now recognized as the site of a variety of biological functions. However, many of these functions occur in different parts of the membrane. For example, agglutination, fusion, and phagocytosis are all properties of specialized regions of the cell membrane surface (Cook and Stoddart, 1975; Hughes, 1975), and the recognition sites for hormones and neurotransmitters are specific protein molecules embedded, to a certain extent, within the cell membrane. Further still, the trans-membrane movement of ions via the various energy-consuming 'pumps' or the channels involved in the transmission of electrical impulses, require molecular entities which are capable of allowing ions to pass from one side of the membrane to the other.

Such processes can best be understood by analysing the relationship between membrane structure and any of the specific functions mentioned above. The cell membrane can be regarded as a complex heterogeneous system whose behaviour is largely determined by the nature and distribution of its constituents, and by the external environment, of which temperature is clearly a parameter of major importance. Most biochemical processes are chemical or physico-chemical rea-actions and, as such, they are influenced by their immediate microenvironment which includes the nature of the solvent and the operating temperature. Temperature determines not only the rate of these processes but also whether they occur at all (Franks, this volume). In this context, the definition

of low temperature is a relative one and whereas the range
1 to 4°C may be quite normal for hibernating or ectothermic
marine animals, it is clearly well below the optimal 37°C
required by most non-hibernating mammals.

In the first part of this chapter we shall describe the
nature and arrangement of the basic elements common to all
cell membranes, and in the second part we shall discuss the
way in which temperature may modify the structure of biomembranes thereby affecting the activity of membrane-bound
enzymes. This should provide a framework in which to view
the structural modifications to cell membranes which are
involved in low temperature adaptation.

BIOMEMBRANE STRUCTURE

Lipid Composition

Lipids are a primary component of all cell membranes,
ranging from 20 percent of the total dry membrane weight in
the case of inner mitochondrial membranes to 80 percent in
the case of myelin plasma membranes (Quinn, 1976). Of these
the glycerol phosphatides represent the predominant structural class (Ansell and Hawthorne, 1964); these usually consist of fatty acid esters of *sn*-glycerol-3-phosphate. The
attachment of various substituents such as choline, ethanolamine, serine, or inositol to the phosphate group, provides
a further subdivision. In a similar way, the fatty acid
esterification and substituent linkage of sphingosine gives
rise to the sphingolipids such as sphingomyelin and the
cerebrosides. The fatty acids themselves normally consist of
long, even-numbered, hydrocarbon chains which can be either
fully saturated, or contain a varying number of *cis*-double
bonds (structures of fatty acids, sterols and phospholipids
are given in the section on Nomenclature). The large number
of possible combinations of head group and fatty acid chain
means that most cell membranes contain a wide spectrum of
different lipids. The functional significance of this
heterogeneity will become apparent later.

In addition, many cell membranes contain the sterol cholesterol and the ratio of cholesterol to phospholipid varies
from one cell type to another. Thus mitochondrial membranes
contain very little sterol, whereas in the erythrocyte membrane, 50 percent of the non-protein material is cholesterol
(Quinn, 1976).

Bilayer Arrangement

Within the membrane, the phospholipid molecules are arranged
in the form of a bilayer with polar groups at the intra- and
extracellular surfaces. These groups are thus able to interact with the aqueous phases. The fatty acyl chains, on the
other hand, are stacked in parallel fashion at right angles
to the plane of the membrane, with the terminal methyl groups
situated in the interior of the bilayer. This arrangement
of polar surfaces with an apolar interior provides both a
permeability barrier and affords a degree of electrical resistance to the membrane, since the insertion of a small
polar molecule or ion into the hydrophobic interior requires
a large activation energy.

Cholesterol, when it is present, is randomly distributed
with the β-hydroxyl group interacting with the polar head-
groups of the lipids, and the sterol hydrocarbon skeleton
intercalated among the lipid acyl chains. Van der Waal's
interactions between lipids and cholesterol are therefore
possible down to the ninth or tenth carbon atom of the lipid
chains.

Proteins

The protein moieties of the cell membrane are responsible
for much of its biological activity, although, of course,
many proteins serve to maintain the membrane's structural
integrity (*i.e.* those comprising the cytoskeleton). Membrane
proteins are incorporated into the lipid matrix such that
they can either lie along part of the membrane surface, or
penetrate, to a greater or lesser extent, the width of the
bilayer. Some proteins such as bacteriorhodopsin (Henderson
and Unwin, 1975), rhodopsin (Jan and Revel, 1974) and those
comprising the ion channels of excitable membranes, can
actually span the entire bilayer. Proteins can broadly be
classified according to the ease with which they may be
removed from their native membranes. Thus, extrinsic or
peripheral proteins can be removed by relatively mild procedures such as altering the ionic strength of the medium,
while intrinsic or integral proteins usually require treatment with detergents to solubilize them.

Membrane Asymmetry

Protein distribution across the membrane is asymmetric and
this asymmetry is an absolute requirement. Proteins will be
exposed on one side of the membrane or the other (usually

the cytoplasmic side), and even when they span the membrane
they do so asymmetrically (Rothman and Lenard, 1977).
   Membrane lipids are also arranged asymmetrically with
respect to the two halves of the bilayer, although the nature
and extent of the asymmetry varies from one membrane type to
another. Thus, in erythrocyte membranes, amino phospholipids
such as phosphatidylserine and phosphatidylethanolamine are
located primarily in the cytoplasmic half of the bilayer,
while phosphatidylcholine and sphingomyelin are found mainly
in the outer half (Bretscher, 1973). On the other hand, in
the membrane of the influenza virus, the location of phosphatidylethanolamine is reversed and, in addition, phosphatidylinositol is found mainly in the inner half of the bilayer
(Blough, 1975). A corollary of the asymmetric distribution
of phospholipids is that phospholipid exchange from one half
of the bilayer to the other ('flip-flop') is likely to be
extremely slow (Kornberg and McConnell, 1971). The functional significance of membrane asymmetry is still unclear,
particularly in the case of lipids, although it presumably
reflects the different biochemical roles played by each half
of the membrane and the physical constraints imposed by the
close packing of head groups in each half of the bilayer.

Lateral Distribution of Components

The arrangement of membrane constituents within the plane of
the membrane has for many years been the subject of much
speculation. A variety of models have been proposed which
invoke specific phospholipid/cholesterol or phospholipid/
protein complexes leading to an ordered or structured array
within the plane of the membranes. This will be discussed
in more detail with reference to the effects of temperature
on protein/lipid interactions. The most plausible model
consistent with the available experimental data is that of
a random arrangement of phospholipids, cholesterol and proteins. In such an arrangement, it is the fluid nature of the
membrane phospholipids which allows the normal functioning
of membrane proteins. In the second part of this chapter,
we shall show how low temperatures modify the fluidity
characteristics of cell membranes, and, in so doing, change
the lateral distribution of membrane constituents. The
alteration in lateral distribution is a key element in any
resulting loss of membrane function.

THE EFFECTS OF TEMPERATURE

Lipid Fluidity

The behaviour of membrane lipids is usefully described in
terms of fluidity (Chapman, 1966; Chapman *et al.*, 1967), a
term which reflects the degree of order in the acyl hydrocarbon chains of the phospholipid molecules. Such information can be obtained by a variety of spectroscopic probe
techniques (Cossins, this volume). For example, the fluorescence polarization of the molecular probe perylene is
dependent upon the rotational motion of the fluorophore within a phospholipid bilayer, and this parameter is in turn related to the phospholipid microviscosity (Cogan *et al.*, 1973;
Shinitsky *et al.*, 1971). Alternatively, the use of paramagnetic groups (usually nitroxides) covalently attached to
fatty acids, phospholipids, or sterols, allows one to interpret the electron spin resonance (ESR) spectra of the probes
in terms of orientation and degree of order of the membranes
into which they are incorporated (Hubbell and McConnell, 1969,
1971; Jost *et al.*, 1971). Similar information can be obtained
by nuclear magnetic resonance (NMR) techniques and in particular the incorporation of deuterium nuclei at specified
positions along the phospholipid hydrocarbon chain, enables
one to measure the rotational motion of the chains from the
nuclear quadrupole splitting (Seelig and Seelig, 1974).
    From probe studies such as those mentioned above, there is
general agreement that the interior of the phospholipid
bilayer is highly disordered or fluid, in contrast to the
region close to the polar head groups at the membrane surface.
However, ESR and deuterium NMR data suggest different fluidity gradients along the acyl chains. Thus, ESR studies of
egg lecithin/cholesterol liposomes indicate a uniform increase
in disorder from the lipid head groups to the terminal methyl
groups (Hubbell and McConnell, 1971), whereas the deuterium
NMR data suggest that there is little change in order along
the hydrocarbon chains until the last three methylene groups,
where a sudden decrease in order occurs (Horowitz *et al.*,
1973; Seelig and Seelig, 1974). The discrepancy has been
attributed to the perturbing effect of the paramagnetic
nitroxide groups used in ESR (Seelig and Niederberger, 1974).

Thermotropic Behaviour

It is a fundamental property of phospholipid bilayers that
they exhibit thermotropism, *i.e.* pure phospholipids in
aqueous dispersion undergo abrupt changes in phase at characteristic temperatures referred to as phase transition

temperatures. The type of phase present in such systems will
depend upon both the temperature and the degree of hydration
for a particular lipid (lyotropic mesomorphism). For the
purpose of this chapter, we will restrict our discussion to
two main phases, namely the liquid crystalline or mesomorphic
phase and the gel phase (for a detailed review of lipid phases
and transitions see Chapman, 1975). In the gel state, phospholipid molecules are arranged in an orderly crystalline
lattice with a characteristic hexagonal packing. The hydrocarbon chains are predominantly in the fully extended or
all-*trans* conformation. At the main transition temperature
(T_c), there is an abrupt, endothermic phase change and the
phospholipid hydrocarbon chains assume a disordered configuration characterized by a number of *gauche* conformations
along each chain. Since the bilayer structure is maintained
the fluid arrangement of phospholipid molecules above T_c is
known as the liquid crystalline state.

Thermotropic Behaviour of Single Phospholipids

For pure phospholipids containing a single type of fatty
acid chain, the transition process is highly cooperative,
i.e. the thermal energy supplied to the all-*trans* chains is
efficiently transmitted from one chain to another throughout
the bilayer. In such systems the gel to liquid crystalline
transition therefore occurs over a very narrow temperature
range and can be treated, to a good approximation, as a
first order process. The technique of differential scanning
calorimetry (DSC) provides a direct measurement of the heat
changes involved in phospholipid transitions (Hinz and
Sturtevant, 1972b; Ladbrooke and Chapman, 1969), although
the transition temperature itself can be examined by a number
of spectroscopic probe techniques (Williams and Chapman, 1970).
Early calorimetric studies on liposomes consisting of 1,2-
diacyl-L-phosphatidylcholine, have established that T_c increases with increasing acyl chain length but decreases
dramatically in the case of unsaturated fatty acid chains
(Chapman *et al.*, 1967). Thus dioleoyl phosphatidylcholine
undergoes the transition at -22°C, whereas its saturated
analogue distearoyl phosphatidylcholine remains in the gel
state until 54°C (Table 1). Polar head groups also exert
a modulating influence on the thermotropic behaviour of
phospholipids, *i.e.* for a fixed acyl chain length, the T_cs of
phosphatidylethanolamines are much higher than those of the
corresponding phosphatidylcholines (Chapman, 1975). Presumably the relatively lower stability of choline-containing
phospholipids is a reflection of the steric properties of the

Table 1

Transition temperatures (T_c) for 1,2-Diacyl-L-phosphatidyl-choline / Water systems

1,2-Diacyl-L-Phosphatidylcholine	Acyl Chain	T_c ($^{\circ}$C)
Diarachidoyl	20:0	75
Distearoyl	18:0	54
Dioleoyl	18:1	−22
Dipalmitoyl	16:0	41
Dimyristoyl	14:0	23
Dilauroyl	12:0	0
Egg yolk lecithin	Mixed	−7 to −15 (broad)

Adapted from Chapman et al., (1967)

head group.

Thermotropic Behaviour of Mixed Phospholipids

Natural membranes contain a wide spectrum of phospholipids and much of our understanding of the effects of temperature on membranes derives from studies on bilayers of mixed phospholipids where variations in chain length, degree of unsaturation and head-group can be experimentally controlled. As an example, calorimetrically obtained phase diagrams for the binary systems distearoyl/dipalmitoyl phosphatidylcholine and distearoyl/dimyristoyl phosphatidylcholine, shows that in the former case, there is a continuous series of solid solutions below the T_c line whereas in the latter case, upon cooling there is a migration of phospholipid molecules within the bilayer to give crystalline regions corresponding to each component (Ladbrooke and Chapman, 1969). This phenomenon is of fundamental importance because it demonstrates that in a mixed phospholipid membrane, where there is appreciable chain mis-matching, a lateral phase separation occurs within the plane of the bilayer. Subsequent experiments on mixed phospholipid classes established that in a variety of two phase systems, lateral phase separation occurs at low temperatures (Chapman *et al.*, 1974; Clowes *et al.*, 1971; Oldfield and Chapman, 1972). These results have also been confirmed by spin label studies (Butler *et al.*, 1974; Shimshick and McConnell, 1973a).

Thermotropic Behaviour of Cholesterol/Phospholipid Systems

Cholesterol is a major component of many eukaryote cell membranes and in the last decade, much work has been directed towards establishing its role in modulating membrane function. In particular, the effect of cholesterol on the thermotropic behaviour of phospholipid bilayers has been studied by a variety of techniques including differential scanning calorimetry (Hinz and Sturtevant, 1972a; Ladbrooke *et al.*, 1968a), proton magnetic resonance (Darke *et al.*, 1972), electron spin resonance (Schreier-Muccillo *et al.*, 1973; Shimshick and McConnell, 1973b), and fluorescence polarization (Papahadjopoulos *et al.*, 1973). Taken together these studies have established that in a fluid bilayer (that is, above the transition temperature), the effect of cholesterol is to order the acyl chains and thereby reduce the fluidity of the bilayer. This can be rationalised in terms of the rigidity of the sterol hydrocarbon skeleton and, in fact, NMR experiments show that when cholesterol is incorporated into a phospholipid bilayer it reduces the rotational mobility of the first ten methylene groups in the phospholipid hydrocarbon chain (Phillips and Finer, 1974). Below T_c however, cholesterol reduces the transition endotherm. Thus, when cholesterol is added to dipalmitoyl phosphatidylcholine bilayers, the temperature range of the main transition is increased, *i.e.* the cooperativity of the process is reduced. When the cholesterol concentration becomes equimolar with that of the phospholipid, there is no phase transition at all and crystallization to the gel state is completely prevented (Ladbrooke *et al.*, 1968a). A later study by Hinz and Sturtevant (1972a) suggested that the phase transition is removed at a 2:1 phospholipid to cholesterol molar ratio, although more recent work from the same laboratory (Mabrey *et al.*, 1978) confirmed the original data of Ladbrooke *et al.* (1968a). It has also been shown that as increasing amounts of cholesterol are added to liposomes of phosphatidylcholine, the sharp increase in water permeability which occurs at the transition temperature is abolished (Blok *et al.*, 1977).

One can summarise the effects of cholesterol on lipid thermotropic behaviour as follows: above the transition temperature the hydrocarbon chains of phosphatidylcholines in aqueous dispersion have less thermal motion in the presence of cholesterol than in its absence. On the other hand, below the transition temperature the chains have more thermal motion than in the absence of cholesterol.

The molar ratios at which cholesterol was found to abolish lipid crystallization, prompted the idea that there exist stoichiometric complexes between the phospholipid and the sterol. Thus 1:1 complexes were suggested by Darke *et al.* (1972) and Phillips and Finer, (1974), whereas 2:1 complexes were proposed by Hinz and Sturtevant (1972a) and Engelman and Rothman (1972). The latter authors elaborated their hypothesis to include a model for 2:1 stoichiometry which involved each phospholipid molecule being surrounded by an annulus of cholesterol molecules. However, recent studies using NMR and DSC are inconsistent with the existence of any form of phospholipid /cholesterol complex (Estep *et al.*, 1978; Mabrey *et al.*, 1978; Opella *et al.*, 1976). Thus, in the absence of any unequivocal evidence for a highly ordered structure such as that proposed by Engelman and Rothman (1972), the random distribution model of Chapman *et al.* (1978) would seem most consistent with experimental data.

Thermotropic Behaviour of Biomembranes

It is now accepted that the bilayer structure, discussed above in relation to aqueous dispersions of phospholipid (liposomes), is the predominant structure of cellular membranes (Wilkins *et al.*, 1971). Many of these membranes also exhibit thermotropic behaviour, but because of the wide variety of phospholipids comprising most cell membranes, it is necessary to think in terms of lateral phase separations rather than the simple, sharp phase transitions. Cholesterol will also play a critical role in modulating the thermotropic phase changes in biomembranes. The situation is further complicated by the presence of membrane proteins for not only do intrinsic proteins affect the thermotropic behaviour of membrane lipids, but the phase behaviour itself has a marked influence on protein distribution and function. Thus a calorimetric study of erythrocyte membranes showed that the intact membrane does not display thermotropism, but upon removal of cholesterol, a broad endotherm is exhibited (Ladbrooke *et al.*, 1968b). The functional consequences of cholesterol removal are illustrated by the recent experiments of Bottomley *et al.* (1980) who showed that rat thymocytes which have been depleted of cholesterol are unable to undergo mitogen-induced transformation.

A number of prokaryotes have been found to exhibit broad (weakly cooperative) thermotropic transitions in their membranes. For example, cells of *Acholeplasma laidlawii* display a broad transition over a 30° range, as do the extracted lipids (Stein *et al.*, 1969); this transition encompasses the growth temperature of $37^\circ C$. With an oleic acid (18:1) supplement in the growth medium, the transition occurs well below the growth temperature. Growth is severely inhibited when stearic acid (18:0) is included in the medium since the transition temperature for distearoyl phosphatidylcholine is several degrees higher than the normal growth temperature (Chapman and Urbina, 1971; Oldfield *et al.*, 1972). In general, cell growth occurs at or above the membrane phase transition temperature but is restricted or abolished below it. Similar effects have been observed in fatty acid auxotrophs of *Escherichia coli*, and again the growth rate is dependent upon the nature of the fatty acids present in the growth medium (Fox and Tsukagoshi, 1972). Fatty acid supplements also control the transport properties of *E. coli* auxotrophs, and Arrhenius plots of transport exhibit two discontinuities which coincide with the upper and lower limit of lipid phase separation within the membrane, as determined by spin label studies (Linden *et al.*, 1973; Cossins; Herbert, this volume).

PROTEIN-LIPID INTERACTIONS

Reference has already been made to the lateral phase separations which occur in mixed phospholipid bilayers when the temperature falls below the transition temperature of one or more of the components. We have also discussed the relationship between cell growth and thermotropic behaviour of natural membranes. In this section, we shall focus in more detail on the physico-chemical properties of phospholipid protein systems. In particular we shall describe the mechanism which links the physical perturbation produced by protein molecules embedded in lipid membranes, with biochemical processes such as the activity of membrane-bound enzymes.

Packing Faults

The effects of low temperature on membrane function can best be understood in terms of a redistribution of membrane constituents as a direct consequence of the dynamic nature of protein/lipid interactions. If one represents phospholipid and protein molecules as small spheres of appropriate relative size, one can construct a two-dimensional array of hexagonally packed phospholipids and observe the effect of inserting protein molecules into the hexagonal lattice.

Such studies show that when a single protein molecule does not occupy the space of an integral number of phospholipid molecules, the resulting lateral stress gives rise to a perturbation in the form of packing faults radiating outwards from the inserted protein (Chapman *et al.*, 1977a). Such perturbations are rapidly attenuated for very low concentrations of protein. However, as the protein to phospholipid ratio increases, the dislocations become comparable in size to the inter-protein spacing. In such a system, the degree of cooperativity between phospholipid molecules should be reduced. This conclusion, based on model studies, is supported by the experimental observation that the incorporation of proteins into phospholipid bilayers has a similar effect on thermotropic behaviour to that of cholesterol. For example, the addition of the polypeptide ion-channel gramicidin A to dipalmitoyl phosphatidylcholine bilayers (Chapman *et al.*, 1977b), or myelin apoprotein to dimyristoyl phosphatidylcholine bilayers (Curatolo *et al.*, 1978), causes a marked broadening of the phospholipid endotherm and, at sufficiently high concentration, a complete abolition of the phase transition.

Phase Separations

A further consequence of the packing faults described above is that if the protein concentration is not too high, and low temperature crystallization of the phospholipid does occur, then protein molecules will be excluded from the crystal lattice and forced to move along the lines of dislocation into aggregates. In other words, low temperatures can induce in a phospholipid/protein system a lateral phase separation into regions of crystalline phospholipid and regions of high protein concentration, containing small domains of trapped phospholipid. Such arguments are amenable to experimental test and, in fact, freeze fracture electron micrographs of the intrinsic protein Ca^{2+}-ATPase, reconstituted into vesicles of dipalmitoyl phosphatidylcholine show that at temperatures above the transition temperature there is a random distribution of particles (assumed to be the enzyme molecules), whereas below the transition temperature there are discrete areas of crystalline lipid and clusters of proteins containing trapped phospholipid domains (Plate 1). At this point, it should be made clear that because of the random distribution of lipids and proteins in cell membranes at physiological temperatures, and the fact that the components are in dynamic equilibrium, there will be transient protein-rich areas containing small lipid domains even

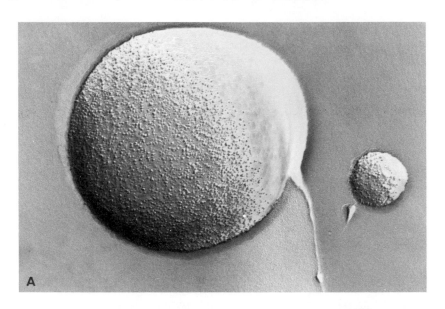

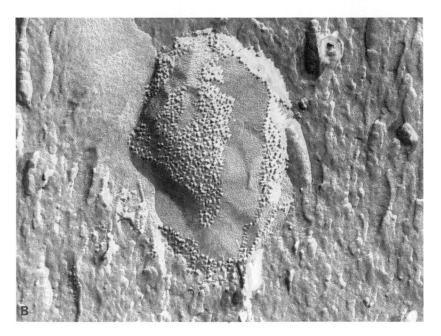

above the membrane transition temperature, especially when the overall protein to phospholipid ratio is high. This suggests that the spin label evidence for the existence of immobile boundary lipids surrounding intrinsic proteins and controlling their activity (Griffith *et al.*, 1973; Hesketh *et al.*, 1976; Warren *et al.*, 1975) should be reinterpreted in terms of restricted probe mobility in those phospholipid domains trapped within the protein aggregates (Chapman *et al.*, 1979). Indeed, recent deuterium NMR studies indicate that lipids adjacent to membrane-bound enzymes are in fact more fluid than the bulk lipids on a time-scale which, though slower than the ESR time-scale, is still orders of magnitude faster than the enzyme turnover rate (Kang *et al.*, 1979; Oldfield *et al.*, 1978).

Protein Diffusion

One of the central problems in elucidating the molecular behaviour of lipid/protein systems as a function of temperature has been to find a suitable physical property of a membrane-bound protein which can be correlated with membrane function. This is in stark contrast to the wealth of information which has accumulated for phospholipids in planar bilayers, multilayers, liposomes and natural membranes. However, in recent years, techniques have been developed for measuring both lateral and rotational diffusion of proteins within a membrane (for a recent review see Cherry, 1979). Thus, the lateral diffusion of rhodopsin in the disc membrane of rod outer segments has been measured by Poo and Cone (1974) and Liebman and Entine (1974) using the technique of Fluorescence Recovery After Photobleaching (FRAP), while a modified FRAP technique has been devised for proteins with relatively slow rates of diffusion (Smith and McConnell, 1978). Protein rotational motion has been measured by saturation transfer electron spin resonance (Hidalgo *et al.*, 1978; Thomas *et al.*, 1976) and by laser-induced dichroism (Cone, 1972; Hoffman *et al.*, 1979; Razi-Naqvi *et al.*, 1973). The limitations of space preclude a detailed discussion of these techniques but the important point is that by studying membrane-bound enzymes both in native membranes and in re-

Plate 1 (opposite). *Freeze-fracture micrographs of a dipalmitoyl phosphatidylcholine/Ca^{2+}-ATPase recombinant with a phospholipid to protein molar ratio of 60:1. (A) frozen from 41°C (magnification X 53,000); (B) frozen from 24°C (magnification X 68,000) (After Hoffman et al., 1980).*

constituted systems, one can correlate functional properties
such as enzyme activity and rotational diffusion with membrane composition and protein/lipid distribution. In the
final section below, we shall describe recent experiments on
the intrinsic protein Ca^{2+}-ATPase where, the above properties
have been explored in relation to the thermotropic behaviour
of the membranes containing the enzyme.

It should be noted that the degree to which integral
proteins are free to undergo lateral motion within the membrane varies from one cell type to another. For example,
Poo and Cone (1974) calculated a lateral diffusion coefficient of approximately $4 \times 10^{-9} cm^2 s^{-1}$ for rhodopsin in retinal
rod outer segments, whereas in the erythrocyte membrane there
is little, if any movement of intrinsic proteins in the plane
of the membrane (Peters *et al*., 1974). Furthermore, erythrocyte membranes do not exhibit thermotropism upon cooling
to $0^{\circ}C$ (Ladbrooke *et al*., 1968b) and freeze-fracture micrographs of these membranes as well as those of lymphocyte
plasma membranes do not show low temperature phase separation
whereas lymphocyte nuclear membranes do (Wunderlich *et al*.,
1974a,b). However, phase separations have been observed in
erythrocyte membranes cooled slowly to $-196^{\circ}C$ (Fujikawa,
this volume). Although lateral phase separations have been
observed in prokaryotes, even at the growth temperature
(Cossins; Herbert, this volume), one possibility for the lack
of protein/phospholipid redistribution at low temperatures
in some mammalian plasma membranes is the high cholesterol
to phospholipid ratio (Ferber *et al*., 1972). It is also
possible that in many membranes the cytoskeletal framework
may prevent protein diffusion and aggregation. Thus the
extrinsic protein spectrin, in association with actin,
comprises the cytoskeletal network of the erythrocyte membrane (Marchesi *et al*., 1976) and this network has been
shown to prevent the aggregation of the intrinsic proteins,
band 3 and glycophorin (Elgsaeter and Branton, 1974). This
is consistent with the more recent measurements of band 3
lateral diffusion by Fowler and Branton (1977) which suggests that the diffusion coefficient at $37^{\circ}C$ is two orders
of magnitude lower than that of rhodopsin. Whether, in the
case of the erythrocyte membrane, the association between
intrinsic and extrinsic proteins is in the form of a direct
linkage (Pinto da Silva and Nicolson, 1974), or merely a
physical entrapment (Cherry *et al*., 1976; Fowler and Branton,
1977), it is clear that structural proteins can have a
marked influence on the protein/phospholipid distribution
and thermotropic behaviour of many types of cell membrane.

Enzyme Activity in Membranes

Arrhenius plots for Ca^{2+}-ATPase activity in native sarcoplasmic reticulum vesicles show a discontinuity at about $15^{\circ}C$
which has been ascribed to a phase change in the membrane
lipids (Inesi *et al.*, 1973; Madeira *et al.*, 1974). However,
since Dean and Tanford (1978) observed the same break in
activity after the enzyme had been isolated from the membrane
and reconstituted in detergents, it would seem more likely
that the discontinuity in the Arrhenius plots of enzymatic
activity is due to a conformational change in the membrane
protein. The latter interpretation is consistent with recent
experiments using laser-induced dichroism where it has been
shown that the rotational motion of the enzyme in the plane
of the membrane also exhibits a discontinuity at $15^{\circ}C$ despite
the fact that the microviscosity of the membrane lipids shows
no such temperature-dependent break (Hoffman *et al.*, 1979).
In the same study it was shown that from $4-15^{\circ}C$, the activation energies for both enzymatic activity and protein
rotational motion are equal, which suggests that, in the
temperature range up to the discontinuity, these processes
are interdependent. This interdependence may well be mediated
by membrane fluidity.

Enzyme Activity in Reconstituted Systems

To gain further insight into the temperature-dependence of
protein movement and function, experiments on reconstituted
systems have proved invaluable. Ca^{2+}-ATPase can be relatively
easily isolated from the sarcoplasmic reticulum and reconstituted into vesicles consisting of defined phospholipids.
Thus with a dipalmitoyl phosphatidylcholine to protein molar
ratio of 60:1, the enzyme displays very little hydrolytic
activity at low temperature. However, at about $28-30^{\circ}C$ there
is a sudden increase in activity, which continues steadily
up to about $40^{\circ}C$ at which point the activity begins to
plateau (Hoffman *et al.*, 1980). The rotational motion of the
enzyme displays similar behaviour. Below $30^{\circ}C$ the enzyme is
completely immobile but at this temperature it exhibits a
sudden onset of rotational motion. Both phenomena, therefore,
occur at the same temperature, and, perhaps more importantly,
this temperature is about $10^{\circ}C$ below the transition temperature for dipalmitoyl phosphatidylcholine. Others workers
have observed the same onset in enzymatic activity below the
lipid transition temperature for these recombinants, and
they have interpreted the effect in terms of a melting of the
lipid annulus. It has been suggested that the melted annulus

allows the Ca^{2+}-ATPase to function until bulk membrane lipids undergo the characteristic transition from gel to liquid crystal at the normal transition temperature, which is 41°C in the case of dipalmitoyl phosphatidylcholine (Hesketh *et al.*, 1976; Metcalfe and Warren, 1977). From the arguments advanced above, the data for Ca^{2+}-ATPase in dipalmitoyl phosphatidylcholine are here interpreted as indicating that below 30°C there is lateral phase separation giving rise to regions of crystalline phospholipid and protein-rich aggregates containing small domains of trapped phospholipid (Plate 1). In such domains the spin labels used in ESR experiments will undergo restricted motion and appear to exist in a separate, immobile phospholipid component. At 30°C, there is a melting of the small phospholipid domains, which are more disordered than the bulk phospholipid region because of protein/phospholipid interactions (Oldfield *et al.*, 1978). The initial melting should lead to a disaggregation of the enzyme clusters and finally to a melting of the bulk membrane lipid. This has been confirmed by freeze-fracture electron microscopic studies of Ca^{2+}-ATPase in vesicles of dipalmitoyl phosphatidylcholine (Hoffmann *et al.*, 1980; Kleeman and McConnell, 1976) and dimyristoyl phosphatidylcholine (Gomez-Fernandez *et al.*, 1980). In the case of the former recombinants, X-ray diffraction data confirms that the phospholipid melting at 30°C is crystalline, hexagonally packed phospholipid, in no way distinct from the bulk membrane phospholipid (Hoffmann *et al.*, 1980).

CONCLUSION

The picture of a cell membrane we have tried to present in this review is that of a heterogeneous system of lipids and proteins which exist in dynamic equilibrium with each other. Hydrophobic interactions determine the basic bilayer arrangement and confer thermodynamic stability to the gross structure, and the lateral distribution of membrane components within the plane of the bilayer plays a crucial role in the normal functioning of the membrane. Lipid composition and environmental temperature determine the degree of fluidity within the membrane and this, in turn, affects both lateral and rotational diffusion of intrinsic membrane proteins. We have described how cell growth is inhibited below the membrane transition temperature and we have discussed the inhibition in terms of protein-lipid interactions.

ACKNOWLEDGEMENTS

We gratefully acknowledge the support of the Cancer Research Campaign during the preparation of this work, and we would also like to thank the Wellcome Trust.

ADAPTATION TO LOW TEMPERATURE

LOW TEMPERATURE ADAPTATION IN BACTERIA

R.A. Herbert

*Department of Biological Sciences, The University,
Dundee, Scotland.*

INTRODUCTION

Microbiologists have long appreciated that temperature affects
the activities of microorganisms. Temperature influences
microorganisms through its effect on solute solubility,
growth rate, nutritional requirements, enzyme activity, cell
composition, ion transport and diffusion. Since microorganisms
lack thermoregulatory mechanisms and can be grown readily
under laboratory conditions, they provide a convenient method for studying the effects of temperature on the living cell.
An additional advantage is that microorganisms can be grown
in continuous culture with relative ease. During growth of
microorganisms in batch culture several environmental factors
such as nutrient concentration, pH and dissolved oxygen
content may change, often unpredictably, and influence significantly the metabolic function of the cells, thereby
masking effects due solely to temperature (Hunter and Rose,
1972). In chemostat culture these variables can be maintained
at constant values, thus permitting a rigorous examination
of the effects of temperature on living cells.

Many natural environments are at low temperatures and in
addition to the polar regions which comprise some 14 percent
of the earth's surface, 90 to 95 percent by volume of ocean
waters are at temperatures of $5^{\circ}C$ or less (Morita, 1966).
Further, refrigerated storage is now a commonly used method
of food preservation and as a consequence the applied micro-
biologist has a considerable interest in the survival and
growth of spoilage microorganisms, including pathogens, at
low temperatures. Forster, in 1887, was the first to demonstrate the existence of bacteria capable of growing at low

temperatures when he reported that phosphorescent bacteria
isolated from fish preserved in ice grew well at 0°C. Subsequent studies have shown that microorganisms capable of
growth at low temperatures are common in soil (Lockhead,
1926), marine and freshwater (Herbert and Bhakoo, 1979;
Morita, 1966; Witter, 1961). Nor are they confined solely
to polar regions for bacteria capable of growth at low temperatures can be isolated in considerable numbers from a wide
variety of temperature environments (Sieburth, 1967; Stokes
and Redmond, 1966).

Traditionally, microorganisms have been classified into
thermophiles, mesophiles and psychrophiles according to their
ability to grow at high, medium and low temperatures, respectively (Simon, this volume). The first two groups can be
readily distinguished by their growth temperature optima.
Microorganisms which can grow at low temperatures (0 to $+5^{\circ}$C)
are more difficult to define precisely and it is only during
the last decade that some agreement has been achieved
(Ingraham and Stokes 1959; Morita, 1966, 1975). For the
purpose of this paper psychrophilic bacteria are defined as
those which have an optimal growth temperature < 16°C, a
maximal growth temperature of approximately 20°C and a minimum
growth temperature < 1°C (Morita, 1975). Microorganisms which
will grow at 0 to $+5^{\circ}$C and at maximum temperatures exceeding
25°C are considered to be psychrotrophic. The majority of
the isolates obtained by the early investigators were psychrotrophic and the occurrence of psychrophilic microorganisms
was not unequivocally proven until 1964 when Morita and Haight
isolated *Vibrio marinus* MP-1. Subsequently, psychrophilic
bacteria have been isolated from a range of natural habitats
(Harder and Veldkamp, 1967; Herbert and Bhakoo, 1979; Moiroud
and Gounod, 1969; Stanley and Rose, 1967). The majority of
psychrophilic bacteria that have been isolated are Gram
negative, aerobic, asporogenous, rod shaped bacteria which
have usually been assigned to the genera *Pseudomonas,*
Flavobacterium, Alcaligenes, Vibrio and *Achromobacter* (Herbert
and Bhakoo, 1979; Morita, 1966; Witter, 1961). Gram positive
psychrophilic bacteria have been reported, although much less
frequently, and include representatives of the genera
Micrococcus (McLean *et al.,* 1951), *Microbacterium* (Brownlie,
1966), *Brevibacterium* (Baker and Smith, 1972) and *Bacillus*
(Shehata and Collins, 1971). There have been few reports of
psychrophilic anaerobes, although psychrophilic *Clostridium*
spp. have been isolated from marine sediments (Finnes and
Matches 1974; Liston *et al.,* 1969).

Bacteria which grow at low temperatures thus form two
distinct groups, those that are specifically adapted to growth

at low temperatures (psychrophiles) and those that are merely
cold tolerant (psychrotrophs). Comparative studies of the
effects of temperature on the physiology and composition of
psychrophilic and psychrotrophic bacteria have greatly contributed to our understanding of microbial growth at low
temperatures.

Significance of Psychrophilic Bacteria in Natural Environments

Although bacteria capable of growing at low temperatures have
been isolated from a variety of natural environments few
studies have been made to assess their ecological significance
in aquatic or terrestrial environments. Sieburth (1967) made
a detailed seasonal study of the effect of water temperature
on the heterotrophic bacterial flora of Narragansett Bay,
Rhode Island. In winter, when water temperatures reach a
minimum of $-2^{\circ}C$ psychrophilic bacteria become the principal
component of the microflora, whereas during the summer,
when water temperatures reach $+23^{\circ}C$, psychrotrophic and
mesophilic types become dominant. Laboratory studies on
selected isolates obtained during the year confirmed these
field data. Furthermore, there was no evidence for the
enhancement or suppression of any single taxonomic group as
a result of temperature. Similar findings have been reported
by Tajima *et al*. (1974) for offshore coastal waters of
Hakodota Bay. The results from both these studies indicate
that significant populations of psychrophilic bacteria develop
at particular times of the year in temperate waters. However,
these data do not provide evidence of either actual or
potential microbial activity *in situ*.

Harder and Veldkamp (1971) demonstrated with a series of
elegant competition experiments at several temperatures using
mixed populations of psychrophilic and psychrotrophic bacteria
that at all dilution rates the psychrophile outgrew the
psychrotroph at $-2^{\circ}C$, whilst at $16^{\circ}C$ the reverse situation
occurred. At intermediate temperatures (4 and $10^{\circ}C$) only
the dilution rate influenced the outcome of the competition.
This laboratory study confirms that low temperatures apply a
selective pressure and the bacteria which predominate under
these conditions have growth characteristics closely fitting
the environmental conditions. It is therefore reasonable to
assume that in permanently cold environments such as ocean
waters psychrophilic bacteria are primarily responsible for
nutrient cycling. This has been demonstrated in coastal
waters surrounding Signy Island in the maritime Antarctic
with a mean annual temperature of $-1^{\circ}C$ where nutrient regen-
eration due to the activities of psychrophilic bacteria is

rapid following the collapse of the annual phytoplankton
bloom (Tanner and Herbert, 1981).

EFFECT OF TEMPERATURE ON SUBSTRATE UPTAKE

The hypothesis that substrate uptake may play a central role
in governing the ability of a microorganism to grow at low
temperatures has gained much credence in recent years. The
first experimental evidence for this came from a comparison
of the effects of temperature on the respiratory activity of
a psychrotrophic *Candida* sp. and a mesophilic strain of
C. lipolytica (Baxter and Gibbons, 1962). The psychrotroph
oxidised exogenous glucose at an appreciable rate at $0^{\circ}C$
whilst virtually no exogenous substrate was metabolised by
the mesophile at temperatures $< 5^{\circ}C$. Endogenous substrates
were, however, metabolised by *C. lipolytica* at near-zero
temperature. It was therefore concluded that the psychrotroph,
but not the mesophile was able to transport sugars into the
cell at temperatures $< 5^{\circ}C$. Similar results were obtained
using mesophilic strains of *C. utilis* and *Arthrobacter* sp.
(Rose and Evison, 1965). However, even in psychrotrophic
bacteria substrate uptake can be significantly reduced at
low temperatures (Bhakoo and Herbert, 1980; Paul and Morita,
1971).
     In contrast, in a psychrophilic *Candida* sp. sugar transport
was found to be largely independent of temperature and subsequent work confirmed these findings for a number of Gram
positive and negative bacteria (Wilkins, 1973). Russell
(1971) demonstrated in *Micrococcus cryophilus* that lysine
uptake occurred at the same rate when the cells were grown
at $0^{\circ}C$ as at $20^{\circ}C$, whilst in *Vibrio* AF-1 maximum uptake of
^{14}C-glucose and lactose occurred at $0^{\circ}C$ and decreased with
increasing temperature (Herbert and Bell, 1977). If
psychrophiles and psychrotrophs differ from mesophiles in
being able to transport solutes into the cell at near-zero
temperatures, the problem is to explain how transport processes differ in these groups of microorganisms. The ways in
which solute uptake could be inactivated at low temperatures
in mesophiles have been proposed by Farrell and Rose (1967),
namely:

1) Solute carrier molecules in the cytoplasmic membranes of
   psychrophiles and psychrotrophs were less susceptible to
   low temperature inactivation than their mesophilic
   counterparts.
2) Solute carrier molecules in mesophilic microorganisms were
   not abnormally cold-sensitive but that due to changes in
   the molecular configuration of the membrane, solute

molecules were unable to combine with their respective
carrier proteins.
3) At low temperatures there was a shortage of energy to
   support active transport in mesophiles.

There is no convincing evidence to indicate that individual
carrier proteins are cold-labile, nor has the lack of ATP for
active solute uptake been demonstrated. Indeed, endogenous
respiration can proceed at temperatures below which growth of
the microorganisms ceases (Baxter and Gibbons, 1962). As a
consequence, most studies have concentrated on the manner in
which low temperatures affect the cytoplasmic membranes.

EFFECT OF TEMPERATURE ON MEMBRANE COMPOSITION

It has been known for many years that the fatty acid composition of many microorganisms changes in response to temperature (Bhakoo and Herbert, 1979; Cullen *et al.*, 1971;
Kates and Baxter, 1962; Kates and Hagen, 1964; Marr and
Ingraham, 1962). Considerable interest has been shown to
those changes which may result in altered membrane function
(Pringle and Chapman, this volume).

Membrane Lipid Composition

As with higher organisms polar lipids in bacteria are built
around a *sn*-glycerol-3-phosphate skeleton (see section on
Nomenclature). Phosphatidylethanolamine is distributed
widely amongst bacteria and in Gram negative species it is
frequently the major phospholipid. Both phosphatidylglycerol
and its derivative diphosphatidylglycerol (cardiolipin) are
commonly found in bacteria, but being difficult to separate,
are frequently expressed as the sum of the two components.
Phosphatidylserine is usually only found in trace quantities
since in microbial cells it is rapidly decarboxylated to
yield phosphatidylethanolamine. Phosphatidylcholine and
phosphatidylinositol, although they are both important components in eukaryotes (Goldfine, 1972), are rarely found in
bacteria.
   Bacteria rarely contain sterols, or if present the quantities synthesised are much smaller than those found in the
tissues of higher organisms. The absence, or the low levels
of sterols in prokaryote cells have proved extremely useful
in studying the effects of temperature on liquid-crystalline
phase transitions in cell membranes.
   The fatty acids of bacteria are generally 10-20 carbon
atoms in length, with C_{15} to C_{18} acids predominating.
Bacterial fatty acids may be either straight chain saturated,

straight chain unsaturated, branched chain (*iso* or *anteiso*)
or cyclopropane acids. The long chain polyunsaturated fatty
acids commonly found in eukaryotes are normally absent in
bacteria, as are C_{18} di- and tri- unsaturated acids. Conversely
cyclopropane acids are, with the exception of higher plants,
rarely found in eukaryotes. Cyclopropane fatty acids are
widely distributed amongst Gram positive and negative bacteria
whilst *iso* and *anteiso* branched acids predominate amongst
Bacillus and *Micrococcus* spp.

Effects of Temperature on Fatty Acid Composition

Numerous studies have been made of changes in fatty acid
composition with temperature in psychrophilic, psychrotrophic
and mesophilic microorganisms. The interpretation of certain
of these data is difficult since the microorganisms were
grown in batch culture and, as previously described, changes
in growth conditions can greatly influence fatty acid
composition.

Comparative studies of mesophilic and psychrophilic yeasts
have shown that the psychrophiles are generally endowed with
a higher proportion of unsaturated fatty acids, in particular
hexadecenoic (16:1) and octadecenoic (18:1) acids, than the
mesophiles (Brown and Rose, 1969; Kates and Baxter, 1962).
In bacteria the effects of low temperature on fatty acid
composition are more complex (Table 1). Mesophilic bacteria,
such as *Escherichia coli* ML 30 respond to decreased temperatures by synthesising increased proportions of the unsaturated fatty acids, 16:1 and 18:1 at the expense of the
corresponding saturated acids (Marr and Ingraham, 1962).
Sinensky (1974) proposed that these alterations in fatty acid
composition enabled *E. coli* to maintain a constant fluidity
of its membrane lipids, a process he termed homeoviscous
adaptation. Similar alterations have been reported by Gill
(1975) for the psychrotroph *Pseudomonas fluorescens*. Fatty
acid profiles of the psychrophile *Vibrio* AM-1 show the predominance of C_{14}, C_{15} and C_{16} saturated and mono-unsaturated
components notably 15:1 and 16:1. Significant quantities of
heptadecenoic acid (17:1) were synthesised only at 0 and 8^{0}C.
The response to growth at low temperatures in *Vibrio* BM-2 was
to synthesise increased proportion of the shorter chain fatty
acids 9:1 and 12:1. The levels of 16:1 and 17:1 acids in this
bacterium were relatively low at all temperatures compared
with the other psychrophilic vibrios studied (Bhakoo and
Herbert, 1979). Reductions in fatty acid chain length occur
in *Micrococcus cryophilus* when grown at low temperature and
again there was no change in the degree of unsaturation with

Table 1

Effect of growth temperature on the fatty acid composition (percent total fatty acids) of some Gram-negative bacteria

Fatty acid	Escherichia coli[1] growth temperature			Pseudomonas fluorescens[2] growth temperature			Pseudomonas T-6[3] growth temperature			Vibrio AM-1[4] growth temperature			Vibrio BM-2[4] growth temperature		
	10°C	20°C	30°C	10°C	20°C	30°C	0°C	8°C	20°C	0°C	8°C	15°C	0°C	8°C	15°C
9:1	–	–	–	–	–	–	–	–	–	–	–	–	11.4	6.6	5.4
12:0	–	–	–	–	–	–	1.5	2.3	2.1	–	–	0.8	0	6.4	9.0
12:1	–	–	–	–	–	–	–	2.6	–	8.2	10.7	1.0	13.7	1.0	–
13:0	–	–	–	–	–	–	2.2	2.6	–	–	–	–	–	–	–
13:1	–	–	–	–	–	–	4.4	8.1	9.9	–	–	–	–	–	–
14:0	3.9	4.1	3.8	–	tr	tr	5.1	6.1	9.9	1.2	1.7	21.8	–	1.0	1.2
14:1	–	–	–	–	–	–	4.3	1.7	–	5.5	–	–	20.6	7.5	12.0
2-OH14:0	12.6	10.4	10.1	–	–	–	–	–	–	–	–	–	–	–	–
15:0	–	–	–	–	–	–	–	–	–	–	–	–	0.7	0.9	1.1
15:1	–	–	–	–	–	–	1.1	1.2	–	8.2	1.0	18.2	1.5	2.5	2.9
16:0	18.2	25.4	28.9	20.4	29.5	34.4	1.1	1.6	1.4	5.1	8.2	17.8	21.6	46.5	38.5
16 cyc	1.3	1.5	3.4	–	–	–	–	–	–	–	–	–	–	–	–
16:1	26.0	24.4	23.3	52.7	46.2	30.9	56.6	54.6	59.0	38.6	39.6	–	20.3	17.7	22.0
17:0	–	–	–	–	–	–	–	–	–	8.6	11.1	13.5	0.4	0.4	0.4
17 cyc	–	–	–	–	5.7	18.1	–	–	–	–	–	–	–	–	–
17:1	–	–	–	–	–	–	7.1	5.3	2.4	20.8	21.9	–	1.3	1.4	1.2
18:0	–	–	–	–	–	–	0.4	1.0	1.3	–	–	–	–	–	–
18:1	37.9	34.2	30.3	26.9	18.6	16.2	15.9	15.1	14.2	–	–	–	–	–	–

[1] from Marr and Ingraham, 1962; [2]Gill, 1975; [3]Bhakoo and Herbert, 1979; [4]Bhakoo and Herbert, 1980.

decreasing temperature (Russell, 1971).

Not all bacteria, however, alter their fatty acid composition in response to a decrease in growth temperature. No significant differences were found in the fatty acid composition of *Staphylococcus aureus* when grown at 25 and 37°C (Joyce *et al.*, 1970), or in either the fatty acid or phospholipid composition of 4 psychrotrophic marine pseudomonads grown at temperatures ranging from 2 to 20°C (Brown and Minnikin, 1973). In these bacteria, 16:1 comprised some 67 to 74 percent of the fatty acids in the cell membrane. These data, together with those obtained by Shaw and Ingraham (1965), show that there is no absolute prerequisite for changes in fatty acid composition to enable bacteria to grow at low temperatures.

Studies using fatty acid auxotrophs have shown in *E. coli* that there are minimum requirements for both saturated and unsaturated fatty acids. The minimum requirement for unsaturated fatty acids in *E. coli* is about 20 percent total fatty acids at 37°C and this increases with decreasing temperature (Cronan and Gelman, 1973) if the total unsaturated fatty acid content falls below this minimum requirement growth ceases. In *E. coli*, the psychrotrophic pseudomonads and the psychrophilic vibrios the levels of unsaturated fatty acids are considerably in excess of those required for growth (at least in *E. coli*). There would thus appear to be no obvious requirement for changes in fatty acid composition with temperature.

Effects of Temperature on Phospholipid Composition

Data on the effects of temperature on phospholipid composition in bacteria are also equivocal. In 4 psychrophilic *Vibrio* spp. the principal effect of decreasing the temperature was to increase significantly the total quantity of phospholipids synthesised (Bhakoo and Herbert, 1979). In *Vibrio* BM-4 there is a 42.9 percent increase in total phospholipids synthesised at 0°C (Table 2) and this correlates with the increase in cell size observed at this temperature (Herbert, unpublished data). There is also some evidence that in this bacterium phosphatidylethanolamine decarboxylase and diphosphatidyl-glycerol synthetase are temperature sensitive, since at 15°C (the isolate's maximum growth temperature) there is a marked decline in both phosphatidylethanolamine and diphosphatidyl-glycerol and an accumulation of their immediate precursors. In contrast, the psychrotroph *Pseudomonas* T-6 shows an almost constant phospholipid composition over the temperature range 0 to 20°C and there was no observed increase in total

Table 2

Effect of temperature on the phospholipid composition (μg phospholipid/mg dry weight) of the psychrophile Vibrio BM-4 and the psychrotroph Pseudomonas T-6 grown in continuous culture under glucose limitation

Phospholipid	Vibrio BM-4[1]			Pseudomonas T-6[2]		
	0°C	8°C	15°C	0°C	8°C	20°C
Phosphatidylserine	4.5	7.0	10.6	7.2	6.2	5.8
Phosphatidylglycerol	53.3	22.0	55.9	18.5	20.6	22.4
Phosphatidylethanolamine	25.6	21.2	13.3	112.2	114.2	94.2
Diphosphatidylglycerol	86.8	80.0	38.7	39.6	38.6	32.6
Total phospholipids	169.4	131.0	118.5	177.5	179.6	155.2

[1] data from Bhakoo and Herbert (1979)

[2] data from Bhakoo and Herbert (1980)

phospholipids synthesised at low temperatures (Table 2).
Similar results have also been reported for *Pseudomonas
fluorescens* (Cullen *et al.*, 1971; Gill, 1975) and 4 marine
psychrotrophic pseudomonads (Brown and Minnikin, 1973).

COLD SHOCK INJURY

In certain species of bacteria, cells at the exponential phase
of growth transferred rapidly to a low temperature show a loss
of viability within minutes. This is termed cold-shock
(Simon, this volume) and the phenomenon is analogous to the
thermal-shock of mammalian spermatozoa (Watson, this volume).
P. aeruginosa, a mesophile, is extremely sensitive to cold-
shock when grown at 30°C, whereas when grown at 10°C the
proportion of unsaturated fatty acids in the cell is increased
and the cells are no longer susceptible (Farrell and Rose,
1968).   Furthermore, a psychrotrophic *Pseudomonas* sp., which
had high unsaturated fatty acid levels at 30°C, was less
susceptible to cold-shock than the mesophile grown at this
temperature. These data provide evidence that an increase
in the degree of unsaturation of the membrane lipids makes
the cells less susceptible to cold-shock.

EFFECT OF TEMPERATURE ON THE PHYSICAL STATE OF THE BACTERIAL
MEMBRANE

The liquid-crystalline state of the membrane lipids is
essential for membrane function, including solute transport
(Overath *et al.*, 1970), the assembly of transport proteins
(Tsukagoshi and Fox, 1973) and the activity of membrane-
bound enzymes (Kimelberg and Papahadjopoulos, 1972). As
prokaryote membranes do not contain appreciable quantities
of sterols they make a useful model system to study the effect
of temperature on the physical state of the membrane.
Acholeplasma (previously *Mycoplasma*) *laidlawii* B lacks a cell
wall and has no internal membrane systems and so is an ideal
model organism to investigate prokaryote membrane structure
and function (McElhaney, 1974). Also, *A. laidlawii* B can
synthesise only saturated fatty acids with chain lengths
ranging from C_{12} to C_{18} (Pollock and Tourtellotte, 1967).
Thus by supplying unsaturated fatty acids in the growth medium
the fatty acid composition of the membrane lipids can be
altered in a controlled and predictable manner (McElhaney and
Tourtellotte, 1969).
    The effects of phospholipid head-group composition and
phospholipid fatty composition on the thermotropic behaviour
of membranes is described elsewhere (Cossins; Pringle and

Chapman, this volume). Studies with model systems and
prokaryotes such as *Acholeplasma* have shown that the principal
effect of sterol in a phospholipid bilayer is to smooth out
the phase transition. In *Acholesplasma* there is a discrete
phase transition whilst in eukaryotic membranes there is a
broad phase separation which occurs over a wide range of
temperatures.

One feature of prokaryotic lipids is the high incidence,
in some species, of branched chain fatty acids. The introduction of a methyl group at the *iso* position has little
effect on the melting point of that fatty acid compared with
its straight chain counterpoint, but the addition of a methyl
group at the *anteiso* position results in a fatty acid with a
significantly decreased melting point (McElhaney, 1976).
Anteiso methylene groups thus appear to have a fluidising
effect similar to that of introducing a *cis* double bond

Studies with a variety of physical techniques have shown
that the cytoplasmic membrane of prokaryotes must be in a
region above the phase transition temperature for correct
membrane function (and hence solute transport and growth).
An apparent exception to this is the outer membrane of
certain Gram-negative bacteria such as *E. coli* where ESR data
suggests that this membrane is in a state of lateral phase
separation at the growth temperature (see below). However,
the difficulties of interpretation of spectra from mixed
protein-phospholipid systems must be considered (Pringle and
Chapman, this volume).

*Effect of Temperature on the Membrane Lipids in Gram-Negative
Bacteria*

Whilst the effects of temperature on the structure of lipid
bilayers have been much studied, the changes that occur *in
vivo* in prokaryote cells, and in particular Gram-negative
bacteria are still only imperfectly understood.

The structure of the Gram negative bacterial cell envelope
is complex due to the presence of an outer membrane as well
as the cytoplasmic membrane between which is sandwiched a
peptidoglycan matrix (Costerton *et al.*, 1974). The outer
membrane differs from the cytoplasmic membrane in that it
contains substantial amounts of lipopolysaccharide in addition
to protein and phospholipid (Muhlradt and Golecki, 1975).
The temperature range over which *E. coli* K12 can maintain the
outer membrane phospholipids in a mixed gel/liquid crystalline
state correlates well with the range of growth temperatures
(Janoff *et al.*, 1979). The current view is that the outer
membrane of Gram negative bacteria functions as a molecular

sieve, allowing the entry of small molecular weight hydrophilic
molecules (ions, sugars, amino acids, *etc.*) but not larger
molecules. It has been proposed that this selectivity is
caused by the spatial organisation of the outer membrane
proteins to form pores or transmembrane channels (Nakae, 1976).
The maintenance of this spatial organisation may thus require
a state of lateral phase separation in the outer membrane
and a bulk phase transition may restrict transport processes
and hence growth. Also it has been suggested that a temperature induced decrease in the fluidity of the outer membrane
lipids will affect the processing and assembly of outer membrane proteins (Di Rienzo and Inouye, 1979). Clearly, there
is a need to determine whether the changes shown to occur in
the outer membrane lipids in *E. coli* and *Salmonella typhimurium*
(Osborne *et al.*, 1972) also occur in the Gram-negative
psychrophilic and psychrotrophic bacteria isolated from
permanently cold habitats.

Modification of Fatty Acid Composition in Bacteria

The processes by which bacteria vary their fatty acid composition in response to temperature have been studied in
E. coli (Sinensky, 1971), *Bacillus megaterium* (Fulco, 1970),
and *Micrococcus cryophilus* (Sandercock and Russell, 1980).
     In *E. coli* the ratio of 18:1 to 18:0 incorporated into
membrane lipids increased with decreasing growth temperature
(Sinensky, 1971). The ratio of saturated to unsaturated
fatty acid synthesis in *E. coli* is partially controlled by
the activity of the enzyme β-hydroxydeconoyl thioester
dehydrase (Cronan, 1975). There is also evidence to suggest
that β-ketoacyl ACP synthetases I and II play a role in this
process (Cronan, 1978). It would thus appear that in *E. coli*
temperature regulates the specificity of enzyme activity
rather than absolute enzyme levels.
     In contrast, the desaturation enzymes in *B. megaterium*
appear to be induced by a decrease in temperature and repressed
at elevated temperatures (Fulco, 1970, 1974). *B. megaterium*
does not synthesise unsaturated fatty acids at 30°C but does
desaturate 16:0 at 20°C. The necessary desaturating enzyme(s)
were absent from cells grown at 30°C, but were rapidly induced
at 20°C; addition of chloramphenicol to cell suspensions
showed that this enzyme synthesis was *de novo* (Fulco, 1970).
The desaturating enzyme(s) induced at low temperatures were
rapidly and irreversibly inactivated *in vivo* at 30°C.
     In addition to changes in degree of unsaturation the chain
length of the fatty acid acyl moieties of the phospholipids
are important in determining the thermotropic behaviour of
the membrane. In *M. cryophilus* phospholipid acyl chain length

may be controlled by a membrane bound elongase enzyme, which
interconverts C_{16} and C_{18} fatty acids *via* a C_{14} intermediate
(Sandercock and Russell 1980). The activity of this elongase
enzyme may be regulated by membrane lipid fluidity. Whilst
the volume of data is as yet small it is probable that the
direct effects of temperature on unsaturated fatty acid biosynthesis, as reported for *E. coli*, is the more common means
of altering the membrane lipid composition (Cronan, 1975).

SUMMARY

The data presented in this review show that in certain
instances the lowest temperature at which a microorganism can
grow is related to the physical state of its membrane lipids.
In bacteria such as *E. coli*, *Pseudomonas fluorescens*, *Vibrio*
AF-1 and *Vibrio* AM-1 the mechanism may be mediated by the
production of increased proportions of unsaturated fatty acids
as the growth temperature is lowered. Alternatively, the
average length of the acyl side chain moieties of the membrane
phospholipids is reduced, *e.g. Vibrio* BM-2 and *M. cryophilus*.
These temperature induced changes in fatty acid composition
are assumed to allow the bacteria to control the fluidity of
their membrane lipids at any prescribed temperature and has
been termed 'homeoviscous adaptation' (Sinensky, 1974).
However unequivocal evidence that such adaptation of fluidity
occurs is scanty (Cossins, this volume), and numerous examples
exist where the degree of unsaturation of the fatty acids does
not change with temperature (Bhakoo and Herbert, 1980; Brown
and Minnikin, 1973). There thus does not seem to be any
absolute requirement for changes in fatty acid composition to
occur with decreasing growth temperature, especially when
high levels of unsaturated fatty acids are already present.
Indeed, it is worth reiterating the statement made by
McElhaney (1976) that merely altering the physical state of
the membrane lipids of a bacterium such as *E. coli* will not
convert it into a psychrophile. It may well be that at
temperatures approaching the minimum for growth one or more
key proteins associated with the membrane lipids become cold-
denatured. At present little is known regarding the interactions between the proteins and lipids within membranes,
although ATPase activity in *A. laidlawii* B is closely
correlated with the physical state of the membrane lipids
(De Kruyff *et al.*, 1973). Studies such as these underline
the need to investigate the effect of membrane thermotropic
behaviour on the activity of membrane associated proteins at
low temperatures, hopefully leading eventually to a synthesis
of data on solute transport phenomena, membrane structure and
growth characteristics in bacteria.

EFFECTS OF TEMPERATURE ON THE LIPID COMPOSITION OF TETRAHYMENA

A. Clarke

British Antarctic Survey, Cambridge, England

INTRODUCTION

The way in which organisms respond at the molecular level to
alterations in environmental temperature is a key problem in
biochemistry. This area has received a great deal of attention in recent years and research has relied upon the use of a
small number of easily cultured and experimentally resilient
organisms. Particularly favoured have been *Escherichia coli*
and the goldfish (*Carassius auratus*). Occupying an intermediate position in terms of cellular complexity and lipid
composition is the ciliate protozoan *Tetrahymena pyriformis*;
the relation of the lipid composition of *Tetrahymena* to
temperature forms the subject of this review.
    Concentration of research on a small number of organisms
is useful in terms of depth of treatment, but may blind an
investigator to how typical or not those organisms may be.
In common with other model organisms, *Tetrahymena* has its
advantages and disadvantages.
    Tetrahymena is a ciliate protozoan, 40-60 µm in length
whose natural habitat is fresh water, particularly where decomposition has resulted in a rich bacterial flora. *Tetrahymena* is eurythermal and tolerant of a wide range of environmental conditions (Table 1). Different strains of *T. pyriformis*
have been used by various laboratories (Table 2) and a related
thermotolerant species, *T. thermophila*, has also been investigated. *Tetrahymena pyriformis* is, however, a morphotypic
species consisting of multiple genetic species, all very similar in size, architecture and karyotype (Nanney, 1980). It
must therefore be borne in mind when comparing results from
different strains of *T. pyriformis* that these may be as divergent

Table 1

Reported extremes of environmental variables which will support growth of Tetrahymena pyriformis (data from Bick, 1972).

Temperature	1 to $40^{\circ}C$
pH	7.0 to 8.9
dissolved O_2	0 to 10 mg L^{-1} (this represents virtually 0 to 100 percent saturation)
free CO_2	0 to 200 mg L^{-1}
NH_4^+	0 to 250 mg L^{-1}
free NH_3	0 to 25 mg L^{-1}
H_2S	0 to 2 mg L^{-1}

genetically as different species of other organisms. A further complication is that genetic drift may occur during long-term maintenance of certain strains by continual sub-culturing (Borden *et al.*, 1973).

Tetrahymena has many of the advantages of bacteria for experimental work on temperature acclimation. It is easily cultured in lipid-free media, it can be maintained axenically, and the biochemical effects of an alteration in temperature appear swiftly. Unlike prokaryotes, most protozoan membranes contain sterol, which renders experimental results of more relevance to membranes of higher organisms (Herbert, this volume). In *T. pyriformis*, however, there are no sterols; instead membranes contain the pentacyclic triterpenoid tetra-hymenol, although it is likely that the effects of this compound on membrane structure and function resemble those of sterols.

A disadvantage of *Tetrahymena* is that the lipid composition varies between the various cellular membranes. A bulk extraction thus contains lipid from many different membrane systems, not all of which will react in the same way, or at the same rate, to a change in temperature.

A final, subtle, disadvantage of *Tetrahymena* for temperature adaptation research is one common to all eurythermal organisms. That is in being eurythermal, their mechanisms of temperature acclimation may give only a limited insight into mechanisms of evolutionary temperature adaptation. A complete picture of low temperature adaptation can only emerge by also studying adaptations in more stenothermal organisms, and how these

Table 2

Strains of Tetrahymena pyriformis used in temperature acclimation experiments

T. pyriformis	WH-14		Conner and Stewart, 1976
			Nozawa et al., 1974
T. pyriformis	NT-1	a thermotolerant strain capable of growth at 40°C	Fukushima et al., 1976
			Kasai et al., 1976, 1977
			Kitajima and Thompson, 1977a,b
			Martin et al., 1976
			Nandini-Kishore et al., 1979
T. pyriformis	GL	an amicronucleate strain	Nägel and Wunderlich, 1977
			Speth and Wunderlich, 1973
			Wunderlich and Ronai, 1975
			Wunderlich et al., 1973, 1974, 1975
T. pyriformis	1630/1W		Morris et al., 1981
T. pyriformis	W		Berger et al., 1972
			Schick et al., 1979
			Shipiro et al., 1978
T. pyriformis	E		Thompson et al., 1972

break down under thermal stress. The subject of the failure
of adaptation to low temperature (*i.e.* cold shock and chilling
injury) forms a major section of this volume.

THE LIPID COMPOSITION OF TETRAHYMENA

As is the case with most prokaryotes and also other unicellular
eukaryotes, there is no such thing as a definitive lipid
composition for *Tetrahymena*.
     The various cellular membranes will all react differently
to stimuli such as pH, ionic strength, osmolarity, exogenous
compounds and temperature, and so the lipid composition of
Tetrahymena isolated from different habitats (or cultured in
different laboratories) may vary greatly. Nevertheless, it
is possible to make some general comments.

Neutral Lipids

Neutral lipids in *Tetrahymena* can vary greatly in amount and
composition, depending on the nutrient status on the cells
(Everhart and Ronkin, 1966; Table 3). Thus, total lipid
extracts of *Tetrahymena* at different stages of the growth
cycle in batch culture will contain different proportions of
neutral lipids. Although much of the increase in lipid content during logarithmic phase growth is due to storage of
triacylglycerol, there is also an increase in the amount of
lipid phosphorus (Table 3). As nutrients become limiting,
there is a decrease in both neutral and polar lipids, due
primarily to a decrease in cell size although there is also
an increase in the tetrahymenol/polar lipid ratio in the
various membranes (Nozawa and Thompson, 1979).
     Neutral lipid classes which have been reported in *Tetra-*
hymena include hydrocarbons (including squalene), wax ester,
terpenes, ubiquinone, free fatty acid, free fatty alcohol,
triacylglycerol and partial glycerides (Hill, 1972; Holz and
Conner, 1973; Nozawa and Thompson, 1979).

Tetrahymenol

This pentacyclic triterpenoid has a sterol-like structure
(see introductory section on Nomenclature) and appears
functionally to replace sterol in the membrane of *T. pyriformis*.
Tetrahymenol has not been found in other animals, though it
has been reported from at least one higher plant (Zander *et*
al., 1969). *T. pyriformis* also contains small amounts of
diplopterol, an analogue of tetrahymenol (Nozawa and Thompson,
1979). Tetrahymenol appears to be an essential component of

Table 3

Lipid content and lipid composition of batch cultured Tetrahymena pyriformis 1630/1W. Cells were grown at 20°C in 5L of proteose peptone + yeast extract (PY) medium (Morris et al., 1981). Lipid composition was determined by TLC and scanning densitometry, unsaturation by capillary gas chromatography (Morris et al., 1979).

Age of culture (days)	Number of cells ml^{-1} x 10^{-5}	Total lipid $pg\ cell^{-1}$	Phospholipid $pg\ cell^{-1}$	Triacylglycerol $pg\ cell^{-1}$	Lipid unsaturation $db\ mol^{-1}$
Cultured in x 0.2 strength medium.					
2	0.18	12.59	4.11	2.26	1.11
3	4.6	16.04	2.56	3.60	1.11
4	3.5	33.89	4.74	5.83	-
6	3.25	51.63	6.93	12.27	1.20
8	4.5	23.07	2.62	9.80	1.24
10	4.75	1.20	0.15	0.20	1.40
14	< 0.10				
Cultured in x 2.0 strength medium.					
3	0.17	-	-	-	-
5	0.46	35.02	6.48	12.94	1.18
7	3.25	22.55	4.24	7.63	1.38
9	4.87	26.41	4.31	8.02	1.29
10	5.87	23.56	3.03	7.09	1.19
12	6.75	23.99	5.90	4.75	1.11
14	13.0	10.97	2.11	1.88	1.20

$db\ mol^{-1}$: mean number of double bonds per molecule of phospholipid fatty acid.

T. pyriformis membranes, but the proportion varies from membrane to membrane (Table 4) and also with stage of culture (Thompson *et al.*, 1972).

Although *T. pyriformis* does not require sterols, cells will readily absorb and metabolise exogenously supplied sterols. Many of these, including cholesterol, are converted to 7,22-*bis*dehydrocholesterol and this blocks tetrahymenol synthesis. *T. pyriformis* is unusual in having the ability to desaturate the sterol nucleus at the Δ^5, Δ^7 and Δ^{22} positions. Thus cholesterol (Δ^5-cholesta-1-en-3β-ol) is converted to cholesta-5,7-*trans*-22-trien-3β-ol. These pathways have been reviewed by Holz and Conner (1973).

Ergosterol is taken up by *T. pyriformis* and incorporated into membranes without modification, although there is a concomitant alteration in membrane fatty acid composition (Nozawa and Thompson, 1979). Some species of *Tetrahymena* appear to have a specific growth requirement for small amounts of sterol when grown on chemically defined media (Hill, 1972; Holz and Conner, 1973).

Phospholipids

Tetrahymena contains only a small number of dominant phospholipids, but a large number of minor components. These phospholipids contain a high proportion of alkyl linked fatty acyl chains and phosphonolipids, as well as traces of alk-1-enyl (vinyl ether or plasmalogen) phospholipids. *Tetrahymena* was one of the first organisms in which plasmalogens were identified (by staining techniques) but despite several contrary reports, it appears that alk-1-enyl linkages are only a minor component of *Tetrahymena* lipids.

Ether lipids can cause problems in the analysis and interpretation of fatty acid composition. Alk-1-enyl linkages are cleaved at low pH so acidic methods of producing fatty acid methyl esters (such as methanolic HCl) will also produce fatty aldehyde dimethyl acetals (DMAs). If these are not first removed they will decompose during gas chromatography on several of the more usual polar phases and produce interfering peaks (Mahadevan *et al.*, 1967). This problem can be circumvented by mild alkaline hydrolysis of phospholipids (Dawson *et al.*, 1962), or an intermediate clean-up procedure, but in *Tetrahymena* contamination of fatty acid methyl esters with aldehyde DMAs is not usually a problem. Alkyl linkages are stable to both acid and mild alkaline hydrolysis, and so when these are present in significant amounts, fatty acid analyses give only a very incomplete picture of acyl chain composition. In *Tetrahymena* about 30 percent of plasma mem-

Table 4

Tetrahymenol content and phospholipid composition (as percent total phospholipid P) of Tetrahymena pyriformis NT-1 grown at $24°C$ (Thomson and Nozawa, 1977).

	Whole cells	Cilia	Pellicles	Microsomes
Tetrahymenol/phospholipid molar ratio	0.078	0.260	0.090	0.055
Diphosphatidylglycerol (cardiolipin)	5.6	0.4	0.3	4.8
2-aminoethylphosphonolipid	25.4	37.0	32.9	20.8
Ethanolamine phosphoglyceride	34.4	21.4	36.2	35.5
Lysophosphatidylethanolamine, lyso-2-aminoethylphosphonolipid and ceramide-aminoethylphosphonate	5.4	14.0	4.6	3.7
Choline phosphoglyceride	26.2	18.1	21.4	27.6
Lysophosphatidylcholine	2.6	1.5	2.7	3.8

brane phospholipids and 50 percent of ciliary membrane phospholipids may contain alkyl linkages (Fukushima *et al.*, 1976;
Nozawa *et al.*, 1974), and thus functional interpretations based
on fatty acids alone may lead to error.

A variety of phospholipids have been reported in *Tetrahymena*,
though in most analyses the three major components choline
phosphoglyceride, ethanolamine phosphoglyceride and its phosphonolipid analogue 2-aminoethylphosphonolipid account for 76
to 90 percent of the total (Table 4). In smooth microsomal
membrane and ciliary membrane, the phosphonolipid analogue
of sphingomyelin, ceramide 2-aminoethylphosphonolipid, can
be as much as 14 percent of the total phospholipid (Wunderlich
and Ronai, 1975; Table 4) and 53 percent in cilia phospholipid
(Nozawa *et al.*, 1974). In addition to those phospholipids
listed in Table 4 there have been reports of small amounts of
phosphatidylserine (Erwin and Bloch, 1963), phosphatidylinositol and monomethyl-aminoethylphosphonolipid (Wunderlich and
Ronai, 1975). As with reported differences in proportion it
is difficult to decide whether these represent strain variations or differences in technique and interpretation. The
precise phospholipid composition varies from membrane to
membrane in *Tetrahymena*, as is commonly found in other organisms. Since the various head groups have differing ionic
charge and size, their packing behaviour will vary and the
response of each membrane to an alteration in temperature
will also differ.

The spectrum of phospholipid types observed in *Tetrahymena*
would not appear to be essential on a structural basis, for
a membrane could apparently function with only 2 or 3 types
of head group. The wide variety observed presumably reflects
as yet unknown functional requirements, a knowledge of which
is nonetheless essential for a complete understanding of
membrane adaptation to temperature.

Fatty Acid Composition

In comparison with vertebrates and many higher invertebrates,
the fatty acid composition of *Tetrahymena* is simple. Nevertheless, high resolution capillary gas chromatography does
reveal a substantial number of fatty acids (Table 5).

Reported fatty acid compositions vary from lab. to lab.,
reflecting differences in strain and culture conditions. In
particular, strain 1630/1W (Table 5) contains only small
amounts of 18:3ω6 in comparison with NT-1 (Thompson and
Nozawa, 1977), WH-14 (Nozawa *et al.*, 1974) and GL (Wunderlich
and Ronai, 1975). It is, however, possible to make a number
of generalisations. Chain length is predominantly C_{12} to C_{18},

Table 5

Total phospholipid fatty acid composition of Tetrahymena pyriformis 1630/1W. Cells were grown in PY medium at $20^\circ C$ and then transferred in bulk to either 15 or $30^\circ C$ (Morris et al., 1981). Phospholipids were separated as previously described (Morris et al., 1979) and fatty acid methyl esters analysed on a 25m SP1000 WCOT glass capillary column with 1 ml min^{-1} He carrier gas. Data are presented as percent total fatty acids in range C_{10} to C_{22}. Structures were assigned following precision isothermal chromatography with and without standards (where available).

Fatty acid	$20^\circ C$ day 3	$15^\circ C$ day 2 after transfer	$30^\circ C$ day 3 after transfer
< 12:0[a]	0.85	0.31	< 0.20
12:0	3.36	3.08	2.47
12:1	0.17	0.35	0.29
iso 14	0.15	< 0.01	< 0.01
14:0	7.36	8.71	8.17
14:1 (4 isomers combined)	1.41	3.01	2.98
15:0	0.40	0.15	0.13
15:1	0.19	0.06	0.29
iso 16	0.41	0.28	0.78
16:0	3.43	3.57	4.98
16:1 (4 isomers combined)	8.35	12.89	6.41
$16:2\omega4$	0.66	2.43	2.59
$16:2\omega7$	1.12	1.08	0.35
16:2 structure unknown	0.23	0.47	0.20
17:0	0.90	0.63	0.14
17:1 (2 isomers combined)	0.91	0.57	0.38
18:0	0.59	0.55	0.72
18:1 (3 isomers combined)	32.42	29.58	25.33
$18:2\omega7$	3.79	4.90	7.27
$18:2\omega6$	21.55	23.09	27.24
18:2 structure unknown	0.44	0.70	0.55
$18:3\omega6$	1.94	1.71	3.65
18:3[b]	0.82	1.20	1.78
db mol^{-1}	1.17	1.44	1.27
unsaturated/saturated ratio	4.46	4.83	4.46

a — total fatty acids of chain length < 12:0. b — total fatty acids of retention time $< 18:3\omega6$ (C_{20} to C_{22}). db mol^{-1}: mean number of double bonds per fatty acid molecule. In addition, 24 minor components detected at levels < 0.01% total; these will include branched chain and α-hydroxy fatty acids (Ferguson et al., 1972).

and unsaturation varies from 0 to 3 double bonds per fatty acid. Most reports show significant amounts of the unusual octadecadienoic acid 18:2ω7 (cilienic acid), although this is not always resolved from 18:2ω6. Even in comparison with algae and other organisms with fatty acids of similar chain length distribution, *T. pyriformis* has a markedly high unsaturated/saturated ratio (Table 5; Morris *et al.*, this volume). This high ratio is a feature of all strains of *T. pyriformis* (Nozawa *et al.*, 1974; Wunderlich *et al.*, 1973).

Each phospholipid class has a distinct fatty acid composition (Berger *et al.*, 1972; Fukushima *et al.*, 1976), and this may vary from membrane to membrane and also between strains (Nozawa and Thompson, 1971; Thompson *et al.*, 1971). Since different membranes may react differently in both phospholipid head group composition and acyl chain composition of each phospholipid, it is obvious that analyses of response to temperature in terms of total cell, or even total phospholipid, fatty acid composition alone is a very insensitive approach.

There are relatively few analyses of triacylglycerol fatty acids in *Tetrahymena*, although in *T. pyriformis* this fraction contains mostly saturated and monoenoic acids (Erwin and Bloch, 1963).

Variations in Lipid Composition with Batch Culture

It is well known that in batch culture of both prokaryote and eukaryote microorganisms, changes in stage of the cell cycle, cell density and medium composition during culture result in an alteration in the lipid composition of the cultured organisms (Herbert, this volume). Almost all work on *Tetrahymena* has involved batch culture, and this makes comparison of different sets of results difficult.

Data for *T. pyriformis* grown in batch culture are given in Table 3. Although when grown in dilute medium there is no net increase in cell number after day 3, the cells are not in a steady state in terms of lipid synthesis. The amount of lipid per cell at first increases and then decreases, and by day 10 the cells contain almost no lipid suggesting utilisation of reserves. A decrease in cell size is typical of cells grown in media depleted of nutrients (Cameron, 1967).

In double strength medium the cells show typical logarithmic growth, and during this period the lipid content of the ciliates is constant at about 24 pg cell^{-1}. This consists of approximately 4.4 pg phospholipid, 9 pg tetrahymenol and 8 pg triacylglycerol, although the amount of the latter two

components falls by day 12. By day 14 the total cell lipid
has dropped to 11 pg, suggesting that either the cells have
decreased in size or that some membranes have been catabolised.
It is clear that cells 'taken from logarithmically growing
cultures' or 'from stable cultures' are neither comparable,
nor necessarily stable.

A further problem is that samples from cultures at the
logarithmic phase of growth contain cilicates at all stages
of the cell cycle. Since it is known that chemical composition varies with the stage of the cell cycle (Holz and
Conner, 1973), only an average composition will be obtained
Tetrahymena cultures can be synchronised by heat shock but
relatively few studies have utilised such cultures.

Synthetic Pathways

Since fatty acid composition is perhaps the simplest parameter
of lipid composition to measure, alterations in fatty acids
in relation to temperature have been well studied.

T. pyriformis grows well in axenic lipid-free media and so
must be capable of the synthesis of all necessary fatty acids
de novo; it thus lacks the typical 'animal' requirement for
dietary 18:2 and 18:3 (Hill, 1972). Palmitic acid (16:0) is
the usual end product of *de novo* fatty acid synthesis, and
studies with $1-^{14}C$-palmitic acid (Erwin and Bloch, 1963) have
suggested the following desaturation pathway:

$$16:0 \rightarrow 18:0 \rightarrow 18:1\omega9 \rightarrow 18:2\omega6 \rightarrow 18:3\omega6$$

$18:3\omega6$ is γ-linolenic acid, a typical 'animal' fatty acid,
with the *cis* double bonds in the methylene-interrupted posit-
ions $\omega12$, $\omega9$ and $\omega6$ ($\Delta^{6,9,12}$). No $18:2\omega9$ was observed when
$1-^{14}C$-stearic acid (18:0) was supplied (Erwin and Bloch, 1963).
In higher animals $18:3\omega6$ is produced by Δ^6 desaturation of
dietary $18:2\omega6$, and then elongated and desaturated to produce
the polyunsaturated acids $20:4\omega6$, $20:5\omega3$ and $22:6\omega3$ characteristic of animal phospholipids (Gurr and James, 1971).

Like *Tetrahymena*, yeasts and green plants are capable of
synthesising *de novo* 18:3, linolenic acid. The structure of
this fatty acid is however different, the third double bond
being introduced into the $\omega3$ (Δ^{15}) position, to yield α-
linolenic acid, $18:3\omega3$. With the exception of yeasts and
fungi, $18:3\omega3$ (α-linolenic acid) is found in eukaryotes as
a component of chloroplast glycolipids and $18:3\omega6$ (γ-linol-
enic acid) as a component of mitochondrial phospholipids.
Monoenoic acid synthesis in bacteria such as *E. coli* is by
an entirely different, anaerobic, pathway.

Tetrahymena contains a variety of C_{16} and C_{18} di- and

trienoic fatty acids (Table 5), and several other synthetic
pathways have been proposed. These include:

$$16:0 \rightarrow 16:1\omega7 \rightarrow 18:1\omega7 \rightarrow 18:2\omega7 \ (\Delta^{6,11}, \text{ cilienic}$$
$$16:1\omega7 \rightarrow 16:2\omega7 \quad (\Delta^{6,9}) \qquad \text{acid})$$
$$16:1\omega7 \rightarrow 16:2\omega4 \quad (\Delta^{9,12})$$

(Conner and Koroly, 1972; Ferguson *et al.*, 1975). It has
been proposed that temperature affects fatty acid composition
by altering the relative activity of these various synthetic
pathways (Conner and Stewart, 1976).

A preliminary characterisation of the Δ^9 desaturase of
T. pyriformis W showed that this enzyme was localised in the
microsomal fraction (Shipiro *et al.*, 1978). It required molecular oxygen and a reduced pyridine nucleotide, the K_m for
NADPH, NADH and FAD being respectively 14.3, 4.5 and 1.4 μM.
The optimal temperature was 30°C and a shift of culture
temperature to 16°C did not affect the relative specificity
of the enzyme to 16:0-CoA and 18:0-CoA.

THE EFFECTS OF ALTERING THE GROWTH TEMPERATURE ON LIPID COMPOSITION

Methods of Altering Culture Temperature

Different workers have used various ways of altering the
growth temperature of *Tetrahymena* in order to examine the
changes in lipid and fatty acid composition. These differing
techniques are not necessarily comparable in their effect on
the cells, and it is not always clear from the published
reports exactly which method was used. The commoner methods
are:

1) Allowing the cells to grow to a certain stage (often early
   log phase, day 3), and then transferring the whole culture to a new ambient temperature.
2) Innoculating a small volume of cells into a large volume
   of fresh medium, and then transferring the bulk culture
   to a new temperature.
3) Innoculating a small volume of cells into a large volume
   of precooled medium.

When a whole culture is transferred to a new temperature,
the rate of change of temperature of the cells can be very
slow (2°C h^{-1}for a 5 litre culture transferred from 25 to
10°C; Clarke, unpublished observations). Often, but not
always, when decreasing culture temperature the rate of
change of temperature is increased by the use of an ice-water

bath (*e.g.* Nozawa and Kasai, 1978). A slow rate is comparable to the rate of acclimation of the cells (see below) and this effect can be complicated by the effects on cell growth rate of sudden dilution (Hill, 1972). Transfer of a small volume of logarithmic stage culture to cold culture medium will subject the cells to a cold-shock. If the drop in temperature is sufficiently great this cold shock may have severe effects on membrane structure and function (see below), albeit temporarily. This is obviously a very different treatment from bulk transfer of the culture, at least as far as the cells are concerned.

Alterations in Lipid Class Composition

Since the fatty acid composition of the various phospholipid classes varies, changes observed in overall phospholipid fatty acid composition on altering the growth temperature may result from either changes in the fatty acid composition of the various phospholipids, a change in phospholipid composition, or both.

That both processes operate in *Tetrahymena* has been clearly shown (Nozawa and Kasai, 1978). *T. pyriformis* NT-1 was grown at 39.5°C and then the temperature of the whole culture was dropped over a 30 min period to 15°C. For 10 h following this shift no cell division occurred, and there was no detectable change in phospholipid composition (Fig. 1). Once cell division had started once more, however, marked changes in phospholipid composition were observed. The major feature was a switch in the relative importance of ethanolamine phosphoglyceride and its phosphonolipid analogue 2-aminoethylphosphonolipid. In contrast, alterations in fatty acid composition were detected within the period of no cell division. This suggests strongly that there are at least two mechanisms operating during temperature acclimation in *Tetrahymena*, one altering phospholipid composition and the other regulating fatty acid composition.

Alterations in Fatty Acid Composition

Tetrahymena is one of the most intensively studied of experimental organisms in terms of temperature adaptation. Most experiments demonstrate the well described increase in unsaturation of phospholipid fatty acids at lower growth temperatures, though this is not true of all. Thus Conner and Stewart (1976) found no alteration in the mean number of double bonds per fatty acid molecule over the range 35 to 15°C. This result appears, however, to be atypical for

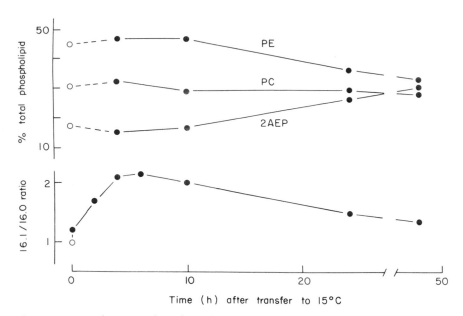

Figure 1. *Changes in phospholipid composition and 16:1/16:0 ratio in Tetrahymena pyriformis NT-1 following a temperature change from 39.5 to 15°C over a 30 min period. PC: choline phosphoglycerides, PE: ethanolamine phosphoglycerides, 2AEP: 2-aminoethylphosphonolipid.* ● *cells cooled to 15°C* ○ *control cells grown isothermally at 39.5 C. Redrawn, with permission, from data in Nozawa and Kasai (1978).*

Tetrahymena, although there are several reports of a lowered temperature having no effect on the fatty acid composition of bacteria (Gill and Suisted, 1978; Hunter *et al.,* 1981).

Wunderlich *et al.* (1973) chilled *T. pyriformis* GL from 28 to 10°C, and found that cell division was inhibited for 16 hours. This contrasts with the 10 hour inhibition observed in NT-1 cells for a 25°C drop in temperature by Nozawa and Kasai (1978), although it is difficult to be certain that the cells were cooled at the same rate in both cases and different strains of *T. pyriformis* were used. Nonetheless it is interesting that the major changes observed in the proportion of saturated and monoenoic fatty acids would sugg- est a pattern of the 16:1/16:0 ratio with time similar to that found by Nozawa and Kasai (1978; Fig. 1). Fatty acid alterations were observed immediately, but in this study phospholipids were not separated from other cell lipids (Fig. 2).

As would be expected, changes in fatty acid composition

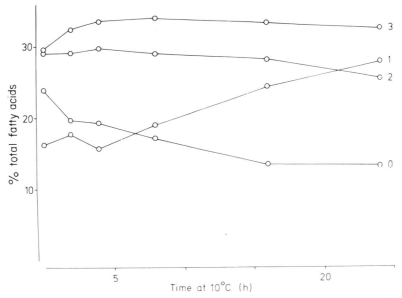

Figure 2. *Changes in total cell fatty acid composition in Tetrahymena pyriformis GL following a rapid temperature change from 28 to 10°C. 0: saturated, 1: monenoic, 2: dienoic, 3: trienoic fatty acids. Redrawn, with permission, from Wunderlich et al. (1973).*

affect all membrane fractions. Nozawa *et al.* (1974) grew *T. pyriformis* WH-14 at 25°C for 24 h before transferring the cells to 15 or 34°C. In whole cells, cilia, pellicles, mitochondria, microsomes and 'post mitochondrial supernatant' total lipid fatty acids were more unsaturated at 15°C, and less unsaturated at 34°C than at 25°C.

Few studies have looked specifically at phospholipids, although relatively pure membrane preparations will presumably contain little neutral lipid. When isolated phospholipids have been examined, the pattern has generally been similar to that reported for whole cells and membrane fractions, namely an increase in unsaturation at low temperatures (*e.g.* Nozawa and Kasai, 1978). In a recent study, however, it was found that in *T. pyriformis* 1630/1W unsaturation of phospholipids increased not only at temperatures below the optimal growth temperature, but also above the optimum (Morris *et al.*, 1981). This result would suggest that the relationship between temperature and lipid composition is not directly causal.

Lipid Composition and Growth Rate

The lowering of temperature has a wide variety of effects on
an organism's metabolism, and one major consequence is usually
a slowing of the growth rate. It is, however, difficult to
design an experiment which controls for the effect of growth
rate when temperature is lowered, and several other variables
also suffer from this drawback including oxygen content of
the medium, viscosity and pH (Franks, this volume).

Although tissue oxygen tension has been implicated as a
regulatory factor in plant desaturase activity (Harris and
James, 1969), in *Tetrahymena* desaturase activity does not
appear to be related to the oxygen tension in the medium
(Skriver and Thompson, 1976). Since palmitoyl-CoA desaturase
at least shows a requirement for molecular oxygen (Shipiro
et al., 1978), this is presumably only true if oxygen tension
remains above a threshold level.

There are several reports that growth rate affects lipid
composition in microorganisms. A survey of 4 mesophilic and
3 psychrophilic bacteria showed that these species reacted
differently to a lowered temperature, but all showed variations
in unsaturation with growth rate, independent of the effect
of temperature (Gill and Suisted, 1978).

The yeast *Saccharomyces cerevisiae* NCYC 366 showed little
change in fatty acid composition in both batch and chemostat
cultured cells when the growth temperature was altered.
However chemostat culture suggested that changes in lipid
class composition were caused by both a lowering of the growth
temperature and a decrease in growth rate (Hunter and Rose,
1972).

In order to separate the effects of temperature and growth
rate on the lipid composition of *Tetrahymena*, cells were
cultured at different temperatures, and also isothermally at
a variety of growth rates obtained by the addition to the
medium of NaCl or glucose, or by varying the strength of the
medium.

Optimal growth occurred at about $15^{\circ}C$, with slower growth
at both higher and lower temperatures. Phospholipid fatty
acid composition showed an inverse pattern; unsaturation was
minimal at the optimum growth temperature but higher both
above and below this temperature (Fig. 3). When cells were
cultured isothermally at $20^{\circ}C$ but with their growth rates
modified by varying the composition of the medium, unsaturation was again inversely proportional to the growth rate
(Fig. 4). There was a significant relationship between growth
rate and unsaturation (Kendall's $\tau = 0.539$, $S = -24$, $0.046 >$
$P > 0.028$).

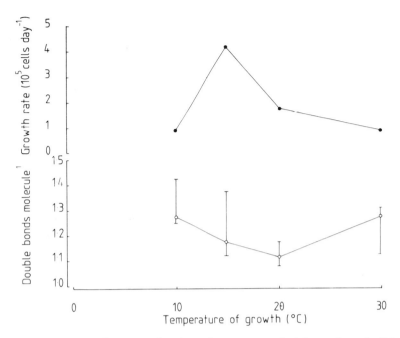

Figure 3. *Growth rate (●) and representative phospholipid fatty acid unsaturation values (○) for Tetrahymena pyriformis 1630/1W cultured at different temperatures in normal strength PY medium. Growth rates were estimated by fitting least squares regression lines to the cell number data. The slope was considered a realistic expression of growth rate for comparative purposes when 4 < n < 10 and r^2 > 0.80. Representative phospholipid unsaturation values were taken from the middle of this range (often day 3) and the height of the vertical bars indicates the total observed range of unsaturation at all stages of culture. (Reproduced, with permission, from Morris et al., 1981).*

These data suggest that in *Tetrahymena* the relationship between temperature and lipid unsaturation may not be a direct one, but may not be a direct one, but may be mediated through growth rate. If this is so, then the effects on membrane lipid unsaturation of compounds such as alcohols and anaesthetics (Nozawa and Thompson, 1979) could conceivably also be mediated via growth rate, rather than by a direct effect on the membrane. The evidence from the effects of exogenous unsaturated fatty acid supply or ergosterol replacement of tetrahymenol in membranes, however, suggest that such mediation via growth rate is unlikely.

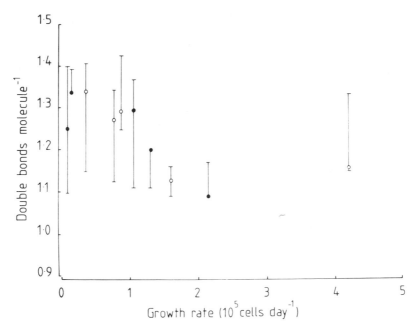

Figure 4.   *Relationship between growth rate and phospholipid*
fatty acid unsaturation for Tetrahymena pyriformis 1630/1W.
(○) growth rates obtained in normal PY medium at various
temperatures.   (●) growth rates obtained at 20°C in modified
growth media.   Representative unsaturation values and total
observed range as for Figure 3.   (Reproduced, with permission,
from Morris et al., 1981).

EFFECTS OF TEMPERATURE ON MEMBRANE STRUCTURE AND FUNCTION

Membrane Structure

The physical appearance of *Tetrahymena* membranes has been
extensively studied by freeze fracture electron microscopy
of glutaraldehyde fixed material.
    Several studies have shown that when *Tetrahymena* membranes
are cooled, the membrane intercalated particles visible on
freeze fracture electron micrographs alter in ditribution
from the more usual random scatter, to a clumped pattern,
leaving large areas of particle-free membrane faces.   The
temperature at which this segregation occurs varies with
culture temperature.   This effect has been quantified by
Martin *et al.* (1976) by use of a particle density index (PDI).
This index is essentially a measure of how close to the normal (PDI=0) or the most clumped distribution (PDI=100), are

the particles visible in selected areas of any given fracture face. It should be noted that this PDI refers only to areas where particles are visible; it thus says nothing about membrane particle distribution as a whole and it is thus not possible to deduce whether the overall number of particles has changed (Robards *et al.*, this volume).

In the outer alveolar membrane of *T. pyriformis* NT-1 the PDI for any given fixation temperature was found to be higher in cells grown at $39.5^{\circ}C$ than those grown at $15^{\circ}C$ (Fig. 5). The translation of the PDI/fixation temperature curve to the right (a lower fixation temperature) is about 24° for a $25^{\circ}C$ decrease in temperature. Compensation thus appears to be perfect (Cossins, this volume), although it must be recognised that the PDI (*sensu* Martin *et al.*, 1976) is a very restricted measure of membrane structure.

When cells were cooled from 39.5 to $15^{\circ}C$, compensation was achieved within about 8 hours. This agrees well with data for *T. pyriformis* microsomal membranes, where the time course of changes in fluorescence polarisation, phospholipid unsaturation and the temperature of incipient appearance of particle-free areas were similar (Fig. 6; Martin and Thompson, 1978). Different membranes, however, have been shown to react at different rates (Kitajima and Thompson, 1977b).

If freeze fracture electron micrographs of the outer alveolar membranes are examined during the course of acclimation of *Tetrahymena* cells to a lower temperature then it can be seen that immediately upon chilling large particle-free areas are visible on the fracture faces. However these gradually reduce in size and eventually an apparently random distribution of particles is once again achieved (Nozawa and Kasai, 1978). The particles counted were 11 to 13 nm in diameter and cells were apparently unable to divide until these particle-free areas had vanished (Nozawa and Thompson, 1979).

Not all membranes in *Tetrahymena* show this behaviour, however. When *T. pyriformis* cells were grown at $28^{\circ}C$ and chilled to $10^{\circ}C$, marked structural alterations were found in the membranes of the pellicle, alveolar sacs, endoplasmic reticulum (microsomes) and nuclei, whereas mitochondria peroxisomes, mucocysts, vacuoles and smaller vesicles showed only occasional structural alterations (Wunderlich *et al.*, 1973). These alterations were reversible, aggregated particles becoming apparently random once more on returning the cells to $28^{\circ}C$ (Speth and Wunderlich, 1973).

It is tempting to speculate that the particles observed represent protein molecules, but recent research on membrane particles indicates that this may not be so. Intramembranous particles visible on freeze fracture electron micrographs

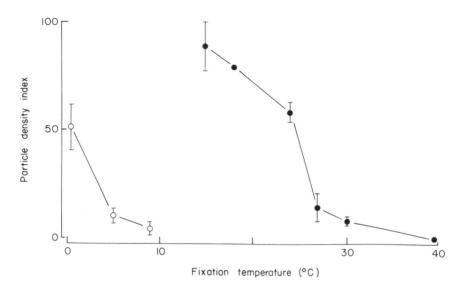

Figure 5. *Relationships between particle aggregation and fixation temperature in the outer alveolar membrane of Tetrahymena pyriformis NT-1. Particle density index was determined from freeze fracture electron micrographs, and is plotted against the temperature at which the cells were fixed with glutaraldehyde following chilling from growth temperature over a 4 minute period. (O) cells grown isothermally at 15°C, (●) cells grown isothermally at 39.5°C. Redrawn, with permission, from data in Martin et al. (1976).*

have been demonstrated in lipid bilayers containing no protein (Miller, 1980; Verkleij *et al.*, 1979). These particles may represent inverted micelles within the bilayer, incidence of which would be related to the phospholipid composition of the membrane and in particular the proportion of phosphatidyl-ethanolamine (Cullis and de Kruijff, 1979). Also glutaraldhyde fixation has been shown to alter the appearance of freeze fracture faces (Breathnach *et al.*, 1976; Parish, 1975).

Whether the observed particles represent proteins or some form of lipid micelle within the membrane, the observed patterns of behaviour of these particles on chilling suggest strongly that lateral phase separations (Pringle and Chapman, this volume) are responsible for the production of particle-free areas. In addition the observed compensation associated with changes in fatty acid composition of phospholipids, together with the similar time scales of adjustments to composition freeze-fracture appearance and cell division, suggest strongly that these changes do indeed serve a functional

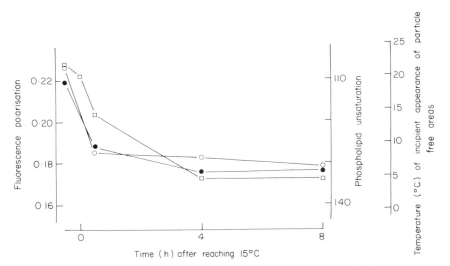

Figure 6. *Changes in fluorescene polarisation (●), phospho-lipid unsaturation, as mean number of double bonds per 100 molecules of fatty acid (o) and temperature of incipient appearance of particle-free areas (□) in microsomes of Tetrahymena pyriformis, following a shift from 39.5 to 15°C over a 30 min period. Redrawn, with permission, from Martin and Thompson (1978).*

role in maintaining membrane integrity.

Spectroscopic Studies

Several authors have examined *Tetrahymena* membranes with spectroscopic probes. As would be predicted from the diff-erences in lipid composition the fluidity of membranes as estimated by electron spin resonance (ESR) using the probe 5-nitroxystearate increased in the order cilia < pellicles < microsomes (Nozawa *et al.*, 1974). This order is paralleled by a decreasing tetrahymenol content; pellicles are also richest in phosphonolipids.

Measurement of microviscosity by fluorescence polarisation with 1,6-diphenyl-1,3,5-hexatriene (DPH) as probe also sugg-ests an increase in fluidity in the order cilia < pellicles < microsomes (Shimonaka *et al.*, 1978).

Observations of microsomes with the fluorescent probe 8-anilino-1-naphthalenesulphonic acid revealed a discontin-uity in the Arrhenius plot between about 15 and 20°C; a similar break was found with proton NMR (Wunderlich *et al.*, 1975). This discontinuity was not interpreted as a phase

transition, as has usually been the case with such breaks,
but as a phase separation, perhaps that revealed by freeze
fracture electron microscopy.

Functional Effects

That these observed structural changes may have functional
effects is illustrated by a study of RNA transport through
the nuclear membrane. Pulse labelling studies of *T. pyriformis*
GL with ^{14}C-uridine suggest that above about 18°C the rate
limiting step for the appearance of ribosomal RNA in the
cytoplasm is rRNA synthesis. Below about 18°C the rate
limiting step is transport from the nucleus to the cytoplasm
(Wunderlich *et al.*, 1974). This marked decrease in rRNA
transport across the nuclear membrane at about 20 to 16°C
coincides with a morphological change in the macronuclear
envelope, in particular the appearance of smooth areas free
of nucleopores. The total number of nucleopores appears to
remain constant, however, suggesting that a lateral phase
separation has occurred, moving the nucleopores and thus
affecting nucleocytoplasmic rRNA transport (Nägel and Wunderlich, 1977). This inhibition is reversible for on rewarming
a normal macronuclear envelope is once more visible.

   The pellicle shows high activities of adenylate cyclase
compared with other membrane fractions, and this enzyme
preparation shows an Arrhenius discontinuity at 28°C, which
is the temperature of the onset of phase separation visible
in freeze fracture faces (Nozawa and Thompson, 1979; Shimonkak
and Nozawa, 1977). In contrast however, 5'-nucleotidase
activity in pellicular membranes showed no Arrhenius discontinuity (Nozawa and Thompson, 1979).

   Glucose-6-phosphatase activity in microsomes shows a discontinuity at 12°C in *T. pyriformis* GL grown at 18°C, the
same temperature at which particle-free areas appear in
microsome freeze fracture faces (Wunderlich and Ronai, 1975).
Furthermore, smooth ribosomes from cells grown at 28°C show
Arrhenius discontinuities in glucose-6-phosphatase activity
at about 17°C. This coincides with discontinuities in the
Arrhenius plots of fluorescence intensity of 8-anilino-1-
naphthalene sulphonic acid, outer hyperfine splitting of
5-doxylstearate and phase partition of 4-doxyldecane
(Wunderlich *et al.*, 1975).

   These observations would appear to offer strong evidence
for a direct relationship between membrane structure and
membrane function. Such deductions, however, must be treated
with care. The interpretation of apparent discontinuities
in Arrhenius plots is equivocal and sometimes plain erroneous

(Franks, this volume), for soluble enzymes can also show such patterns.

Nevertheless, the weight of evidence suggests that membrane structure and function are indeed closely related in *Tetrahymena*.

MECHANISMS

Although there is now a considerable body of data on the effects of temperature on lipid composition in *Tetrahymena*, little is known of the mechanisms by which these changes are affected.

Despite the demonstration that adaptation to temperature involves alterations in phospholipid head group composition, nothing is known about how these changes occur. Time course studies of the incorporation of the non-biological head group analogue isopropyl ethanolamine into monolayers of L-M fibroblasts suggest that this process is slow (maximum reached after 24 h: Snyder *et al.*, 1980). Culturing *T. pyriformis* NT-1 in media containing 8.1 mM choline analogues alters the phospholipid head group composition, but not the fatty acid composition (Nozawa and Thompson, 1979). Preliminary experiments indicate that exchange of phospholipids between membranes occurs in *Tetrahymena*, mediated by a carrier protein molecule. If the exchange involves the same head group in each case, however, this mechanism cannot be invoked to explain alterations in composition. Transfer of phospholipid molecules seems to be governed by the fluidity of the receptor membrane (Iida *et al.*, 1978; Maeda *et al.*, 1978).

Alterations in Fatty Acid Composition

One of the problems in investigating alterations in fatty acid composition associated with temperature change is that alteration can be affected at several different levels, for example:

1) At the level of transcription by *de novo* synthesis of new desaturase enzymes.
2) At the level of enzyme activity by a temperature induced modification of the activity of desaturase enzymes.
3) At the level of the membrane by modification of the activity or specifity of acyl transferases.
4) At the level of overall metabolic control by altering the relative balance of synthesis and desaturation of fatty acids available to the membrane.

A temperature induced synthesis of desaturase has been

demonstrated in the bacterium *Bacillus megaterium* 14589
(Fulco, 1969, 1972; Fulco and Fujii, 1980). This organism
contains only one desaturase enzyme which is membrane bound,
shows a requirement for oxygen and desaturates only at the
Δ^5 position of acyl chains (C_{15} to C_{19}) attached to intact
membrane phospholipids, and not those present as CoA or ACP
thioesters. In bacteria grown at $35^{\circ}C$ the enzyme is absent,
but on lowering the temperature the enzyme is synthesized
de novo; this synthesis can be prevented by blocking protein
synthesis with cycloheximide before lowering the temperature.
In *T pyriformis* the activity of the enzyme which converts
16:0 to 16:1ω9 (palmitoyl-CoA desaturase) increases after a
shift of temperature from 39.5 to $15^{\circ}C$, and this increase is
prevented by cycloheximide indicating that synthesis of new
enzyme is involved (Fukushima *et al.*, 1979; Nozawa and Kasai,
1978).

Another possible mechanism for modifying the ratio of
saturated to unsaturated fatty acids is a temperature induced
alteration in the specificity of synthetase and desaturase
enzymes, has has been demonstrated in *E. coli* (Cronan, 1978).
There is no evidence, however, that a lowering of the growth
temperature of *Tetrahymena* alters the temperature characteristics of the palmitoyl-CoA desaturase (Shipiro *et al.*, 1978).

The prokaryote *Brevibacterium ammoniagens* contains a multi-
enzyme complex which produces, *de novo*, both 18:0 and 18:1ω9
fatty acids. The ratio of these two products is dependent
on the temperature *in vitro* (*i.e.* the assay temperature),
but not the growth temperature. This suggests a direct influence of temperature on the fatty acid synthetase/desaturase
complex (Kawaguchi *et al.*, 1980).

There is as yet no evidence of such a mechanism operating
in *Tetrahymena*, and experiments using ^{14}C labelled 18:3ω6
and 18:2ω7 (cilienic acid) indicate that alterations in membrane fatty acid composition in *Tetrahymena* are not caused
by a differential turnover of fatty acids (Schick *et al.*,
1979).

Much experimental work on desaturase activity in *Tetrahymena*
has centred on the regulation of activity of membrane bound
desaturase enzymes. These are associated with the endoplasmic
reticulum and their activity appears to be regulated by membrane fluidity.

Comparison of the incorporation rate of ^{14}C-acetate into
fatty acids of *Tetrahymena* grown isothermally at $39.5^{\circ}C$ and
cells cooled to $15^{\circ}C$ showed that at the lower temperature
there was an increase in the rate of desaturase activity
compared with the rate of synthesis *de novo*. This change in
the balance of pathways was quite rapid (Martin *et al.*, 1976).

Supplementing ciliates with 18:2 and 18:3, acids which
are rapidly incorporated into membranes, indicated that this
shift in the balance of synthesis and desaturation was not
due to a differential response to temperature. Rather it was
due to a direct effect of membrane fluidity on the activity
of the desaturase enzyme (Martin *et al.*, 1976). This hypothesis has received support from studies where isothermal
cultures of *Tetrahymena* have been fed exogenous 9- and 10-
methoxystearic acid or anaesthetics, which increase the
fluidity of the membrane. In all cases net desaturase activity
was reduced (Kitajima and Thompson, 1977a, Nandini-Kishore
et al., 1977). Supplementation with ergosterol which enters
the membrane and reduces membrane fluidity, as judged by ESR,
increases the conversion of ^{14}C-16:0 to 16:1 and as a result
there is a switch in the relative amounts of 18:3ω6 and 18:2ω7
fatty acids produced (Kasai *et al.*, 1977).

The hypothesis that a decrease in membrane fluidity directly increases desaturase activity offers an elegant mechanism
for rapid adaptation to variations in temperature. Interestingly a similar mechanism has recently been proposed for
control of a phospholipid desaturase in microsomes of the
yeast *Candida lipolytica* (Kates and Pugh, 1980) and ATPase
in the plasma membrane of rat liver (Riordon, 1980).

DISCUSSION

It is clear that *Tetrahymena* shows the classic response to a
lowered temperature by increasing the unsaturation of its
membrane phospholipids and altering the proportion of the
various phospholipid head groups. Investigations of the
mechanisms by which these adjustments are made suggest that
both fast and slow changes are involved. A direct control of
desaturase activity by membrane fluidity will obviously allow
a rapid adjustment of fatty acid composition, and such is
indeed observed (Fig. 1,2). The longer term adaptation, by
alteration of phospholipid head group composition, is presumably attained by other, slower, enzymatic pathways. Since
alterations in fatty acid composition occur so rapidly following a change in temperature, Martin *et al.* (1978) have questioned whether the alterations in phospholipids are actually
important in temperature adaptation. However since changes
in phospholipid composition are also seen in *T. pyriformis*
cultured isothermally at different temperatures (Nozawa and
Kasai, 1978) it seems realistic to assume these changes do
have functional consequences and some adaptive value.

The possible influence of temperature-induced alterations
in growth rate in this process are not at all clear. Certain-

ly a slowing in growth rate *per se* seems to have an affect
on fatty acid unsaturation, but the cause of the increased
unsaturation at high temperatures (Fig. 3) is unclear.
Perhaps there is a differential response to temperature in
synthesis and desaturase pathways in this particular strain
of *Tetrahymena*. Certainly it seems unlikely that the postulated direct control of membrane fluidity on desaturase
activity can be invoked to explain these results.

As it is possible that changes in membrane phospholipid
fatty acid unsaturation are not necessarily related directly
to temperature, and in some cases it is known that alterations
in growth temperature have no effect on *Tetrahymena* lipid
composition (Conner and Stewart, 1976), it could be asked
whether adjustments of membrane composition to temperature
are really necessary.

Certainly in some prokaryotes there is evidence that membranes can tolerate quite wide variations in composition
without affecting function, to the extent that Gill and
Suisted (1978) have questioned whether there is any need for
prokaryotes to adapt their membranes to temperature at all.
A similar view has been forwarded by Silvius *et al.* (1980)
and Cronan and Gelman (1975) who demonstrated that fatty
acid composition in bacteria can vary widely before membrane
function is affected, and there is thus no apparent need
for adaptation to temperature. Prokaryote membranes are,
however, unusual in a number of ways, notably in the absence
of sterol and the apparent requirement for phase separation
to be present (rather than avoided as in eukaryotes) in some
membranes such as the outer membrane of *E. coli* (Janoff *et
al.*, 1979). Also some prokaryote membranes, particularly
those of halophiles and thermophiles, have very unusual lipid
compositions, which often show little variation with temperature (Hunter *et al.*, 1981; Kates, 1978; Langworthy, 1977).
It is therefore unlikely that the behaviour of prokaryote
membranes is directly relevant to that of eukaryotes.

However evidence from freeze-fracture electron microscopy
indicates that in *Tetrahymena* alteration of lipid composition
in response to a lowered temperature is necessary to avoid
lateral phase separation. Should such phase separation occur
cell division stops and many membrane functions are impaired.
Thus *Tetrahymena* is capable of making both rapid and longer-
term adjustments to compensate for alterations in temperature.

It is difficult to quantify how good this compensation is,
but it is at least good enough to allow *Tetrahymena* to exploit
a very wide range of habitats and a thermotolerant strain,
T. pyriformis NT-1, capable of growth at 40°C is well documented. It has been proposed that at least part of this

tolerance to varied external conditions is due to the unusual
lipid composition of those membranes exposed to, or close to,
the external environment. Thus ciliary and pellicle membranes
are rich in both alkyl linkages and phosphonolipids. Although
this may make these mambranes more tolerant of changes in the
external medium they do not render the pellicle at least less
likely to undergo phase separation at low temperatures
(Wunderlich et al., 1973).

Since few, if any, experimental studies of temperature
adaptation in Tetrahymena have examined changes in the alkyl-
linked acyl chains, any picture of alterations in response to
temperature will be incomplete. In most organisms, however,
the range of variety in alkyl-linked acyl chains is much less
than acyl-linked chains, and this is also true in Tetrahymena
where most of the alkyl chains appear to be 16:0 (Fukushima
et al., 1976; Nozawa and Thompson, 1979).

A further aspect of temperature adaptation of fatty acids
which has become apparent in fish but has received little
attention in Tetrahymena is positional alteration in the acyl
chains on the phospho-glycerol backbone. In goldfish acclimated to low temperatures there is an increase in the ethanolamine phosphoglycerides in intestinal microsomal membranes,
as well as an increase in the proportion of the typical fish
polyunsaturated fatty acids 22:6ω3, and 20:4ω6 at the sn-1
and sn-2, and the sn-2 position respectively. There is also
an apparent switch of 18:1 and 20:1 from the sn-2 to the sn-1
position (Miller et al., 1976). Possibly the high proportion
of alkyl linkages at the sn-1 position in Tetrahymena phospholipids limits turnover and exchange of acyl chains at this
position.

There have been few studies of the efficacy of temperature
adaptation in Tetrahymena (Cossins, this volume) in terms of
lipid composition. Reports merely demonstrate that changes
occur, and that these are associated with structural changes
visible in freeze-fracture electron micrographs, and in some
cases physiological consequences. It is now becoming clear,
however, that proteins and protein-lipid interactions are
important in the adaptation of many membranes to temperature
(Pringle and Chapman, this volume). No studies have yet taken
into account the possible role of membrane proteins in
adaptation of Tetrahymena to temperature. Indeed Kitajima
and Thompson (1977a) stated that 'lipids are the only structural elements of living cells known to participate in the
temperature acclimation phenomenon'. The PDI data of Martin
et al. (1976), however, although based on a somewhat arbitrary
measure of membrane appearance, does suggest that compensation
approaches the perfect state (Cossins, this volume).

Although the value of *Tetrahymena* as an experimental organism for the investigation of the effects of temperature on
the biochemistry and physiology of membranes is clear, it
should be borne in mind that this is an extremely tolerant
and adaptive organism. Like bacteria, *Tetrahymena* shows a
remarkable ability to cope with varying environmental
conditions. As such it is an invaluable model for examining
such responses in a complex eukaryotic cell. However, just
as the very different membrane of prokaryotes limits their
usefulness as models from which to extrapolate to eukaryotes,
it is possible that concentration on a supremely adaptive
organism may blind us to the mechanism of evolutionary
adaptation to low temperature. The latter may be very diff-
erant from the very rapid short-term adjustments to temperature shown by organisms such as *Tetrahymena pyriformis*.

THE ADAPTATION OF MEMBRANE DYNAMIC STRUCTURE TO TEMPERATURE

A.R. Cossins

Department of Zoology, University of Liverpool, England.

INTRODUCTION

The alteration of temperature can produce as dramatic effects upon the physiology of organisms as almost any environmental manipulation, and yet living organisms have invaded virtually all thermal habitats on this planet. The evolutionary adaptation of organisms to diverse thermal environments and the acclimatisation displayed by many organisms to seasonal variations in temperature involve a great many adjustments at the cellular and molecular levels of organisation (Hazel and Prosser, 1974). These adaptations may be of two distinct types; those that permit survival in transient conditions which otherwise would be lethal (resistance adaptations) and those that compensate for the direct effects of temperature on biological processes (capacity adaptations).

Perhaps the best studied and most ubiquitous cellular response to altered temperature is the compensatory adjustment of the dynamic structure or 'fluidity' of cellular membranes for the direct effects of the temperature change. This process has been called 'homeoviscous adaptation' by Sinensky (1974), a term that implies firstly that membrane fluidity is maintained in a more or less constant and presumably optimal condition, and secondly, that the maintenance of this optimal state has some adaptive value for the species. Of course, the integrity and functional properties of cellular membranes are vital to numerous cellular processes and all of them will be affected to varying extents by temperature variations. Thus, the strategy of maintaining membrane function independent of temperature changes by modulating the effective viscosity of the hydrophobic compartment of mem-

branes would seem to have obvious survival value both during
rare crisis periods and also considerable importance for the
competitive fitness of species at more normal temperatures.
    Concerning the effectiveness of the homeoviscous response,
most studies of membrane adaptation were based, until recently,
upon the compositional analysis of membrane lipids of different species or of experimentally manipulated individuals.
Generally, these studies have demonstrated an increase in the
unsaturation of the fatty acids of membrane phospholipids at
reduced temperatures (Cossins, 1976; Hazel, 1973; Hazel and
Prosser, 1974). The functional and structural significance
of these analyses were usually interpreted on the basis of
studies of model systems, such as liposomes (De Gier *et al*.,
1968; Klein *et al*., 1971). These indicated that the compositional adjustments would, indeed, serve to compensate membrane fluidity and associated properties for the effects of
the temperature change. However, our current understanding
of the exact contribution of acyl chains to fluidity of complex cellular membranes is rudimentary (Lands, 1980) and the
comparison of membrane lipid composition cannot provide
quantitative information on the effectiveness of the proposed
homeoviscous response. It is, for example, relatively easy
to conceive of homologous membrane-types with quite different
acyl group compositions and similar membrane fluidities
(Cossins *et al*., 1978). The unequivocal demonstration of
homeoviscous adaptation and its effectiveness demands a more
direct and quantitative means of assessing the degree of
molecular motion or of order within the membrane. Only then
can the modification of membrane composition be linked to
functional or structural adaptations of cellular membranes.
The demonstration of structural membrane adaptations is the
subject of this contribution.
    The significance of homeoviscous responses can only be
fully appreciated from a thorough understanding of the effects
of temperature on membrane structure. Temperature-induced
changes in bulk fluidity and order are therefore discussed in
some detail, together with a critical assessment of some
experimental techniques for the quantitative estimation of
these effects. The adaptive value of homeoviscous responses
depends upon the magnitude of the compensation, and the
occurrence and effectiveness of some described adaptations
are assessed.

MEMBRANE FLUIDITY AND ITS MEASUREMENT

The fluid mosaic model for the structure of biological membranes lays great emphasis upon the molecular motion of its

constituent molecules (Singer, 1974; Singer and Nicholson, 1972; Pringle and Chapman, this volume). This motion is manifest in a number of distinct ways such as the lateral movement of lipids and proteins in the plane of the membrane, the 'flip-flop' of lipids from one monolayer to the other, the wobbling and rotation of entire molecules about a plane normal to the membrane and the flexing of acyl chains and polypeptides by rotation about their carbon-carbon bonds. Each type of motion occurs over a distinct time-course and each may have distinct and important roles in the various functional properties of the membrane. Nevertheless, the resultant dynamic state is commonly described by the all-inclusive term 'fluidity', and a major problem in membrane biology is the quantitative interpretation of such a complex and heterogeneous property in terms of the quantitative techniques currently available. Different spectroscopic techniques may be sensitive to molecular motion of different types and time-scales, so that it should be borne in mind in the following discussion that any measure of membrane fluidity refers only to the types of motion sensed by that technique and is heavily biased to only one or a few components of the overall dynamic state. In general, most spectroscopic techniques furnish information about the time-averaged orientation of the molecules involved (*i.e.* order parameters, limiting anisotropy) or on the rate of motion of those molecules (relaxation times, correlation times, diffusion coefficient, microviscosity). The correlation between membrane order on one hand and molecular mobility on the other is not well understood and it is necessary to keep these two concepts apart (Schreier *et al.*, 1978; Seelig and Seelig, 1980). For this reason the following discussion is divided into a consideration of average membrane fluidity and its compensation, followed by thermotropic structural rearrangements and their modification during temperature acclimation.

Average Membrane Fluidity

Perhaps, the most widely used methods for the measurement of membrane fluidity are electron spin resonance (ESR) spectroscopy and fluorescence polarisation spectroscopy. In both cases molecules (probes) are introduced into membranes by virtue of their hydrophobic character and information regarding their rotational characteristics may be gleaned from measurements of their spectroscopic properties. A full account of these techniques may be found in Pesce *et al.*, (1971), Shinitzky and Barenholz (1978) and Swartz *et al.*, (1972). The basic assumption of all such methods is that the motion of the probe reflects that of its environment.

When interpreting information derived from these techniques, it is important to be aware of their limitations. The intro- duction of foreign molecules such as spectroscopic probes into membranes may itself cause some disruption of the mem- brane structure. This effect is of two types, a longer-range, cooperative perturbation of bulk bilayer properties and a short-range disruption of the immediate microenvironment of the probe. The first effect can be circumvented simply by reducing the proportion of probe molecules to membrane lipid molecules to 0.1-1.0 percent, whilst the second and potent- ially more serious effect can be minimised by the use of probes whose molecular structure mimics as closely as possible that of the surrounding hydrocarbon chains. Nevertheless, even small local perturbations may seriously affect the absolute nature of the measurements since it is this perturbed solvent environment that is sensed and reported by the probe (Cadenhead *et al.*, 1975). Provided experiments are performed on a comparative basis, in such a way that deviations from absolute measurements apply equally to all preparations, then the order and fluidity of the different preparations can be validly compared.

A further problem is that, probe techniques provide average data for all the hydrophobic environments sampled by the probes, and this lack of resolution makes the interpretation of fluidity rather difficult. Indeed, it is usually not possible to distinguish between an altered average fluidity in a homogeneous fluid bilayer as a result of experimental treatment, and a shift in the proportion of gel phase within a predominantly fluid bilayer. Different probes may partition into different sites within the bilayer or at its surface by virtue of their hydrophobic or polar character and, hence, will report on different aspects of membrane fluidity. This fact can however be used to advantage depending upon the information that is required. Thus, the fluorophore 1,6- diphenyl-1,3,5-hexatriene (DPH) which partitions equally between coexisting gel and liquid-crystalline phases (Andrich and Vanderkooi, 1976; Lentz *et al.*, 1976) is thought to report a weighted average value for membrane fluidity (Shinitsky and Barenholz, 1978), whilst *trans*-parinaric acid partitions preferentially into gel phase lipids and provides information of a more specific nature (Sklar *et al.*, 1979). The fluorophore PRODAN appears to partition between a hydro- phobic site and a site of intermediate hydrophobicity, per- haps near the headgroup region of the bilayer. By the use of appropriate filters, probe motion can be silultaneously estimated at each site (Chong, Cossins and Weber, unpublished observations).

The interpretation of data from spectroscopic measurements is usually based upon semi-empirical models of probe rotation which, strictly speaking, are not generally applicable to the hindered, anisotropic rotations experienced by molecules in a phospholipid bilayer. Such data must be considered as experimentally convenient but less than rigorous. The recent development of more elaborate models of rotational behaviour (Israelachvili *et al.*, 1975; Kinosita *et al.*, 1977) together with the introduction of more sophisticated spectroscopic techniques (Lakowicz *et al.*, 1979) permits a more realistic and detailed quantitative description of membrane fluidity and how it is modified by homeoviscous adaptation (Cossins *et al.*, 1980). It is now clear, for example, that the polarisation of DPH fluorescence is heavily dependent upon the ordering of the probe within its anisotropic environment rather than upon its rate of rotational motion (Hildenbrand and Nicolau, 1979). However, for most purposes the use of the less sophisticated, semi-empirical formulations such as the Perrin equation provide adequate information on the existence and extent of homeoviscous responses. In a recent comparison of membranes from thermally acclimated fish (Cossins, unpublished observations) the steady state fluorescence polarisation technique provided essentially identical differences between samples as did the differential polarised phase technique.

Increased temperature causes a progressive and general increase in the rate of molecular motion within the bilayer and a decrease in order. The temperature dependence of fluidity depends to some extent upon the lipid composition of the bilayer. Phospholipid liposomes have an Arrhenius activation energy for microviscosity of approximately 15 kcal mol^{-1}, whilst the incorporation of equimolar amounts of cholesterol reduced this by half (Shinitzky and Inbar, 1976). As might be expected, increased molecular mobility and decreased order in membranes results in increased rates of the various processes associated with membranes. A large part of the temperature dependence of enzymatic activity is due to changes in the viscosity of the solvent environment (Gavish and Werber, 1979; Franks, this volume) and intrinsic membrane-bound enzymes are no exception (Sinensky *et al.*, 1979).

Thermotropic Transitions of Membrane Structure

A second important effect of temperature upon bilayer structure is the induction of reversible gel to liquid-crystalline phase transitions (Melchior and Stein, 1976). For bilayers

of a single phospholipid such transitions occur over very
narrow temperature ranges, indicating a high degree of cooperativity between the hydrocarbon chains in the interior of
the bilayer. In the gel state these chains exist in an
extended, all-*trans* configuration and are aligned roughly
perpendicular to the plane of the membrane (Chapman, 1969).
Flexing motion and wobbling of the hydrocarbon chain is
severely restricted by the semi-crystalline alignment of its
neighbours. In the liquid-crystalline state, however, the
hydrocarbon chains exhibit a considerable amount of flexing
motion through rotations about their carbon-carbon bonds.
This motion creates free space which permits the wobbling and
flexing of adjacent chains (Chapman 1969). A phase transition
in a simple artificial membrane, therefore, results in a
dramatic change to the dynamic structure of the membrane,
although the general bilayer configuration remains undisturbed
(Pringle and Chapman, this volume).

Cellular membranes are usually composed of a bewildering
array of different phospholipid head groups and hydrocarbon
chains as well as cholesterol and intrinsic membrane-bound
proteins. Phase transitions in such heterogeneous mixtures
usually occur over a fairly wide range of temperatures, within which segregated areas of gel and liquid-crystalline
phases coexist to give what has been termed a 'phase separation' (Shimshick and McConnell, 1973a).

More subtle effects may also be induced by temperature
changes. Lee *et al.*, (1974) have suggested that the formation of "clusters" of quasi-crystalline lipids in the bilayer mark the beginning of a phase transition. These
clusters were thought to be small densely-packed arrangements
of relatively ordered lipid that exist over a short lifetime
within an area of fluid, dispersed lipid. One might further
suppose that as temperature was progressively reduced the
magnitude and lifetime of these clusters increased to produce two distinct phases, and ultimately a single gel phase.
As the clusters grow in size and permanence the degree of
order within the cluster increases due to co-operative effects
(Franks, this volume). Wu and McConnell (1975) discuss the
phase diagrams of phospholipid mixtures and show how immisc-
ibility of fluid domains may occur under specified conditions
of temperature and phospholipid composition. This immiscib-
ility can give rise to lateral phase separations and may
contribute to the generation of asymmetrical bilayers.

A great many adaptive responses of membranes have been
demonstrated by observing shifts in transition temperature
during acclimation, so for the following discussion it will
be helpful to examine critically the techniques which help

to define the temperatures over which transition phenomena
occur. Perhaps the least equivocal evidence is that provided
by calorimetric measurements in which the uptake or evolution
of heat by a sample provides direct evidence of a bulk phase
transition (Chapman, 1975; Ladbrooke and Chapman, 1969).
Spectroscopic probes in artificial membranes show very dramatic changes in rotational behaviour as the membrane undergoes a phase-transition, reflecting the markedly different
motional environment provided by the two phases (Papahadjopoulos *et al.*, 1973). In cellular membranes with more gradual
transitions such dramatic effects usually do not occur.
Instead, discontinuities in the Arrhenius plots of motional
parameters of membrane-bound probes have been observed and
interpreted as evidence of a structural rearrangement such
as a phase transition (Kumamoto *et al.*, 1971; Raison *et al.*,
1971a). The temperatures of Arrhenius discontinuities of
probe motion have sometimes been related to similar discontinuities in the Arrhenius plots of activity of membrane-
bound enzymes (Raison *et al.*, 1971b; Raison and McMurchie,
1974), to phase transitions detected by calorimetric techniques and to phase separations observed by freeze-fracture
studies (Grant and McConnell, 1974; Kleeman *et al.*, 1974).

Finally, freeze-fracture electron microscopy provides
useful correlative evidence of phase separations in cellular
membranes. When samples are rapidly cooled from temperatures
within a calorimetric phase transitions, the fraction surfaces
display two distinct areas, smooth particle-free patches and
areas containing large numbers of intramembranous particles
e.g. Hochli and Hackenbrock, (1979). Evidence has been presented that these particles are membrane-bound proteins,
though recently similar structures have been observed in
pure lipid bilayers and alternative interpretations have been
offered (De Kruijff *et al.*, 1979; Verkleij *et al.*, 1980).
When quenched from above the calorimetric phase transition,
these intramembranous particles are usually dispersed randomly across the fracture face. These observations are
believed to illustrate the formation of a gel bilayer which
by its crystalline arrangement tends to squeeze-out the
proteins into the remaining areas of fluid lipid, much as
salt is excluded from ice crystals (Verkleij *et al.*, 1972;
Pringle and Chapman, this volume). This technique is unique
in providing a means of observing transitions *in situ* and in
a variety of different membrane-types in the same preparation.
The development of quantitative techniques would make this a
very powerful tool, provided that it can be shown that
fracture faces are produced randomly (Robards *et al.*, this
volume).

Arrhenius Plots

Mainly for reasons of experimental convenience, most workers
have used spectroscopic probes and membrane-bound enzymes to
study thermotropic lipid transitions. In cases where sudden
and dramatic changes in the slope of Arrhenius plots are
observed, it is relatively easy to estimate the transition
temperature. However, many published graphs show rather
small changes in slope, which are often spread over a range
of temperatures. Many such graphs may be described equally
well by a curvilinear plot, with consequent implications for
their interpretation as phase transitions. The subjective
assignment of two straight lines where a single curvilinear
plot is equally appropriate is a problem that bedevilled
Crozier in the 1920's and the recent use of statistical
techniques to assign the most likely transition temperature(s)
really do not provide a complete solution. It is quite clear
that some correlations between discontinuity temperatures of
probe rotation and enzymatic activity are due to 'brute
force' graphical methods (Franks; Wilson and McMurdo, this
volume).

In other instances convincing biphasic or multiphasic
Arrhenius plots have been presented. But before it is
concluded that they result directly from lipid transitions,
it should be remembered that discontinuities may be produced
by other effects, including thermotropic changes in protein
structure (Lee and Gear, 1974; Franks; Pringle and Chapman,
this volume) and temperature-induced alterations in substrate-
binding kinetics (Silvius *et al.*, 1978; Silvius and McElhaney,
1980). Indeed, transitions in the structure of membrane-
bound proteins may be influenced by bulk membrane fluidity
of an entirely fluid membrane since the conformational
flexibility of polypeptide chains are sensitive to the order
and motion of their immediate environment (Careri *et al.*,
1979). Thus, protein transitions may be modified, in principle, by lipid supplementation procedures even when no lipid
phase transition has occurred. The use of inappropriate
assumptions and approximations in the methods used to calculate motional rates and order parameters for spin probes may
themselves lead to artefactual discontinuities (*e.g.* Schreier
et al., 1978). In some cases discontinuities have been
observed in the Arrhenius plot of polarisation of fluorescence, which upon correction for variations in fluorescence
lifetime become entirely straight (Cossins, 1977, 1981).

In view of these limitations, it is clearly unsatisfactory
to interpret Arrhenius breaks of probe motion or membrane-
bound enzyme activity as bulk phase transitions without

confirmatory evidence from calorimetric, X-ray diffraction
or freeze-fracture studies. As discussed below, calorimetric
phase transitions have been observed in mammalian membranes
at temperatures well below those of Arrhenius breaks of probe
motion. At present, the meaning of Arrhenius breaks that
are undetected by calorimetry is unclear but probably represents a more subtle and gradual structural rearrangement of
limited proportions of the membrane rather than a bulk phase
transition. Possible molecular arrangements that may account
for such phenomena are discussed by Lee (1975), Lee *et al.*
(1974) and Wu and McConnell (1975).

THE EFFICACY OF FLUIDITY COMPENSATIONS

The fluidity of mitochondrial membranes isolated from 5°C
and 25°C acclimated green sunfish (*Lepomis cyanellus*) illustrate some general features of homeoviscous adaptation in
fish (Fig. 1a). In this study membrane fluidity was estimated using the steady state fluorescence polarisation technique with the probe 1,6-diphenyl-1,3,5-hexatriene (DPH).
The results were interpreted using the Perrin equation to
provide a rotational diffusion coefficient ($\bar{R}$) as a measure
of probe motion during its fluorescence lifetime (Cossins,
1977). The Arrhenius plots of $\bar{R}$ for both cold- and warm-
acclimated fish were linear, suggesting that no major
structural rearrangements occurred over the range 0 to 40°C.
The graph for cold acclimated fish was displaced to a lower
temperature compared with the corresponding graph for warm-
acclimated fish, so that at any measurement temperature $\bar{R}$
was greater in the cold than the warm-acclimated fish. This
is taken to indicate that the mitochondrial membranes of cold-
acclimated fish constrained probe motion somewhat less than
the membrane of warm-acclimated fish and hence were more
fluid (Fig. 1b). Since the Arrhenius plots in Figure 1a
were parallel, the observed shift of the graph along the
temperature axis conforms to what temperature biologists have
termed a 'translation' (Prosser, 1973). However, this
response was somewhat less than 'ideal' or 'perfect' in the
sense proposed by Precht *et al.* (1973) because the membrane
fluidity was not identical at the respective acclimation
temperatures. For an 'ideal' response the Arrhenius plots
for cold- and warm-acclimated fish would have to be separated
by approximately 20°C, to give a ratio of translation to
difference in acclimation temperatures (hereafter termed
'homeoviscous efficacy') of 1. The translation illustrated
in Figure 1a was approximately 8°C, only 40 percent of that
required for perfect compensation of fluidity.

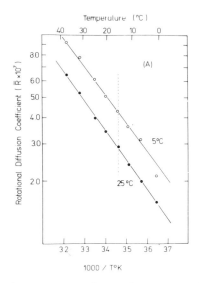
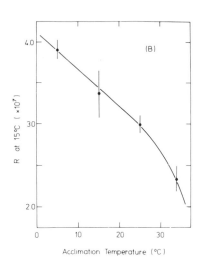

Figure 1. *The effects of thermal acclimation of green sunfish
(Lepomis cyanellus) upon the fluidity of liver mitochondrial
membranes. Membrane fluidity was estimated by the steady
state fluorescence polarisation technique.
(A) Arrhenius plots of the rotational diffusion coefficient,
R, for mitochondria of $5°C$ and $25°C$ acclimated fish. A high
value of R indicates a high membrane fluidity and vice versa.
(B) Membrane fluidity (R at $15°C$) against acclimation
temperature. If the membranes of the differently acclimated
fish were identical then this graph would be horizontal.
The negative slope indicates that the fluidity of the membranes of cold-acclimated fish was greater than the fluidity
of warm-acclimated fish (after Cossins et al., 1980).*

The partial compensation of membrane fluidity as a result
of altered temperature appears to be a general feature of
homeoviscous compensation in fish (Table I). There appear
to be differences in the homeoviscous efficacy of different
membranes from the same organisms. In thermally acclimated
goldfish (*Carassius auratus*) mitochondria show the largest
compensatory response and sarcoplasmic reticulum none at all.
Thompson and Nozawa (1977) have demonstrated by ESR spectroscopy that homeoviscous responses of a similar magnitude
occur in the various cellular membranes of the ciliate
Tetrahymena pyriformis. The homeoviscous efficacy of some
prokaryotes, however, appears to be more impressive (Table
I). For example, Esser and Souza (1974) have shown by ESR
spectroscopy that the fluidity of spheroplasts of *Bacillus
stearothermophilus* was approximately constant when grown

over the range $42°$ to $65°C$, whilst Sinensky (1974) found that
the estimated fluidity of phospholipid extracts of *Escherichia
coli* was identical at growth temperatures between $15°C$ and
$43°C$. Huang *et al.* (1973) demonstrated relatively constant
membrane fluidities at different growth temperatures for
Acholeplasma laidlawii.

However, incomplete compensations have also been observed
in prokaryotes. Fluidity as measured by ESR spectroscopy was
compensated by only 10-20 percent in the outer membrane of
E. coli and by 20 percent in the cytoplasmic membrane (Janoff
et al., 1979) which is in contrast to the results of Sinensky
(1974). The data of Rottem *et al.* (1978) show a homeoviscous
efficacy of 100 percent for phospholipid extracts of *Proteus
mirabilis* grown at different temperatures using the spin
probe 5-doxylstearate, but only 30-50 percent using 12-doxyl-
stearate as probe. These particular probes differ in the
depth of the spectroscopic doxyl moeity within the bilayer
and because of their different rotational characteristics
rather different procedures were used in the interpretation
of their spectra. This observation provides a salutory lesson
in exercising great caution in the quantitative description
of fluidity. It also raises the possibility that fluidity
adjustments during adaptive responses may occur at specific
points in the so-called fluidity gradient (Seelig and Seelig,
1980) that exists across the bilayer.

ADJUSTMENTS OF TRANSITION TEMPERATURES

Many prokaryotes have membranes which display phase transitions or separations at their growth temperature (Melchior and
Stein, 1976). By contrast, most eukaryote membrane systems
are predominantly fluid at physiological temperatures so that
thermotropic phase transitions usually occur at temperatures
well below body or acclimation temperature. In both cases,
however, the profound changes in membrane structure that
result from such transitional effects suggests that adjustments of transition temperatures to preserve the *status quo*
may be of greater importance than the compensation of fluidity
per se. Indeed, a number of studies have shown that adjustments of transition temperatures during thermal acclimation
are generally more complete than compensations of fluidity.
Lipid extracts of *E. coli* display a single phase transition
approximately $15°C$ below their growth temperature over the
range 15 to $43°C$ (Sinensky, 1974). Esser and Souza (1974)
detected a similar ESR discontinuity in membrane preparations
of *B. stearothermophilus* at growth temperatures between 42
and $65°C$.

Table 1

Homeoviscous responses in thermal acclimation

Organism	Membrane	Tech-nique[*]	Range of growth temperatures (°C)	Efficacy[+]	Reference
Escherichia coli	Outer	b	12–43	0.1–0.2	Janoff *et al.* (1979)
	Cytoplasmic	b	12–43	0.2	"
E. coli	Outer	a	20–37	0.25–0.4	Janoff *et al.* (1980)
E. coli	Phospholipid extracts	d	15–43	1.0	Sinensky (1974)
Proteus mirabilis		b	15–43	1.0	Rottem *et al.* (1978)
		c	15–43	0.3–0.5	"
Bacillus stearothermophilus		b	42–58	1.0	Esser and Souza (1974)
Tetrahymena pyriformis	Cilia	b	15–34	0.25	Nozawa *et al.* (1974)
	Pellicle	b	15–34	0.2–0.5	"
	Microsomes	b	15–34	0.25	"

Table 1 (Continued)

Organism	Membrane	Technique [*]	Range of growth temperatures (°C)	Efficacy [+]	Reference
Carassius auratus (goldfish)	Synaptosomal	a	5-25	0.3	Cossins (1977)
	Sarcoplasmic Reticulum	a	5-25	0.0	Cossins *et al.* (1978)
	Brain Synaptic	a	5-25	0.4	Cossins (unpub. obs.)
	Brain Mitochondrial	a	5-25	0.5	"
	Brain Myelin	a	5-25	0.25	"
Lepomis cyanellus (green sunfish)	Liver mitochondrial	a	5-25	0.5	Cossins *et al.* (1980)
	"	e	5-25	0.75	"
	Liver Microsomal	e	5-25	0.4	"

+ *Efficacy was calculated as the ratio of the translation of the fluidity/temperature graph along the temperature axis during acclimation or hibernation, to the difference in acclimation or body temperature. All values were calculated from data supplied in original references.*

* *Techniques (and probes) as follows:- a Steady state fluorescence polarisation (1,6-diphenyl-1,3,5-hexatriene), b Electron spin resonance (5-doxyl stearate), c ESR (15-doxyl stearate or 16-doxyl stearate), d ESR (methyl-12-doxyl stearate), e Differential polarised phase fluorimetry.*

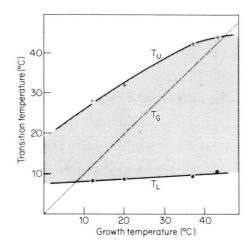

Figure 2. *Adjustments in the lipid phase transition temperatures for the outer membrane fraction of Escherichia coli
when grown at different temperatures. The upper (T_U) and
lower (T_L) limits of the phase transition were determined by
ESR spectroscopy. The shaded area indicates the temperatures
over which a phase separation exists and the dashed line
indicates growth temperature (T_G). (Data from Janoff et al.
1979).*

In a more comprehensive study of *E. coli*, Janoff *et al.*
(1979) observed a gradual transition over a $10-20^{\circ}$C interval.
Two discontinuities in Arrhenius plots of ESR parameters
were detected in the cytoplasmic and outer membranes and were
interpreted as the upper and lower limits for the phase
separation. The temperature for the lower discontinuity was
relatively constant when growth temperature was varied. The
temperature for the upper discontinuity was increased as
growth temperature increased (Fig. 2), with the net result
that the outer membranes always existed within the limits for
the proposed transitions. The authors concluded that the
temperature range over which growth occurs in *E. coli* correlates closely with the temperature range over which mixed
phases exist in the outer membrane.

Nakayama *et al.* (1980) have also studied the effects of
altered growth temperature on *E. coli* membranes using an X-
ray diffraction technique. A discontinuity was detected
which was interpreted as the temperature at which the membrane
was entirely fluid. This presumably corresponds to the upper
transition detected by ESR spectroscopy by Janoff *et al.*
(1979) except that the transition temperatures observed by
Nakayama *et al.* (1980) were $10-15^{\circ}$C lower. In addition

Nakayama *et al.* (1980) found that the transition for the
cytoplasmic membrane was considerably more mobile when growth
temperature was altered than was the transition for the outer
membrane with the result that at higher growth temperatures
the outer membrane was entirely fluid. Janoff *et al.* (1980)
have recently compared the discontinuities detected by ESR
spectroscopy with those detected by fluorescence polarisation.
The temperatures for the upper discontinuity were in good
agreement whilst the temperatures for the lower discontinuity
were rather different. These discrepancies highlight the
problems of comparing studies using different analytical
techniques and serves to emphasise the point that each technique may report rather different information.

     Abbas and Card (1980) have used differential scanning
calorimetry to determine temperatures for the bulk phase
transition in the bacterium *Yersinia enterocolitica.* The
mid-points for these transitions were -13, -9 and 1°C for
membranes from cells grown at 5, 22 and 37°C, respectively.
Since these transitions occurred well below all growth
temperatures, it may be that this adjustment is directed more
towards providing an appropriate level of fluidity.

     In the eukaryote *Tetrahymena*, Kasai *et al.* (1976) and
Martin *et al.* (1976) have observed phase separations by
freeze-fracture electron microscopy. The degree of particle
aggregation in the various cellular membranes at different
incubation temperatures was used as a semi-quantitative index
(the particle density index, PDI) to describe the onset and
completion of the phase separation. Comparison of alveolar
membranes from *Tetrahymena* grown at 39.5 C and 15°C (Fig. 3)
indicated a shift of the PDI to temperature curve of approximately 25°C, so that in both cases the onset of phase
separation occurred well below growth temperature but at the
same temperature interval. The magnitude of this difference
in temperature for phase separation at the different growth
temperatures contrasts with the comparatively small compensation of membrane fluidity as detected by spectroscopic
probes (Nozawa *et al.*, 1974).

EVOLUTIONARY ADAPTATION OF MEMBRANE FLUIDITY

Goldfish and green sunfish inhabit environments which experience wide seasonal fluctuations in temperature, and these
eurythermal fish are able to adjust the structure of some
cellular membranes to maintain them at least partially independent of seasonal temperature changes. Other fish and
mammals live in relatively extreme and unvarying thermal
environments. For example, some fish of the family

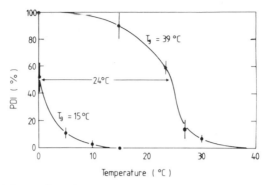

Figure 3. *Shifts in the temperatures that cause lateral phase separations in the outer alveolar membranes of Tetra- hymena pyriformis as a result of altered growth temperature. Cells were grown at 15°C and 39°C and incubated at various temperatures between 0°C and 33°C for 5 minutes. After incubation the cells were fixed with 1% glutaraldehyde. The degree of aggregation of the intramembranous particles at each incubation temperature was visualised on freeze-fracture replicas and quantified as a particle density index (PDI). (Modified after Martin et al., 1976).*

Nototheniidae live exclusively in the Antarctic Ocean at approximately -1.9°C throughout the year, whilst mammals and birds have a core temperature that deviates little from 37- 42°C. If the concept of an optimal membrane state applies to homologous membranes of these various species then one might expect that they would show adaptive differences in membrane fluidities to suit them for their respective thermal environments. Furthermore, it might be expected that sel- ection pressure for adaptation to extreme thermal environments would be severe. When these extreme environments are also thermally stable, it is possible that the potential for seasonal flexibility exhibited by eurythermal species such as goldfish need not be maintained. This may allow a more complete adaptation over evolutionary time than might be possible in a eurythermal species. This hypothesis was examined by comparing the fluidity of the synaptosomal membranes of the eurythermal goldfish and green sunfish with two relatively stenothermal fish species, the arctic sculpin, *Myoxocephalus verrucosus* (whose normal habitat temperature is 0°C) and the desert pupfish, *Cyprinodon nevodensis* (34°C). In these experiments fluidity was estimated from polarisation measurements; a high value indicated low fluidity and *vice versa*.

Arrhenius plots of polarisation for synaptosomal prepar-

ations of the various species (Fig. 4a) show a clear shift
to lower temperatures for species with lower acclimation or
body temperatures. Thus the synaptosomal membranes of the
arctic sculpin were more fluid than those of the 5°C-treated
goldfish, whilst the membranes of the desert pupfish had a
fluidity between those of the 25°C-treated goldfish and the
rat (37°C). The graph for the arctic sculpin was shifted
approximately 5°C with respect to the 5°C-treated goldfish
which corresponds to the difference in acclimation temperatures. For this particular comparison there is thus complete
or ideal fluidity compensation (Fig. 4b). Similarly, the
desert pupfish and rat were shifted 9°C and 12°C with respect
to the 25°C-treated goldfish, again corresponding to their
respective differences in acclimation or body temperature
(Fig. 4b). However, the polarisation values obtained for the
arctic sculpin and 5°C-treated goldfish at their respective
cellular temperatures were different from those obtained for
25°C-treated goldfish, desert pupfish and rat indicating that
all species did not adapt to similar levels of fluidity. The
intermediate polarisation value obtained for goldfish at
15°C suggests that this eurythermal fish can not only adapt
to either extreme, but also to temperatures in between (Fig.
4b). The observation that rat, hamster and 5°C and 25°C-
treated green sunfish fall into the same general sequence
(Cossins and Prosser, 1978) also suggests a general relationship between cellular temperature and synaptosomal membrane
fluidity which crosses broad phylogenetic boundaries.

A major problem of interspecific comparisons such as this,
is the basic assumption that the homologous preparations from
different species are purified to the same extent and contain
the same distribution of membrane-types. The possibility
remains that some of the observed differences were due to
preparative inconsistencies.

In a study of muscle sarcoplasmic reticulum, Morse et al.
(1975) observed that the ESR probe, 2-heptyl-2-hexyl-5,5-
dimethyloxazolidene-N-oxyl, was significantly more immobilised
in vesicles prepared from rabbit than those from lobster.
The Arrhenius plots of an empirical motion parameter for
rabbit vesicles were shifted by $10-14^{\circ}$C with respect to the
plot for lobster vesicles, which is approximately half of the
$20-30^{\circ}$C difference in body temperatures of these animals.
On the other hand, Cossins et al. (1978) were unable to
detect consistent differences in the fluidity of sarcoplasmic
reticulum vesicles of goldfish, arctic sculpin, desert pupfish and rat.

Raison et al. (1971a) have shown Arrhenius plots for a
semi-empirical rotational correlation time for spin probes

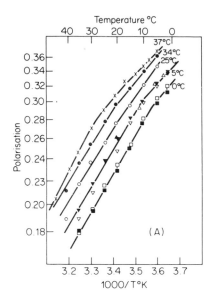

Figure 4a. *Evolutionary adaptation of membrane fluidity to temperature.*
(A) Arrhenius plots of fluorescence polarisation for 1,6-diphenyl-1,3,5-hexatriene in brain synaptosomes of the Arctic sculpin (0°C, □ , ■), 5°C-acclimated goldfish (▼ , △ , ▽), 25°C-acclimated goldfish (O), desert pupfish (34°C, ●) and rat (37°C, X). Each symbol represents a separate preparation. A high value of polarisation indicates a low membrane fluidity.

in liver mitochondria of rat and trout. Although no comparison of fluidity was explicitly made by these authors it is quite clear that the membranes of trout displayed considerably greater fluidity than the corresponding rat membranes. In addition, the Arrhenius plot for rat liver showed a discontinuity at approximately 23°C, whilst the plot for trout liver was entirely straight over the range 0-35°C. On the other hand, McMurchie *et al.* (1973) using similar techniques have demonstrated that toad heart mitochondrial membranes were less fluid than rabbit heart mitochondrial membranes, whilst the reverse was true of heart plasma membranes and endoplasmic reticulum membranes of these animals. Discontinuities were again observed in Arrhenius plots of rabbit membranes but not in toad membranes.

Generally, it seems that mammals display relatively viscous cellular membranes when compared with cold-adapted or cold-acclimated poikilotherms. Arrhenius discontinuities in the activity of membrane-bound enzymes of several endotherms

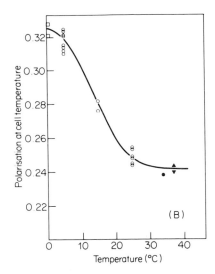

Figure 4b. *Membrane fluidity (as expressed by polarisation) at acclimation or body temperatures against temperature. If the different species were 'ideally' adapted to their respective temperatures then the graph would be horizontal. Note the two horizontal portions at 0-5°C and 25-37°C. Arctic sculpin (□), goldfish (O), pupfish (●), rat (▲) and hamster (▼). Each point represents an individual animal.*

have been interpreted as transitions of lipid order (McMurchie *et al.*, 1973; Raison and McMurchie, 1974; Raison *et al.*, 1971a, 1971b). These discontinuities usually occur between 8°C and 23°C which is well below the body temperature of mammals and birds. However, differential scanning calorimetry has revealed phase transitions at lower temperatures which certainly do not coincide with those in Arrhenius plots. For example, Hackenbrock *et al.* (1976) found bulk transitions in mitochondria of rat liver below 0°C whilst Martonosi (1974) found similar transitions in sarcoplasmic reticulum between 0°C and 5°C. In other instances, calorimetric transitions show good agreement with transitions detected by probe methods (*e.g.* Brasitus *et al.*, 1980). Charnock *et al.* (1980) detected calorimetric events between 18°C and 26°C in phospholipid extracts of hamster myocardial membranes, but these involved less than 2 percent of the membrane lipid and did not appear to influence the Arrhenius discontinuity of the $(Na^{+}+K^{+})$-ATPase. Whatever the case, such bulk thermotropic transitions have little physiological significance for normothermic mammals.

Poikilotherms, on the other hand, usually do not display
Arrhenius discontinuities for probe rotation or for membrane
bound enzyme activity over the biological temperature range,
although there are exceptions (Wodtke, 1976), and it seems
unlikely that they suffer lipid transitions at any temperature
above 0°C. With respect to both dynamic motion and structural
order the membranes of different vertebrate species are
clearly adapted to their respective body and environmental
temperatures, though at present it is not possible to conclude
that these properties are equivalent at their respective cellular temperatures. The presumed optimal condition for any
particular membrane therefore, may not have any absolute significance but may vary between species or, indeed, between
different cell-types of an individual to suit the respective
functional requirements of each particular membrane.

HOMEOVISCOUS ADAPTATION DURING HIBERNATION

The rapid and dramatic reduction in body temperature suffered
by hibernating mammals would appear to be an obvious case in
which an adjustment of membrane fluidity would have considerable adaptive value.
    However there are important differences in the phenomena
of temperature acclimation of fish and mammalian hibernation
which provide reason to question whether a homeoviscous
response *per se* is appropriate during hibernation. Firstly,
the adaptive strategy of fish during cold acclimation is
compensatory so that comparable levels of activity are possible at each acclimated temperature. By contrast, the adaptive strategy of hibernation is exploitative, in that energy
expenditure is dramatically reduced during a period of reduced food availability and inclement conditions (Wang, 1979)
by the relaxation of homeothermy so that body processes occur
at a very small fraction of their normal rate. Provided the
vital functions do not fail there is no reason to look for
compensatory cellular mechanisms for hibernation; hibernation
should be viewed as a resistance adaptation, not a capacity
adaptation. Secondly, the spontaneous and transient arousals
that occur during hibernation result in rapid and large
changes in body temperature (Hudson, 1967). It hardly seems
feasible for an appreciable homeoviscous response to take
effect during these brief transitional periods, so that it
becomes necessary to decide whether it is strategically most
advantageous for the animal to adapt membrane fluidity and
function for hibernating or for normothermic temperatures.
In view of the more demanding functional requirements of
normothermic mammals, it would seem more reasonable to expect

membrane fluidity to be arranged for optimal function at
normal body temperatures. A possible adaptive mechanism which
provides for a satisfactory membrane state at both hibernating
and normal body temperature without a homeoviscous response,
is a reduction in the temperature dependence of membrane
fluidity. At present, there is no experimental support for
such a 'multigrade' membrane viscosity. This topic is
discussed in detail elsewhere (Willis *et al.*, this volume).

CONCLUSIONS

Phenomena such as those described here have supported the
concept of an optimal or most favourable state for membrane
structure. At present we are only beginning to understand
the complexity of organisation of cellular membranes, and so
specific statements concerning the nature of the optimal
condition must await a more complete knowledge of the functional significance of dynamic motion, structural order and
their microheterogeneity. The recent observation of non-
bilayer configurations of phospholipids in both model and
natural membranes opens up exciting new horizons, whose implications for adaptive responses in membranes will take some
time to assess (Cullis and De Kruijff, 1979).
    The overall degree of fluidity in all its diverse expres-
sions is likely to reflect a compromise between the rate-
depressing effects of molecular order on matrix functions,
and the rate-enhancing effects of disorder on barrier properties, although the specific functions of particular membrane-types may introduce other factors. It is possible that
the elaborate microheterogeneity found in some membrame systems
and the so-called 'boundary' lipids of certain membrane-bound
enzymes provides for the specialisation of particular micro-
domains for specific purposes. We must at least consider the
possibility that these different microenvironments are
adapted more or less independently during homeoviscous
adaptation, although the experimental approach to this question is difficult at present.
    The possibility of modulating the effective solvent vis-
cosity for adaptive purposes is one that occurs only in the
hydrophobic compartment of cellular membranes. As an adaptive strategy it is particularly potent because it effects all
functions and processes of membranes that are influenced by
membrane fluidity. In principle, the homeoviscous process
can be very effective over a wide range of temperatures since
the insertion of only one olefinic bond into a saturated acyl
chain produces a reduction of almost $70^{\circ}C$ in the gel to
liquid-crystalline transition temperature. It is hard to

conceive of similar variations in aqueous media without
incurring non-physiological variations in osmotic pressure
or the hydrogen bonding sytem of water (Franks, this volume).
The incorporation of subsequent olefinic bonds into acyl
chains results in progressively smaller effects on the physical properties of phospholipids and there is good reason to
question the structural significance of variations in the
proportion of highly polyunsaturated fatty acids in membrane
phospholipids. Indeed, the adaptive response often correlates
more closely with shifts in the proportion of saturated to
unsaturated fatty acids rather than with changes in the unsaturation index, which is a measure of the average unsaturation of a mixture of acyl chains (Cossins *et al.*, 1977).

In view of the apparently large possibilities for adaptive
responses in membrane systems it is a little discouraging to
observe that in many instances the homeoviscous response is
less than half that which is required to maintain the constancy
of membrane structure. However, even relatively small fluidity
compensations provide an organism with some competitive
advantage which will be favoured by natural selection, and
teleologically the numerous observations of such responses
testifies for its adaptive value. In addition, there are two
reasons why use of the terms 'ideal' or 'perfect' to describe
the extent of the compensation may be inappropriate. Firstly,
it is possible that the optimal membrane condition to which
homeoviscous adjustments are directed may itself vary with
temperature so that the labels 'partial' or 'complete' may
have no absolute significance. Secondly, the homeoviscous
response is probably only one of several adaptive responses,
which together result in a more complete and effective compensation of specific functional properties. For example,
the concentration of electron transport enzymes in mitochondrial membranes is known to increase by up to 50% during cold
acclimation of fish (Sidell, 1977) resulting in an enhanced
cellular catalytic activity. This, combined with the higher
turnover number resulting from a more fluid hydrophobic
environment (Sinensky *et al.*, 1979) may permit comparable
catalytic rates at the different acclimation temperature.

Although many eukaryote membranes exist in a predominantly
fluid condition a reduction in temperature may result not
only in a reduced fluidity but also in structural rearrangements which, though not as profound as a gel to liquid-
crystalline phase transition, still may have deleterious
effects upon membrane function. For membranes whose optimal
condition requires some degree of microheterogeneity of lipid
organisation, such as those of many prokaryotes, a small
increase or decrease in temperature may result in similar

and perhaps more dramatic structural alterations. It is not
surprising, therefore, that the more complete compensations
of fluidity have been recorded in the latter case. However,
adaptive processes in prokaryotes differ generally from those
described in higher animals in one other important respect,
namely that in the former they are induced over many gener-
ations, whilst in the latter they may become expressed within
the lifetime of the individual. The natural selection of
genetic variants with specific attributes from a population
would, in principle, permit greater flexibility of response
than may be expected from the differential expression of a
single genotype. Indeed, selection is known to produce
appreciable effects on the temperature biology of both animals
and microorganisms over a relatively small number of gener-
ations (Morrison and Milkman, 1978) and there is good reason
to belive that phenotype adaptations in microorganisms are of
little importance. The efficacy of evolutionary adaptation
is supported from our studies with vertebrates in which the
more dramatic adaptations of membrane fluidity were noted for
animals that had evolved in the more extreme and stable thermal
habitats (Cossins and Prosser, 1978).

The concept of an optimal membrane structure has also been
extended to consider the long-term effects of other factors
which directly affect membrane fluidity. Homeoviscous responses have been invoked to explain the lack of differences
in membrane fluidity of lipid-supplemented murine fibroblast
LM cells (Schroeder, 1978) and also may be involved during
fatty acid supplementation of *Dictyostelium* (Herring *et al.*,
1980). Another example is provided by many hydrophobic drugs
which partition into cellular membranes and cause a general
fluidisation of the hydrophobic interior. Hill and Bangham
(1975) have suggested that organisms respond to this fluidising effect in much the same way as they respond to the
fluidising effects of increased temperature, by a homeostatic
reduction in fluidity. An attractive feature of this hypothesis
is that it provides possible mechanisms to account for two
commonly observed phenomena of chronic drug treatment, namely
increased tolerance and withdrawal syndrome (Goldstein, 1976).
The adapted state would only be effective during the continued
presence of the drug and its sudden removal would result in a
maladapted membrane which would require re-adaptation to its
original state.

Chin and Goldstein (1977) have recently reported a reduced
fluidising effect of ethanol upon membranes of ethanol-tolerant
mice compared to control mice. Subsequently, these workers
claimed that the principle compositional difference between
these membrane preparations was the increased cholesterol to

phospholipid ratio in the former (Chin *et al.*, 1978) although
there are also differences in the acyl group composition
(littleton and John, 1977).   Nandini-Kishore *et al.* (1977)
have shown that *Tetrahymena* responds to the fluidising
anaesthetic methoxyflurane by decreasing its fatty acid desaturase activity in much the same way as was observed during
growth at high temperature (but see Clarke, this volume).   In
view of the demonstrated adaptive abilities of goldfish membranes we are currently studying the chronic effects of
fluidising drugs on the fluidity and composition of their
brain membranes.

PHASING OUT THE SODIUM PUMP

J.C. Ellory

Department of Physiology, University of Cambridge, England.

J.S. Willis

*Department of Physiology and Biophysics, University of
Illinois, USA.*

INTRODUCTION

There seem to be two main objectives in studying the effect
of changing temperature on the activity of the sodium pump.
In the first case, temperature effects are followed in the
context of biological performance, where the functioning of
cells or organisms over a range of temperatures may critically
depend on their ionic regulation. Special cases like acclimation and hibernation are included in this approach. In
the second case temperature is used as an experimental parameter to investigate the mechanism of sodium pump activity,
looking at the relevance of activation energies, phase transitions and differential effects on partial reactions. In
the present short review we intend to consider both of these
approaches, hoping to show that knowledge from the second
may help us understand the first.

What Does the Sodium Pump do?

In virtually all animal cells the dominant mechanism for
maintaining intracellular ionic composition is the sodium
pump, which transports potassium ions into the cell, and
sodium ions out. Essential functions controlled by this
system include volume regulation (Hodgkin, 1958); the
conservation of high intracellular K^+ levels for maintaining
membrane potential and facilitating essential biochemical

reactions (*e.g.* protein synthesis, glycolysis, glutathione biosynthesis: Kernan, 1980; Willis, 1972); the maintenance of transmembrane Na^+ gradients for electrical events and secondary active transport of sugars, amino acids, Ca^{2+} and Cl^- (*e.g.* Schultz, 1977); and a significant direct electrogenic contribution to the membrane potential in some cell types (Thomas, 1972).

In terms of cellular functioning of the sodium pump with changing temperature, there are three ways in which cell response can be modified, namely:-

1) Direct effects of temperature on the pump mechanism *per se.*
2) The effect of temperature on cell membrane permeability to other ions (this will act not only for the obvious pump ligands Na^+, K^+ and Mg^{2+}, but also other ions which affect the pump including Ca^{2+}, H^+, PO_4^{2-} and VO_3^- and will include the response to changing temperature of other transport systems and cell organelles such as mitochondria).
3) Cell metabolism and in particular the availability of substrate ATP (and ADP and P_i) will be temperature sensitive.

The effect of cooling on metabolism is outside the present scope, and has been touched on recently by Willis (1979) and Behrisch (1978). The interactions of other transport systems affecting sodium pump activity is relevant for considering the cellular response to cooling, and we will therefore look at selected examples where these effects are important.

Mechanism of the Sodium Pump

The detailed reaction mechanism of the sodium pump is complex and the subject of intensive research. Many recent reviews are available (Glynn and Karlish, 1975; Robinson and Flashner, 1979; Stekhoven and Bonting, 1981). Briefly the normal mode of pump function involves the outward transport of three Na ions, the inward transport of two K ions and the hydrolysis of one molecule of ATP. The reaction is specifically inhibited by the cardiac glycoside ouabain. The enzymic correlate of ion transport is thus Na^+, K^+ activated ATPase activity. Under certain conditions the pump will perform modified ion transport with equivalent biochemical events. These functions are called partial reactions, and are particularly relevant to temperature studies since they show a different temperature sensitivity from that of the overall ATPase reaction (Barnett and Palazotto, 1974, and see below).

In the absence of external K^+ and the presence of external and internal Na^+ the pump will perform ouabain-sensitive

Na^+-Na^+ exchange which is dependent on ADP but requires ATP
to be present and is correlated biochemically with ATP-ADP
exchange (Glynn and Hoffman, 1971). Similarly in the absence
of intracellular Na^+ the pump will perform a K^+-K^+ exchange
dependent on intracellular phosphate and ATP (Simons, 1974).
Other important biochemical partial reactions of the sodium
pump include ouabain-binding promoted by either Mg^{2+}, Na^+ and
ATP, or Mg^{2+} and P_i; K^+-dependent phosphatase activity using
a model substrate, for example p-nitrophenyl phosphate
(pNPPase) and phosphorylation which entails reaction of the
enzyme with ATP-^{32}P in the presence of Mg^{2+} and Na^+ to produce
an acid-stable phosphorylated intermediate, which can be
dephosphorylated by K^+ ions. These biochemical approaches,
coupled with fluorescence and ligand binding studies have led
to the development of reaction schemes for the sodium pump
such as that given in Figure 1. Interconversion of the two
forms of the enzyme E_1 and E_2 (Na^+-preferring to K^+-preferring)
is critical for the whole pump sequence, and an important and
obvious consideration to explore in temperature studies is
the possibility that certain steps are more sensitive than
others, affecting partial reactions of the pump differentially.

Technical Problems

It is convenient to mention at the start certain technical
problems which should have influenced experimental design,
and will influence our interpretation of the results. The
principal problems concern changes in pH, ligand affinity and
inhibitor binding. The fact that the neutral point of water

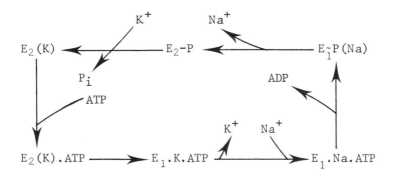

Figure 1. *Diagram of some intermediate steps in the ATPase
reaction sequence, simplified after Glynn and Karlish, 1980.
Reactants inside the square are intracellular, those outside
extracellular.*

(pN) changes with temperature is well known, (Franks, this volume) and has important biological implications (Reeves, 1977). Experimenters are faced with the choice of maintaining either pH or pN constant with temperature, or using a "biological" buffer such as imidazole which may mimic the buffer temperature dependence of the enzyme being studied (Reeves, 1977). The problem has been considered systematically by Park and Hong, 1976, who plotted isopleths for (Na^++K^+)-ATPase activity/pH at different temperatures. For frog skin the pH optima are rather narrow and shift with cooling. Other workers have approached the issue more pragmatically and concluded that the pH isopleth was rather broad, and could be ignored (Sperelakis and Lee, 1971; Willis and Li, 1969).

Changes in affinity for ligands and substrate as a function of temperature may give a clue to reaction mechanisms. This is an area which has been extensively studied (Dixon and Webb, 1964; Hazel and Prosser, 1974; Hochachka and Somero, 1973). Nevertheless, if K_m values change with temperature, as they usually do, experimentally determined enzyme velocities may be at saturating substrate concentrations over part of a temperature range, but not the whole. This can even lead to convincing apparent discontinuities in Arrhenius plots (Silvius et al., 1978). A special case of affinity changes relates to inhibitor binding. Ouabain, as a specific high affinity inhibitor of the pump has a lower apparent affinity and decreased binding rate to (Na^++K^+)-ATPase at low temperatures (Baker and Willis, 1972; Willis, 1969) which limits the usefulness of this probe for certain preparations in the cold.

Although these considerations are easy to control for with isolated ATPase, they are more difficult to maintain with whole cell preparations. When direct comparisons have been made between temperature dependence of sodium pump activity in intact cells and (Na^++K^+)-ATPase activity in a membrane preparation derived from these cells a disparity has been observed (Willis and Li, 1969; Willis et al., 1978), where the cell has consistently performed better than the membrane preparation. Possible explanations for this effect include changes in membrane structure during preparation such as lipid rearrangement in the bilayer, calcium effects, cholesterol loss, partial proteolysis or removal of other modifying proteins.

STUDIES ON CELLS

Although there has been much work on the effects of temperature on cellular Na^+ and K^+ regulation, we are going to

confine the present review to three areas, excitable tissues,
epithelia and erythrocytes. Tissues from ectotherms might be
expected to differ from those of endotherms in terms of temperature adaptation, since they often function over a much
broader temperature range. However, this is not always the
case (Willis, 1978; Willis *et al.*, this volume).

Excitable Tissues

For obvious reasons, repetitively firing excitable cells can
be subjected to relatively large changes in their intracellular sodium concentration. Since certain ectotherms routinely
experience large environmental temperature gradients, thermal
compensatory effects must exist in their nervous systems and
muscle fibres, to accommodate high Na^+ transport rates at
varying temperatures. An important property of the sodium
pump in excitable tissues is its electrogenic component
(Thomas, 1972), which can amount to a 10-20 mV contribution
to the membrane potential (Carpenter and Alving, 1968;
Merickel and Kater, 1974). The electrogenicity arises from
the pump coupling ratio for Na^+:K^+ deviating from unity.
This could arise from the 3:2 stoichiometry shown in erythrocytes (Garrahan and Glynn, 1967; Post and Jolly, 1956) together with a low anion permeability, but could also result
from a contribution from uncoupled Na^+ efflux through the
pump, or a variation in the number of Na and K ions trans-
ported per pump cycle (Sachs, 1977).

   A number of studies have analysed the temperature dependence of membrane potential in molluscan (*Aplysia, Helix,
Anisodoris*) and annelid (*Lumbricus*) neurons. In *Aplysia*
neurons, up to a third of the membrane potential could be
attributed to an electrogenic Na^+ pump at high temperatures
(20-25°C), the pump being inhibited on cooling to 5°C
(Carpenter and Alving, 1968; Marchiafara, 1970). Similar
experiments with *Anisodoris* (Gorman and Marmor, 1970),
Helisoma (Merickel and Kater, 1974) and *Helix* (Zecévic and
Levitan, 1977) confirm the existence of a high temperature
sensitivity of the electrogenic sodium pump. Pump activity
was measured as a current, and specifity established by
adding ouabain or removing K^+, or injecting Na^+ intracellularly.

   In terms of nerve function, detailed analyses confirmed
that the temperature dependence of the other membrane components, including the Hodgkin Huxley conductances, the
transient K^+ conductance (Connor and Stevens, 1971; Partridge
and Connor, 1978) and the passive Na^+ and K^+ permeabilities
(Fishbarg, 1971; Gorman and Marmor, 1970; Zecévic and Levitan,

1977) all interact to compensate temperature effects on the repetitive firing frequency.

The important consequence for Na^+ pump studies is that in these cells the electrogenic component is K^+-dependent and highly temperature-sensitive. Studies on the isolated enzyme indicate that Na^+-ATPase activity (and by inference uncoupled Na^+ efflux) is less temperature sensitive than total (Na^++K^+)-ATPase activity (Blostein, 1970; Neufeld and Levy, 1970) suggesting that uncoupled Na^+ efflux might manifest itself at low temperatures, a prediction not borne out by the electrophysiological studies.

Zecévic and Levitan (1977) examined temperature acclimation in *Helix* and concluded that there was no adaptive change in Na^+ pump activity in their animals, the compensatory change occurring in the relative membrane permeabilities to Na^+ and Cl^-. Similar results were found for *Lumbricus* neurons (Dierolf and McDonald, 1969) although Lagerspetz *et al.* (1973, see also Lagerspetz, 1974) did find evidence for temperature acclimation in earthworms.

The message from excitable tissues seems to be that the electrogenic Na^+ pump operates in the orthodox mode, and fails at low temperatures. Adaptation does not necessarily involve any change in pump activity.

Epithelia

Historically, cooling has been a favoured approach for slowing or abolishing facilitated transport processes. Most of the work on temperature-dependence of epithelial transport has been directed towards using cold as an inhibitor to establish the presence of various transport processes and to characterise them. Since isolated (Na^++K^+)-ATPase usually has a complex, non-linear Arrhenius plot (see below) several groups have proposed that a parallelism between temperature dependence of the short circuit current and (Na^++K^+)-ATPase activity is good evidence for this enzyme driving transepithelial sodium transport. Data of this kind include the work of Kawada *et al.* (1975), Park and Hong (1976) on frog skin; Asano *et al.* (1970) on frog bladder, and Augustus (1976) on salivary gland ductal epithelium.

Temperature studies on transport have sometimes indicated that other mechanisms outside the sodium pump, or differential temperature effects on the asymmetric membranes of epithelial cells are operating. Thus Laugesen *et al.* (1974) working on cat submandibular salivary gland used temperature to dissociate transport rates at the luminal and basolateral membranes. Biagi and Geibisch (1979) investigated transport

in isolated proximal tubules and showed by temperature dependence that coupled Na^+-glucose/amino acid transport was
the rate limiting step. Berthon *et al*. (1980) have shown
that isolated rat hepatocytes undergo transient changes in
K^+ and Na^+ permeabilities consequent on cell swelling on
cooling. Other considerations are an elevation of intracellular Ca^{2+} due to cold sensitivity of the calcium pump,
and of course, direct effects on metabolism.

Temperature acclimation studies in epithelia have produced
the usual conflicting series of results. Kawada *et al*.
(1975) could find no significant acclimation of either short
circuit current or (Na^++K^+)-ATPase activity in frog skin.
By contrast Lagerspetz and Skytta (1979) demonstrated perfect
temperature compensation of short circuit current, although
there was no significant change in (Na^++K^+)-ATPase activity.
This indicated thermal regulation at another site, presumably
the Na^+ entry step on the apical membrane.

In an extensive series of experiments on goldfish intestine
by Smith (1976), three adaptive steps are proposed. The most
rapid response (15-20 h) to thermal transition is a change
in the microvillar permeability to Na^+, and actively-trans-
ported amino acids. A second change (32-48 h later) involves
alterations in membrane phospholipids, but acclimation is not
complete until 2-3 weeks later, when there is a final regulation of sodium pump activity, involving an alteration in
pump turnover rather than the number of sites, and probably
connected with *de novo* synthesis. Obviously again in this
case, regulation is occurring largely through changes in
passive cell permeability.

Our conclusion from this work is that in cold sensitivity
of epithelial transport is governed by changes in passive Na^+
and K^+ permeability through other transport systems outside
the sodium pump.

Erythrocytes

Red cells represent a convenient model system for studying
temperature effects on membrane transport. Cold cation loading of red cells was used historically to investigate active
Na and K transport on subsequent warming (Harris, 1941;
Maizels and Patterson, 1940), giving the best early data on
pump coupling ratios (Post and Jolly, 1957). Early experiments were directed towards comparing Na^+ pump activity at
24 and $37^{\circ}C$ to calculate the activation energy and compare
ion transport and ATPase activity (Glynn, 1957). In common
with other cell types and tissues, erythrocytes have the
problem of the presence of several separate cation transport
systems. Thus, Na^+ crosses the membrane in human erythrocytes

via the Na^+ pump, the Na^++K^+ cotransport system (Dunham *et al.*, 1980; Wiley and Cooper, 1974), the Li^+-Na^+ counter-transporter (Pandey *et al.*, 1979), $NaCO_3^-$ on the Band 3 anion exchanger (Funder and Wieth, 1980), some Na^+-dependent amino acid transport systems (Young *et al.*, 1980) and "passive" or residual permeability. These systems can be dissected with inhibitors, particularly using ouabain and bumetanide (or furosemide). Sodium pump activity has been measured as a function of temperature in erythrocytes of human (Fig. 2), sheep (Joiner and Lauf, 1979), guinea pig (Kimzey and Willis, 1971a) wood chuck (Ellory and Willis, 1978) and ground squirrel (Kimzey and Willis, 1971b; Willis and Ellory, 1982). There are additional data comparing many species at 5 and $37^{\circ}C$ (Willis *et al.*, 1980) which showed the Na^+ pump activity was 0.2-0.7 percent at $5^{\circ}C$ of that at $37^{\circ}C$ for 8 species of hibernator compared with a value of 1.8 percent for hibernator cells. Measurement of apparent K_m for external K^+ and internal Na^+ indicated an increased affinity of 3-fold at the lower temperature.

More detailed data for human erythrocyte Na^+ pump (recalculated from Stewart *et al.*, 1980) are presented as an Arrhenius plot in Figure 2, showing an apparent discontinuity at around $22^{\circ}C$. Similar studies for sheep cells gave different results. Cells from high potassium sheep showed an apparent discontinuity around $29^{\circ}C$, whilst low potassium cells showed linear Arrhenius plots (Joiner and Lauf, 1979).

Stewart *et al.* (1980) also presented data for temperature dependence of the Na^+-K^+ cotransport system, and residual (*i.e.* ouabain and bumetanide insensitive) K^+ influx and Na^+ efflux. The cotransport system was also highly temperature-sensitive and was unmeasurable at temperatures below $8^{\circ}C$. Surprisingly the passive K^+ and Na^+ permeability showed a paradoxical minimum around $12^{\circ}C$, subsequently increasing as temperature was lowered. This effect was demonstrable for both Na^+ and K^+, and similar results have been demonstrated for rat (Friedman *et al.*, 1977) and dog (Elford and Solomon, 1974) erythrocytes, and for human cells whose permeability has been increased with salicylate or thiocyanate (Wieth, 1970). The physical basis of this effect on passive permeability is obscure, since we do not have a clear idea of the way ions cross membranes by "passive permeability".

In terms of cell function, the increasing passive permeability at low temperatures exacerbates the failures of the pump at low temperature, and results in a gain of Na^+ and loss of K^+ at temperatures below $15^{\circ}C$ (see also Kimzey and Willis, 1971a,b).

Thus compensation in erythrocytes is relatively poor, in

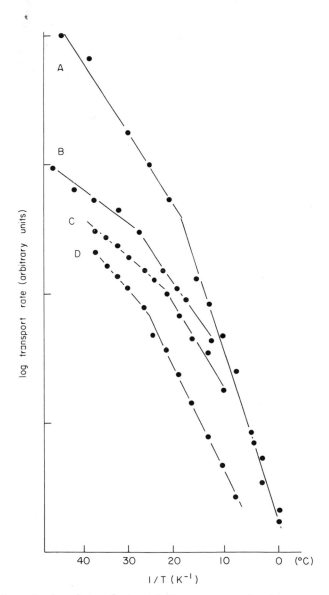

Figure 2. *Arrhenius plots of four separate transport systems in human erythrocytes. A) leucine transport (data redrawn and scaled from Hoare, 1972) B) glucose transport (data redrawn and scaled from Sen and Widdas, 1962) C) Na^+-K^+-co-transport (furosemide-sensitive)(redrawn from Stewart et al., 1980) D) ouabain-sensitive K^+ influx (redrawn from Stewart et al., 1980).*

terms of thermal response. Nevertheless, it is significant
that the sodium pump in erythrocytes performs considerably
better than (Na^++K^+)-ATPase activity measured in isolated
membrane fragments (Willis et al., 1978).

ISOLATED ATPase STUDIES

So far we have been discussing the cellular response to
cooling, pointing out the effects of factors outside the
sodium pump which influence the overall behaviour of the cell.
To look at the effect of temperature on the sodium pump itself we must consider the wealth of literature on (Na^++K^+)-
ATPase, because measurement of ATPase activity in broken
membrane fragments is the most frequent parameter for detailed
temperature studies. These experiments can be considered
under two headings. One concerns the influence of the membrane environment, particularly the lipid, on the ATPase
activity, whilst the second is concerned with the reaction
sequence as a function of temperature; that is, does cooling
stabilise a particular intermediate and slow the overall
ATPase activity via a specific reaction?

Lipid Effects

Almost all thermal studies on (Na^++K^+)-ATPase directed
towards its membrane environment have been presented in
terms of Arrhenius plots (Charnock, 1978), pursuing the
successful concepts of phase transitions or separations
developed in prokaryotic systems, and often correlating results with lipid analyses or physical measurements which
report (with certain reservations) on membrane environment
(Grisham and Barnett, 1973; Kimelberg and Papahadjopoulos,
1974; Shimshick and McConnel, 1973; Wisnieski et al., 1974;
Wu and McConnell, 1973, but see Gordon et al., 1978; Thilo
et al., 1977; Trauble and Eibl, 1974; Cossins, this volume).
The lipid requirements of (Na^++K^+)-ATPase have been the
subject of considerable research and dispute (Fourcans and
Jain, 1974; Stekhoven and Bonting, 1981; Tanaka, 1974).
Part of the problem seems to reside in the methods of de-
lipidation, the role of detergent-replacement and the generation of lyso-derivatives by enzymatic degradation (Stekhoven
and Bonting, 1981). The phospholipid, which probably exists
as an annulus around the (Na^++K^+)-ATPase may differ chemically,
and will certainly differ physically from the bulk membrane
lipid (Brotherus et al., 1980) and will additionally give
rise to domain structure or ordering in other lipids
(Gebhardt et al., 1977; Knowles et al., 1979). Also considering the known membrane lipid asymmetry (Etemadi, 1980)

which effects the temperature dependence of membrane proteins differentially according to their insertion in either half or through the bilayer (Houslay *et al.*, 1979) and the interactions of other membrane proteins, protons, divalent cations, and cholesterol with the system, it is extremely difficult to analyse the involvement of bulk membrane lipid in temperature effects on the pump. Thus although it is clearly possible to obtain marked changes in the temperature dependence of $(Na^+ + K^+)$-ATPase activity by altering membrane lipid composition either *in vitro* (Kimelberg and Papahadjopoulos, 1972; Tanaka, 1974) or *in vivo* (Nemat-Gorgani and Meisanii, 1979; Smith and Kemp, 1969; Solomonson *et al.*, 1976) this may not be readily interpretable. Interestingly some of the most dramatic effects on Na^+-pump activity in erythrocyte can be achieved by manipulating cholesterol levels, where the interaction alters both the ligand affinities for Na^+ and K^+ as well as the rate of transport (Claret *et al.*, 1978). We therefore conclude that although changes in bulk membrane lipids can dramatically affect the temperature-dependence of $(Na^+ + K^+)$-ATPase activity these results are not directly interpretable in terms of phase transition or separation, the involvement of annulus lipids requiring more complex analysis.

Partial Reactions

A wealth of evidence from partial flux measurements in erythrocytes studies on the phosphorylated intermediate, intrinsic and extrinsic fluorescence, K^+ occlusion and binding, proteolytic digestion and various enzymic assays has reinforced the kind of reaction sequence for the sodium pump outlined in Figure 1 (but see Plesner *et al.*, 1981), with the division of the enzyme into a sodium-preferring form (E_1) and a potassium-preferring form (E_2). Although many subtleties of the number of intermediates and their interconversion remain to be settled, it is possible to classify certain reactions, for example Na^+-Na^+ exchange and Na^+-ATPase phosphorylation from ATP+Na as E_1-reactions, whilst K^+-K^+ exchange, K-occlusion, K^+-pNPPase and ouabain binding with Mg^{2+} and P_i involve E_2. Temperature dependence studies of a variety of partial reactions have been carried out, the most comprehensive being the analysis of Barnett and Palazotto, (1974), who attempted to pinpoint the temperature-sensitive step in the reaction sequence. Arguing from the contrast between biphasic Arrhenius plots for $(Na^+ + K^+)$-ATPase, Na^+-ATPase and ouabain inhibition, they concluded that conversion of E_2K^+ to an E_1 form was the most likely temperature-sensitive reaction. Further recent studies (Swann and Albers,

1979, 1981) approach the problem with a different, allosteric model, but confirm that low temperatures favour a form of the enzyme with high K^+ and low Mg^{2+} affinity, consistent with the involvement of $E_2(K^+)$. Very little data is yet available on the temperature dependence of partial fluxes in erythrocytes, but we have recently found that both ouabain-sensitive Na^+-Na^+ exchange and uncoupled Na^+ efflux show a relatively low temperature sensitivity (*i.e.* a high relative activity at low temperatures). Other evidence from the effects of increasing Mg^{2+} concentrations (Kimelberg, 1975) or oligomycin (Gruener and Avidor, 1966) on (Na^++K^+)-ATPase, and the low temperature sensitivity of Na^+-ATPase (Blostein, 1970; Neufeld and Levy, 1970) suggests that perhaps only pump fluxes involving $E_2 \rightarrow E_1$ transitions in either direction will be markedly inhibited on cooling.

In the same series of erythrocyte experiments we compared ouabain-binding at 5 with $37^{\circ}C$, and found the same number of sites at equilibrium, with a simple linear relationship for the plot of percent pump inhibition against number of molecules bound. This is important in terms of the interaction between the sodium pump and its environment, since any differential effects on pump molecules within the population in terms of their turnover would be expected to show up as a non linear pump inhibition/bound ouabain curve. Further the fact that the equilibrium number for binding sites at $5^{\circ}C$ is the same as $37^{\circ}C$ eliminates the possibility that a fraction of the pumps were frozen out and unable to bind ouabain or participate in transport.

CONCLUSIONS AND PROSPECTS

Superficially this selective review of the effects of cooling on ion transport in various cells may seem depressing to those gainfully employed in studying the sodium pump. The significance of alternative membrane transport processes in neuronal temperature compensation, and epithelial transport is clear, and even erythrocyte studies have revealed the significance of paradoxical increased passive permeability at low temperatures. However, pump failure, or the high temperature sensitivity of the overall pump reaction reveals that at least one conformational change involved in ion translocation is highly restricted. Identification of this component as the $E_1 \sim E_2$ reaction already seems plausible, and confirmation should come from thermal studies on K^+-occlusion with the purified enzyme, and measurement of partial fluxes in erythrocytes and enzyme-containing liposomes (perhaps including the trypsinized invalid enzyme:

Anner and Jørgensen, 1979). With regard to the lipid involvement, it is clear that we must look at the specific microenvironment of the pump rather than bulk membrane effects.
As an example of this in Figure 2 we compare Arrhenius plots
for three transport systems in human red cell membranes –
the sodium pump, the glucose carrier and the L-system amino
acid carrier. The differences in breakpoints emphasise that
each system is independent, and not related directly to membrane fluidity. However, correlating sodium pump activity
with its membrane microenvironment by the use of more specific
probes or lipid modification is an attractive possibility
which may lead us further towards understanding the actual
process of translocation, rather than its consequences.

MEMBRANES OF MAMMALIAN HIBERNATORS
AT LOW TEMPERATURES

J.S. Willis

*Department of Physiology and Biophysics, University of
Illinois, USA.*

J.C. Ellory

Department of Physiology, University of Cambridge, England.

A.R. Cossins

Department of Zoology, University of Liverpool, England.

INTRODUCTION

On the face of it the phenomenon of mammalian hibernation
would seem to offer excellent opportunities for examining the
effects of low temperature on mammalian membranes in general.
Mammals capable of hibernation do not differ in other respects from closely related non-hibernators whose membrane
function is (usually) highly sensitive to even modest decreases in temperature, and yet hibernators are able to sustain life for prolonged periods at temperatures between 0
and 5°C. There can be little doubt that this difference in
survival at low body temperature resides largely, if not
entirely, in the failure of membrane function in the sensitive
forms and continuance of function in the resistant (hibernator)
forms.
 While characterization of this difference may be exploitable

* *Unless otherwise indicated, the term hamster in this review refers to the Syrian hamster, and ground squirrel to the
thirteen-lined ground squirrel. Scientific names are listed
at the end of the text.*

experimentally, it is also fraught with hazards, which though
numerous are not entirely obvious. Thus while hibernating
organisms do survive, some specific, presumably non-vital,
tissues may cease functioning in the cold, such as the auditory
nerve in hamsters (Kahana *et al.*, 1950). Conversely, though
non-hibernating mammals die with body temperatures below
about 20°C some tissues of these cold-sensitive species are
remarkably cold-resistant. In some cases this can be related
to low temperature function in peripheral tissues (Willis,
1978), but there are also cases of even core tissues being
very active *in vitro* at 5°C, for example conduction in phrenic
nerve of rabbit (Abbott *et al.*, 1965).

Another issue is the dependence of cold resistance upon
whether the tissue is actually residing within the organism.
While the oldest and most striking demonstrations of intrinsic
cold resistance of hibernators consisted of showing the continued beating of the excised heart, conduction of excised
nerves or transmission at the neuromuscular junction at
temperatures below 10°C (Willis, 1967, 1978, 1979) demonstration of adaptation of brain tissue *in vitro* has been much
less convincing even though it is clearly cold-resistant in
intact hibernators (Goldman and Willis, 1973). Immediate loss
of resistance upon removal of the tissue from the organism
presumably reflects the loss of some form of necessary structural or chemical "support" provided by the higher order of
organisation, such as hormones or cell-cell attachments
Preparations based on longer residence outside the organism,
including static culture, primary and secondary diploid lines
and established cell lines, show progressively less and less
difference in cold sensitivity based on origin of the tissue
(Willis, 1978), although some differences do persist in diploid culture. Such long term changes may be due at least in
part to nutritional factors and genetic selection for more
tolerant cells, or, conceivably, acclimation in the poikilothermic sense (Willis, 1978).

Further complications rise from the fact that "hibernation"
is not a unitary phenomenon and is of polyphyletic origin.
Thus, the nature and degree of adaptation varies from species
to species and there are differences in the extent of cold
tolerance (Willis, 1978). Many species (ground squirrels,
hedgehog) grow fat before hibernation and stop eating during
hibernation whereas others (hamsters) store food and eat
between bouts of hibernation but do not grow exceptionally
fat. In many species the seasonal changes of hormones interact in complex and diverse ways with the seasonal cycles of
eating and hibernating. These differences will affect specific membrane activities in various ways unrelated to their

responsiveness to temperature *per se.*

Finally, the question of whether the organism from which cells are taken is actually in a hibernating condition is sometimes an important variable. The classic demonstrations of differences between species in nerve and cardiac muscle conduction referred to above did not depend on the hibernators actually hibernating; however more subtle differences, or differences between other tissues, are dependent on the physiological state of the organisms.

It is also of relevance to compare the "cold adaptation" of hibernators with that of cold-acclimatised poikilotherms (Cossins, this volume). Most studies have in the latter case been concerned with so-called "capacity adaptation" (Precht *et al.*, 1973) which permits a compensation and presumed optimalization for activity at each sustained temperature. Consequently, there is often a minimal difference in activity or metabolism between, say, fish adapted to two different temperatures. The case with hibernators is different in two important ways. First, the adaptation may be regarded as of the "resistance" type which permits mere survival without sustained general activity at the low temperature. This may be regarded as exploitative in the sense that energy utilization is minimized during periods of reduced food availability. Secondly, hibernation characteristically consists of short "bouts" (a few days in some cases, to many days in others) interspersed with periods of arousal and resumption of normally high body temperature ($37^{\circ}C$). The transitions from one state to the other are rapid (hours) and therefore vital cells must be capable of functioning at both ends of this extreme spectrum ($>30^{\circ}C$) without time for acclimation. This represents a thermal challenge far greater than that tolerated by aquatic poikilotherms.

Scope of this Review

There have been two distinct types of study aimed at analysing the cold resistance of hibernator membranes. The older approach has been to estimate the effect of cooling from temperatures in the normothermic range ($37^{\circ}C$) to those in the deepest hibernation range (about $5^{\circ}C$) on some aspect of membrane activity (*e.g.* permeation, enzymatic function, membrane potential). More recently there have been numerous attempts to determine the composition of membranes and if possible to relate this to some physical measure of membrane fluidity. Occasionally the two approaches merge, as when some enzymatic function is compared with fluidity or composition. The bulk of the studies of the first type has con-

sisted of comparing differences between tissues taken from
hibernating and non-hibernating species, whereas studies of
the second type usually consist of comparisons based on
tissues taken only from members of hibernating species in one
or the other physiological state (awake/summer *vs* hibernating/
winter). Both comparisons are of interest and the segregation
according to approach is reasonable in that the larger functional differences are usually seen between species, whereas
compositional comparisons between species are difficult to
relate simply to temperature adaptation. Although the latter
point might also apply to differences in physiological state
related to diet, hormones, *etc*., one might at least hope to
sort out these effects and in any case with intraspecific
comparisons spurious genetic differences are minimised.

In both approaches the centre of interest has been in one
or the other of two membranes - the plasma membrane or the
mitochondrial inner membrane. These are of course very
different entities - the mitochondrial membrane being very
much richer in protein than the typical plasma membrane but
also lacking entirely in cholesterol, both of which differences affect in complex ways expectations of temperature
transitions (Pringle and Chapman, this volume).

Consequently, we shall deal here first with "descriptive
studies" of both plasma membrane activities and mitochondrial
membrane activities before turning to studies of the "compositional" type.

In discussing either membrane type we have been necessarily
selective. In mitochondria we have avoided a more general
discussion of metabolism and provision of ATP. Important as
this is for cells at low temperature it has been reviewed
elsewhere (Willis, 1979) and we have focussed only on activities related to membrane function. Similarly, for plasma
membrane we have avoided topics in which membrane activity
is expressed as an organ or tissue function (renal absorption,
ECG, CNS activity), which also have been discussed elsewhere
(Willis, 1979). More narrowly still, we have not discussed
the cold adaptation of excitation, conduction, maintenance of
membrane potential and synaptic transmission in nerve and
muscle of hibernators. Even though these are useful starting
points for consideration of the plasma membrane, the literature on this subject has acquired an antique patina and has
not been significantly advanced since it was last reviewed
(Willis, 1967, 1979). In plasma membranes, therefore, we
have dealt only with the subjects of ion pumps, leaks and
permeation by organic molecules. This not only represents
the collective interest of the authors, but is also a subject
in which several groups have made recent advances.

Finally, there have been two types of criterion by which
adaptation at low temperature has been judged. The favourite
diagnostic tool has, of course, long been the venerable but
resilient Arrhenius plot. In older descriptive literature on
hibernation this was prized largely for the fact that it may
be used to give a single numerical value (or at least a small
set of values) to the effect of temperature on measured
activity. More recently much emphasis has been given to
"breaks" in the curve, usually with consequent steeper segments at low temperature (Cossins, this volume). Thus, cold
resistance is often seen as consisting of an Arrhenius plot
with no breaks, or breaks confined to very low temperatures
with less steep slopes prevailing throughout the plot.

The other approach has been to compare activities at only
two temperatures, a "high" and a "low" temperature, repres-
enting the two extremes at which the tissues of the hibernating organism actually reside. This procedure has the
practical advantage of allowing for a more frugal use of
samples or for a wider variety of controls and use of parallel
inhibitors. It also has the virtue of not being wedded to
a set of physical chemical concepts of dubious applicability.
For example, it is possible for the net decline in rate between, say, 37 and 5°C to be less for a function with a steeply
broken Arrhenius curve than for a function with a straight,
unbroken curve of moderate slope. In their enthusiasm for
breaks and slopes at intermediate temperatures, perpetrators
of Arrhenius plots have sometimes failed to extend their
measurements down to the actual temperatures of hibernation.
On the other hand, it would often be naive to infer that a
tissue performing at 37 and 5°C *in vitro* is reflecting the
capacity of that tissue in the intact organism at the same
two temperatures, so that a look over the whole temperature
scale would be more suitable. Thus, both sets of criteria
have their valid uses and their temptations to over-interpret-
ation

It is also well to recognize that owing to arithmetic
considerations the two criteria may in certain circumstances
lead to opposite conclusions. Thus, a doubling of an activity
at 5°C achieved by a simple translation of the Arrhenius
curve would not be seen as an increase in resistance in the
"Arrhenius" sense (slope unchanged) but would appear so by
the other criterion (*i.e.* the $5^{\circ}/37^{\circ}$ activity ratio would
increase).

PERSISTENCE OF MEMBRANE ACTIVITIES AT LOW TEMPERATURES

Mitochondria

Description of mitochondrial activity at low temperature has
dealt mainly with respiratory capacity. Results and comparisons are complicated by the various choices of primary substrate and diversity of conditions related to phosphate
acceptor. Thus, the connection with the truly membrane-
governed processes of hydrogen ion and electron transport and
ATP synthesis are often somewhat indirect, involving various
numbers of soluble-enzyme mediated steps before the cytochrome
chain is engaged. There have apparently been no studies of
mitochondrial solute transport (*e.g.* Ca^{2+}, phosphate,
adenylates) at low temperature in hibernators.

  Most work with hibernators has been on mitochondria of
liver or heart. Brown fat mitochondria of hamsters (a hibernating species) have been extensively investigated, but
with few exceptions (Cannon and Polnaszek, 1976; Liu *et al.*,
1978; Viebke *et al.*, 1978), not with regard to temperature
adaptation.

  With rat as the standard species of comparison, there is
no persuasive evidence of a membrane related cold adaptation
in mitochondria of either liver or heart from either ground
squirrels or hamsters.

  To be sure, early work on rat heart mitochondria (Hannon,
1958) in comparison with hamster (South, 1960) indicated that
the Arrhenius plot for pyruvate respiration with phosphat
acceptor was less steep and less convex in the hamster than
in the rat. In contrast, with succinate as primary substrate
for rat and hamster heart mitochondria no such difference was
observed (Roberts and Chaffee, 1973). An incidental result
which emerged from a study of dietary lipids on cardiac
mitochondria of rats was that the Arrhenius plot of succinate
oxidation by ground squirrel mitochondria was steeper throughout than that of rat mitochondria (Schatte *et al.*, 1977). It
seems likely that the differences found by South (1960) may
be attributable to complex or rate limiting steps prior to
succinate, such as pyruvate dehydrogenase or isocitrate
dehydrogenase.

  For liver mitochondria the situation is more complex. No
systematic differences were observed in cold sensitivity of
succinoxidase activity of mitochondria from either hamster or
ground squirrel compared with rat mitochondria (Liu *et al.*,
1969) or warm-room rats compared with hamsters (Roberts *et al.*,
1972). But when cold-acclimated rats were compared with
those of hibernating hamsters, the Arrhenius function between

20 and 6°C was steeper for the rats. On the other hand, the
reverse was true in the range from 34 to 20°C, so that the
overall difference of sensitivity was not great.

In making such comparisons much may depend on the species
chosen. Thus, mitochondria of both liver and heart of squirrel
monkeys were found to be much more temperature sensitive
(steeper Arrhenius slopes, greater convexity) than were those
of either hamsters or rats (Roberts et al., 1972; Roberts and
Chaffee, 1973). It is not known whether these tissues are
more temperature-sensitive in squirrel monkeys, but it is not
unlikely that the whole organism is less resistant to hypothermia than a rat. Unfortunately for many practical reasons
comparisons seldom involve more than one or two species in
each "camp".

With regard to differences related specifically to the
physiological state of the organism, respiration in heart
mitochondria has generally been observed to be elevated in
hibernation (hamsters: Roberts and Chaffee: 1973; South,
1960) and to be depressed in liver mitochondria (hamsters:
Roberts et al., 1972; ground squirrels, Liu et al., 1969).
The overall slope of the Arrhenius curve of respiration and
its "straightness" have not been observed to change with
hibernation in heart mitochondria (Roberts and Chaffee, 1973;
South, 1960), but it has been reported to be less steep,
owing to a diminished convexity, in liver mitochondria of
both hibernating hamsters and hibernating golden-mantled
ground squirrels (Raison and Lyons, 1971; Roberts et al.,
1972. Liu et al. (1969), however, did not observe any differences in temperature sensitivity with hibernation in liver
mitochondria of hamsters and thirteen-lined ground squirrels.
These results, like those of Roberts et al. (1972) on hamsters,
but unlike those of Raison and Lyons (1971) on golden-mantled
squirrels, were averages of several individuals.

It is difficult to see why, with such doubtful indication
of large differences in membrane behaviour, a lipid-fluidity
hypothesis (see below) for mitochondria in this biological
context should even be called for.

Plasma Membrane: Cation Concentrations in Cells

Maintenance of K^+ and Na^+ gradients across cell membranes is
an essential requirement for survival at low body temperature
for a variety of reasons (Willis, 1972). In ground squirrels
and hamsters K^+ gradient has been shown to be maintained
during prolonged hibernation or even to be elevated in kidney,
brain, muscle and liver (Willis et al., 1971). Some skeletal
muscles lose K^+ during hibernation in these species but

diaphragm, which maintains active contraction, does not.
Erythrocytes lose K^+ slowly (Kimzey and Willis, 1971a).

Comparison with non-hibernators in this capacity is difficult owing to complications of forced hypothermia. One must
therefore resort to *in vitro* comparisons, which generally
have the problem of comparing differences in damaged tissues.
Erythrocytes are perhaps the least damaged by removal from
the organism and those of hamster and ground squirrel lose
K^+ far more slowly than those of guinea pig or human when
held at $5^\circ C$ (Kimzey and Willis, 1971a,b). Kidney slices of
hamsters and ground squirrels, both in short and long term
incubation, show a similarly greater K^+ retention than those
of rat of guinea pig, but there is still great loss due to
the experimental procedure (Willis *et al.*, 1972). Rabbit
kidney slices retain and reaccumulate K^+ at low temperature
as well as those of hibernating species (Willis, 1968).
Kidney cells in primary culture grown from explanted renal
cortex of hamster, ground squirrel or guinea pig are stable
at $37^\circ C$, and those from the two hibernator species lose only
about 10 percent of their K^+ in four hours at $5^\circ C$ whereas
those of guinea pig lose about 40 percent (Zeidler and Willis,
1976).

In a particularly careful and detailed study Kamm *et al.*
(1979) found that aortic smooth muscle for rat and ground
squirrel maintained a stable K^+ content *in vitro* for up to
48 hours at both 37 and $17^\circ C$, but that at $7^\circ C$ the concentration fell progressively in the rat aortas while remaining
unchanged in the ground squirrel aortas. By determining
extracellular space they showed that the cell Na^+ and Cl^-
concentrations were also maintained at normal values, except
in rat preparations at $7^\circ C$ where both values increased progressively while cell water content rose.

Diaphragm muscle of hamster retains K^+ during short term
incubation slightly better than those of rat (Willis, 1967).
Brain cortex of hamster did not show any greater capacity to
retain K than that of rat, except in preparations made from
hibernating hamsters where the difference was still relatively
modest.

Plasma Membrane:Pump Fluxes in Cells

Better retention of K^+ at low temperature can in principle
consist either of greater maintained pumping through the
ouabain-sensitive Na^+ pump or more greatly diminished
leak through any of several pathways. Of these two possibilities the first one, sustained pumping, has received more
attention partly because it is more readily estimated.

Early attempts to measure "pump rate" by net movements in leached kidney slices of guinea pig (a non-hibernator) led to the apparently false conclusion that the rate was less at low temperature than in preparations from hamsters and ground squirrels (Willis, 1966, 1968, 1978). In fact, it appears both from measurement of ouabain-sensitive respiration at low temperature (a measure of pump activity) and from unidirectional ouabain-sensitive K^+ influx in cultured kidney cells, that the rate in guinea pig cells is scarcely more reduced than in hibernators (Willis, 1968; Zeidler and Willis, 1976). The net uptake is low apparently because at low temperature the pump is working against a comparatively large leak. These results still leave open the possibility that in kidneys of other species, which fail to retain or reaccumulate K^+ at low temperature, the failure could be due to a cold-sensitive pump.

The net loss of K^+ in erythrocytes was investigated initially by comparing its rate with that predicted by a theoretical, exponential curve, assuming only passive leak without a pump (Kimzey and Willis, 1971a). This prediction was met in guinea pig cells and rate of loss was not increased in the presence of ouabain, whereas in ground squirrel cells loss of K^+ fell along the theoretical curve only in the presence of ouabain and was much slower in uninhibited cells. Hence it appeared that a still active pump was contributing to K^+ retention in erythrocytes of this species of hibernator, but not in erythrocytes of guinea pig. This interpretation was borne out by determination of unidirectional fluxes in the cells of the two species (Kimzey and Willis, 1971b). More recent studies on the same two preparations have shown that the maximum pump rate in erythrocytes of ground squirrels is reduced by only a quarter as much as that in guinea pigs, and the difference is due entirely to differences in turnover number (enzymatic efficiency) not to alteration of numbers of pump sites (*i.e.* ouabain-binding sites)(Ellory and Willis, 1978).

A survey of a wider selection of species (Willis *et al.*, 1980) showed that ouabain-sensitive K^+ influx was less temperature sensitive (as judged by the criterion of $5^\circ/37^\circ C$ ratio) in 6 out of 7 species of hibernators ($5^\circ/37^\circ C$ ratio greater than 1.0 percent) than in 8 out of 9 species of non-hibernators ($5^\circ/37^\circ C$ ratio less than 1.0). The two "cross-overs" were hamsters ($5^\circ/37^\circ C$ ratio 0.8 percent, less than half that of any other hibernator) and mole ($5^\circ/37^\circ C$ ratio 3 percent, or four fold greater than that of any other non-hibernator).

In the only other study of ion fluxes in intact tissue Kamm *et al.* (1979b) attempted the more difficult procedure of measuring Na^+efflux in an aortic muscle preparation and

determining the active component from the difference with
and without external K^+. Their results show an unbroken
Arrhenius plot for the ground squirrel aortas between 4 and
$37^{\circ}C$, but in rat aortas there was a sharp downward break at
$17^{\circ}C$ with consequently greater net reduction in pumping at
low temperature. K^+-dependent Na^+ efflux was slightly less
temperature sensitive in aortas of hibernating than in
"normothermic" ground squirrels, but it is not clear if the
latter were summer active animals or merely squirrels prevented from hibernating in winter.

Meaningful flux measurements have stringent requirements
for simple morphology and thus have not been widely attempted
in other tissues. Urgently needed are suitable models of
nerve tissue and cardiac muscle. Culture of cardiac muscle
cells on fiber supports (Lieberman, 1975) would seem to offer
a good opportunity to investigate species-related differences
in transport capacity at low temperature correlated with
maintenance of membrane potential.

Plasma Membrane: (Na^++K^+)-ATPase

An alternative to determination of pumping of ions in intact
cells is to measure the effects of temperature on (Na^++K^+)-
ATPase activity and the effect of hibernation state on level
of activity of (Na^++K^+)-ATPase in broken membranes. This
method has been employed in four tissues for which unidirectional flux data are difficult to achieve, namely brain, liver,
kidney cortex and cardiac muslce.

With microsomal preparations from brain of awake (summer)
hedgehog there is a steeply concave Arrhenius plot at low
temperature (Bowler and Duncan, 1969), similar to preparations
made from rat brain (Bowler and Duncan, 1968), but in
preparations made from brain of hibernating hedgehogs the
low slope characteristic of higher temperatures persisted at
low temperatures. Microsomal preparations from hamster brain
cortex were similar to preparations from rat, except for
brains of hibernating hamsters which showed increased activity
at low temperature. The "improvement" with hibernation was
not as marked as in hedgehogs, being only a doubling of
activity at $5^{\circ}C$ in the hibernating compared with the awake
hamster brains. In Richardson's ground squirrel no change
in thermal sensitivity occured with hibernation, but in both
the high and low temperature range the Arrhenius slope at
any time was less than or equal to that computed for hibernating hedgehogs or hibernating hamsters from the above
studies (Charnock and Simonson, 1978a). In the brains of
Richardson's ground squirrels which had hibernated for more

than 100 days Charnock and Simonson (1978a) did find an increase in specific activity of $(Na^+ + K^+)$-ATPase, but Vysochina (1977) found a decrease in brains from russet susliks (another species of ground squirrel).

In $(Na^+ + K^+)$-ATPase of a crude nuclear-mitochondrial fraction of hamster kidney cortex activity was detectable at 10 and $5^\circ C$, whereas in preparations made from rat kidney it was not (Willis and Li, 1969). To be sure, the specific activity of the hamster preparation was higher even at $37^\circ C$ so that the loss of activity in rat could have been simply due to low sensitivity of the method, but on the other hand the greater reduction with cooling was evident even at higher temperature (e.g. $20^\circ C$). In his review of this subject Charnock (1978) recomputed these results and found that the hamster Arrhenius plots revealed the high slopes and breaks characteristic of results based on species presumed to be temperature sensitive. The latter studies were, however, carried out on less noisy, more purified preparations. Charnock did not accord the same treatment to the control rat data of Willis and Li (1969).

$(Na^+ + K^+)$-ATPase of kidneys from Richardson's ground squirrel also exhibits a non-linear Arrhenius plot, and actually had less relative activity at $5^\circ C$ ($5^\circ / 37^\circ C$ ratio of 0.3 percent) than rabbit (a non-hibernator) and it was concluded that the enzyme in this hibernator species exhibited no unusual cold resistance (Charnock and Simonson 1978b). In the sense that rabbit kidneys (as well as those of guinea pig) appear to have fairly vigorous pumping at $5^\circ C$ (Willis, 1968) this conclusion is unexceptionable. The real issue seems to be, however, not whether pumps of kidney cells of the ground squirrel are failing at low temperature, but rather whether kidney cells of any mammal need to be considered to have cold-sensitive Na^+-pumps. Charnock (1978) reported that $(Na^+ + K^+)$-ATPase of sheep kidney was, like that of rabbit, no more cold-sensitive than ground squirrel. Unfortunately, nothing is known about performance of intact sheep kidney cells at low temperature, but with the example of the rabbit and guinea pig (see above), assumptions about cold-sensitivity in non-hibernators should be made with caution.

On the basis of the above cited results on $(Na^+ + K^+)$-ATPase, rat kidney seems at least to offer the possibility of one example of a kidney cell pump failing at low temperature, but the results on tissue K^+ (some metabolically dependent retention at $5^\circ C$) and on Na^+-sensitive respiration (an indirect measure of $(Na^+ + K^+)$-ATPase in intact cells) are somewhat ambivalent on this score (Willis, 1968).

Indeed, further comparison of this sort between effect of temperature on measures of $(Na^+ + K^+)$-ATPase function in intact

cells (fluxes, pump related respiration) and effect of temperature on the same enzyme activity in broken membrane
preparations reveals another, more general difficulty (Ellory
and Willis, 1976; Willis and Li, 1969). This is that the
cell-free preparation consistently shows greater (often far
greater) sensitivity to cooling than does the intact cell.
The presence of the disparity in red blood cells as well as
kidney cells permitted a systematic treatment (Willis *et al*.,
1978) which led to the conclusion that the initial rupture
of the membrane by itself, quite aside from any subsequent
manipulation, increases the temperature vulnerability. It
is not clear whether this disruptive effect would alter the
apparent relative temperature sensitivity of the enzyme from
two sources, so that the more cold-sensitive intact pump
would appear in a purified membrane preparation as the less
cold-sensitive of the two.

With regard to seasonal or physiological state changes in
the (Na^++K^+)-ATPase of the kidney, an increase in activity
was found at every temperature from kidneys of hibernating
hamsters compared with awake hamsters (Fang and Willis, 1974).
This increase was confined to (Na^++K^+)-ATPase and was not
detected for other membrane enzymes. It appeared to represent
an increase in turnover number (efficiency) since there was
no increase in Na^+-dependent phosphate incorporation into the
enzyme with hibernation (*i.e.* no increase in pump sites).
The kidneys of Richardson's ground squirrel appeared to be
very different (Charnock and Simonson, 1978b), exhibiting
more than 60 percent reduction with hibernation, both in
specific activity and ouabain binding of the membrane preparation (*i.e.* turnover number remained constant). Such a
difference between these two species might reflect the
tendency for reduced thyroid activity in the ground squirrels
prior to the hibernation season, as compared with the elevated thyroid levels in cold-exposed prehibernating hamsters,
when the (Na^++K^+)-ATPase is increasing (Hudson and Wang, 1979;
Wang, 1982). On the other hand, this particular species of
ground squirrel, unlike others studied, exhibits a high rate
of thyroid secretion and high titres of free thyroid hormones
during the hibernating season.

The temperature sensitivity of (Na^++K^+)-ATPase of hamster
liver (Houslay and Palmer, 1978) and of ground squirrel
cardiac muscle (Charnock *et al*., 1980) has been examined only
with respect to onset of hibernation without any comparison
between species. Like brain (Na^++K^+)-ATPase of hamster, that
of liver had an altered temperature sensitivity after onset
of hibernation. In awake hamsters the Arrhenius curve showed
a break at $25^{\circ}C$ and another at $12^{\circ}C$. In hibernating hamster

the lower break was at $3.3°C$ and the slope between 3.3 and
$26°C$ was less steep than in the awake hamster preparations.
In ground squirrel the Arrhenius plot of cardiac muscle
resembled that of kidney in showing a cold-sensitive type of
plot. On hibernation there were no changes in slope or breaks,
but a general decrease in specific activity (Charnock *et al.*,
1980).

Summary of Evidence of Pump Adaptation

This confusing array of results emerges from the likely
interaction of both real biological diversity and differential
emphasis created by diverse analytical approaches. To comp-
rehend this, it is perhaps useful to recall that, in general,
hibernator tissues retain K^+ at low temperature and tissues
from non-hibernators usually lose K^+. Retention in hibernators probably depends at least in part on sustained pumping
at low temperature in most cases, but not all. Hamster
erythrocytes are a *bona fide* exception. Before assuming that
broken Arrhenius curves of (Na^++K^+)-ATPase also constitute
exceptions (*e.g.* ground squirrel brain, kidney and cardiac
muscle), one needs to recall, first, that the breaks and
steepness may be due to an artefact of membrane disruption
during preparation (and that changes with hibernation may
likewise be due to changes in the same artefact) and, secondly,
that even with breaks and steepness, sufficient biologically
relevant activity may persist at low temperature.

Loss of K^+ probably results in part from severe inhibition
of pump activity at low temperature, in most cases, but not
all (*e.g.* mole erythrocytes, guinea pig and rabbit kidney).
The same caveats about Arrhenius curves therefore apply with
equal force here.

With onset of hibernation, changes in thermal sensitivity
may occur (hamster and hedgehog brain, hamster liver), or
they may not (ground squirrel tissues in general). Levels of
specific activity may either go up (hamster kidney, ground
squirrel brain) or down (ground squirrel kidney and cardiac
muscle). Not only do these physiological state changes vary
with species and tissue, but the issue is also a difficult
one to assess in terms of advantage to the organism. Thus,
elevated pumping might be advantageous in the kidney which
accumulates K^+ in hibernation and may offset any tendency for
hyperkalemia. On the other hand, energy expenditure for
transport is a major part of the basal energy budget and if
the value of becoming cold is to preserve energy, diminished
transport presumably should add to this. Indeed, it has been
suggested that a shift in pH sensitivity of the (Na^++K^+)-ATPase

with falling body temperature accomplishes just this end
(Malan, 1978).

Kinetics of the Na⁺-pump

In principle, an adequate description of the effect of temperature should be based on more than just intact pumps
functioning under ambient condition or (Na^++K^+)-ATPase functioning under maximal stimulation (*i.e.* V_{max}). From the
standpoint of simple comparison, reliance on either tactic
could either hide real difference between species (or
physiological states) or reveal differences in limiting performance which would not reflect the intact situation. Besides, estimation of changes in affinities of ligands, (K^+,
Na^+, ATP and ouabain) could provide clues as to the cause of
failure in those pumps which do fail at low temperature.

     In cultured kidney cells of both guinea pig and ground
squirrel there was no decrease with temperature in apparent
affinity for the effect of external K^+ concentration on
ouabain-sensitive influx, but the data did not permit an
evaluation of the possibility of an increase in affinity
(Zeidler and Willis, 1976). Erythrocytes of the same two
species exhibited increased affinity both for external K^+ and
internal Na^+ at low temperatures but there was no difference
between the two species in this regard large enough to account
for the four-fold difference in pumping at $5^\circ C$ (Ellory and
Willis, 1978). Alteration of internal cell Na^+ concentration
however, did permit a more accurate estimate of this differ-
ency in pumping rates under V_{max} conditions than had been
achieved earlier.

     Binding of ouabain by plasma membranes is affected by
temperature in various ways according to species and tissue
(Willis and Ellory, 1982), but differences which correlate
with cold-resistance of the Na^+-pump have not yet been observed. From a practical standpoint, the frequent effect of
cold decreasing sensitivity to ouabain can make it difficult
to estimate pump-related activity at low temperature.
Alteration in ouabain-related phenomena in several tissues
of ground squirrel have been observed after long term hibernation, which may reflect alterations in membrane organization.
In particular brain (Na^++K^+)-ATPase becomes less sensitive
to ouabain inhibition (*i.e.* there is a shift in dose-response
curve) during winter hibernation, and there is about a
40 percent decrease in rate and amount of specific ouabain
binding (Charnock and Simonson, 1978a). There was also a
decrease in rate and amount of binding in ground squirrel
kidney ATPase (Charnock and Simonson, 1978b), but the two

situations are different. It will be recalled that brain
ATPase activity increased with hibernation (thus giving a
much larger turnover number) whereas specific activity fell
in kidney (in parallel with binding). The latter effect is
easily explained simply as decreased number of ATPase molecules, but the results in brain are more complex and may
suggest alteration in lipid-protein relationships affecting
ouabain binding.

Affinities of the isolated (Na^++K^+)-ATPase ligands have
not been investigated with specific reference to hibernation;
results with standard preparations are discussed elsewhere
(Ellory and Willis, this volume). Use of isolated (Na^++K^+)-
ATPase, however, permitted investigation of the so-called
"partial reactions" of the pump. K^+-activated pNPPase of the
hamster kidney was found to have the same temperature dependence as that of rat, even though there was a difference between the two species in overall reaction (Fang 1971), and
similarly in brain there was no difference in that activity
between hibernating and awake hamsters (Goldman and Albers,
1975) as there is for (Na^++K^+)-ATPase. In the brain preparation it was also shown that there was no difference in
temperature sensitivity of the Na^+-activated ATP-ADP exchange
reactions (Goldman and Albers, 1975). These "negative"
results are compatible with the interpretation, based also on
other standard preparations as well (Ellory and Willis, this
volume) that blockage at low temperature occurs at the point
of the reaction where there is a change of enzyme conformation,
at $E_2(K^+$-preferring$)\rightarrow E_1(Na^+$-preferring$)$, or possibly at
$E_1\rightarrow E_2$. This problem was investigated in hamster and rat
kidney by means of a pulse-chase experiment which permitted
a comparison of the rate of the "initial", Na^+-dependent
phosphorylation of the enzyme. The peak of labelling at $0^\circ C$
was greater in hamster preparations than in those from rat
(Fang, 1971; Willis *et al.*, 1972). While this was interpreted
at the time only in terms of "Na^+-dependency" (in contrast
to K^+-dependency), the result is also consistent with a
difference between the two species in $E_2\rightarrow E_1$ transormation at
low temperature.

Such a block ought to be expected to lead to difference in
the change of external K^+ and internal Na^+ affinity seen at
low temperature and in the amount of ouabain bound at low
temperature, and such differences were not apparent in the
studies of intact erythrocytes discussed above (Ellory and
Willis, 1978). The issue has yet to be investigated more
directly in intact cells by examination of partial fluxes
(Na^+-Na^+ exchange, ouabain sensitive K^+ efflux, *etc.*) and of
rates of ouabain binding.

Passive Permeation of Inorganic Cations

The other way to achieve retention of ion gradients at low
temperature (*e.g.* high intracellular K^+ concentration) is a
reduction in passive leakage. It was shown that in kidney
slices, prepared by a method which minimized loss of K^+,
retention at 5°C was better in hamster than in guinea pig
preparations even in the presence of metabolic inhibitors
(Willis, 1966). These results suggest a difference in the
passive leak of Na^+ and K^+ between the two species at low
temperature. This was later confirmed for K^+ by undirection-
al flux determination in cultured renal cells of these two
species as well as of thirteen-lined ground squirrels
(Zeidler and Willis, 1976).

In erythrocytes of ground squirrels and guinea pigs a
similar difference in leak fluxes was also demonstrated for
K^+ influx and efflux and for Na^+ influx and efflux (Kimzey
and Willis, 1971b). It was also found that in guinea pig
erythrocytes, stored at 5°C rate of K^+ loss was unaffected by
ouabain, but was accelerated by glucose deprivation, suggesting a persisting metabolic component to the leak even in
these temperature-sensitive cells (Kimzey and Willis, 1971a).

There are of course numerous parallel leak paths for Na^+
and K^+, depending upon the cell type. Kidney cells, for
example, might be expected to have cation carriers coupled
with amino acids, sugars and organic acids, as well as apical
leaks for Na^+ and perhaps a "basic leak" underlying all of
it. Since the time of the studies of hibernator erythrocytes
it has been shown that human erythrocytes possess a co-trans-
port system for Na^+, K^+, and Cl^- (perhaps two systems) which
is furosemide-sensitive (Dunham and Sellers, 1980; Wiley and
Cooper, 1974) as well as a Li^+-Na^+ exchange mechanism (Pandey
et al., 1979). There is also a K^+ channel stimulated by
intracellular Ca^{2+} (the so-called "Gardos" channel) which
can be blocked by quinine (Lew and Ferreira, 1978). A full
description of temperature sensitivity of "leaks" must therefore involve a partitioning amongst all these available
mechanisms.

In an attempt to determine the effect of temperature on
the "basic leak" (*i.e.* furosemide-insensitive, ouabain-in-
sensitive K^+ and Na^+ flux) in human erythrocytes, Stewart *et
al.* (1980) observed that there is a zone of temperature
independence (about 20 to 12°C) and then a rise in influx at
temperatures below 12°C. We have examined this in erythrocytes of guinea pigs and ground squirrels and in the absence
of Ca^{2+} we find only a fall of basic K^+ leak with cooling
between 37 and 20°C and then no further change (*i.e.* no rise).

Under these conditions the leak in the cells of the two species is not greatly different (Ellory, Willis and Hall, unpublished observations).

The results of Kimzey and Willis (1971a,b) both on fluxes and on net changes in cation content, however, were obtained in a medium containing Ca^{2+} (albeit with an excess of phosphate as well). When we re-examined the furosemide/ouabain-insensitive K^+ influx with and without Ca^{2+} we found that the ground squirrel cells were unaffected but that the permeation in guinea pig cells with Ca^{2+} was an order of magnitude higher at $5°C$ and, further, that a rise at low temperature (starting at $12°C$) similar to that seen in human erythrocytes also occurred (Hall and Willis, unpublished results). Most of this effect in guinea pigs is blocked by quinine. Guinea pig erythrocytes stored at $4°C$ with Ca^{2+} lose K^+ more rapidly than those without and this effect is further enhanced in glucose-deprived cells (Hynes, Hall and Willis, unpublished results). These results suggest that in guinea pig erythrocytes most of the greater K^+ leak at low temperature is attributable to the Ca^{2+}-sensitive "Gardos" channel. Whether the absence of such effects in ground squirrels implies the absence of a "Gardos" channel or better retention of Ca^{2+} regulation is under study. The higher passive fluxes of Na^+ at low temperature in guinea pig erythrocytes is also not explained by this interpretation.

Passive Permeation of Amino Acids

Very little has been attempted regarding transport of organic molecules in cells of hibernators at low temperature. We have sought to determine the relative changes of passive fluxes of amino acids through three different carrier systems of erythrocytes of several hibernating and non-hibernating species (Ellory and Willis, 1981). We found that leucine uptake by the "L" system and tryptophan transport by the "T" system were remarkably temperature-insensitive in all species, whereas the "ASC" system (specific for alanine, serine and cysteine and also Na^+-dependent) was more temperature sensitive, but without any correlation as to whether the cells were cold-resistant by other criteria. These results, taken together with those described above showing different temperature sensitivities of active and passive ion movements, strongly suggest that the thermal behaviour of each permeation mechanism is independent and that there is no single, common physical transformation governing the effect of cooling on membrane activity.

COMPOSITION AND FLUIDITY OF HIBERNATOR MEMBRANES

The rapid and dramatic reduction in body temperature tolerated
by many hibernating mammals has attracted interest as a case
in which adjustment of membrane fluidity might have adaptive
value and therefore be observed in mammalian cells. Investigation and discussion of this proposition has hinged mainly
on three issues, the thermal behaviour of membrane-bound
enzymes (*i.e.* Arrhenius plots of mitochondrial respiration,
(Na^++K^+)-ATPase), changes of membrane lipid composition (*i.e.*
mainly in saturation of fatty acids) and the physical analysis
of membrane fluidity (*i.e.* by spectroscopic probes or differential scanning calorimetry).

Change in Arrhenius plots of mitochondrial respiration
from broken to straight with onset of hibernation provided
the earliest stimulus for the proposal that membrane lipid
fluidity was a variable and regulated factor in cold adaptation of hibernator membranes (Raison and Lyons, 1971). The
broken Arrhenius plot was thought to constitute evidence for
a bulk-phase transition which prevented normal function of
low temperatures. Obstacles to easy acceptance of this conclusion have been fully spelled out elsewhere (Charnock, 1978;
Silvius and McElhaney, 1981; Willis, 1978; Cossins, this
volume), namely:-

1) It is not generally agreed that such bulk phase changes do
   uniformly occur either in mitochondrial or plasma membranes.
2) Assuming that these changes do occur at least in certain
   cases, it is not clear that the change in shape of the
   Arrhenius curve necessarily implies a bulk change of state
   of the bilayer, since water-soluble enzymes can exhibit
   similar behaviour.
3) A phase change need not lead to a sharp break in an Arrhenius curve.
4) When change of "fluidity" does occur, it is conceivable
   that it is a result of alteration of the protein component
   as well as, or instead of, the lipid component.

Even with these caveats in mind it is difficult not to be
impressed by the findings of Houslay and Palmer (1978) who
investigated eight membrane associated enzymes in liver plasma
membranes of hibernating and awake hamsters. Three of the
enzymes are associated with the cytoplasmic membrane surface
and showed a break in Arrhenius function at about 25°C, which
did not change with hibernation. One enzyme is associated
with the outer surface of the membrane and showed a break at
13°C, and this "transition temperature" decreased to 4°C in
preparations from hibernating hamsters. Four enzymes penetrate

both halves of the membrane and showed two transition temperatures (25 and 13°C), the lower of which decreased to 4°C
with hibernation. Their conclusion that a change occurs in
the lipid matrix of the outer membrane leaflet allowing for
a lower temperature for a critical phase separation was
bolstered by change in the fluorescence of a membrane probe
(see below).

Compositional studies have been carried out in brain of
hamster (Blaker and Moscattelli, 1978, 1979; Goldman, 1975),
and ground squirrel (Aloia, 1980), kidney of hamsters
(Wilkinson, 1976), heart of ground squirrels (Aloia, 1980;
Schatte et al., 1977), and liver and heart mitochondria of
ground squirrels (Lerner et al., 1972; Platner et al., 1976).
Aloia (1980) has cogently reviewed most of this literature
and has concluded that, although changes in saturation of
specific fatty acids occur, if one takes into account all
such changes and also alterations in mole fractions, then
".... there does not appear to be a consistent trend toward
increased levels of unsaturated fatty acids during hibernation
in all tissues of hibernators". A major problem with the
interpretation of lipid compositional studies in terms of
membrane fluidity is that relatively little is known about
the packing characteristics of polyunsaturated fatty acids
that usually dominate most indices of unsaturation. It is
clear, however, that membrane properties of phospholipids
are not a linear function of the overall unsaturation (Demel
et al., 1972; Ghosh et al., 1971), since the most dramatic
change in physical properties comes with the insertion of one
olefinic bond into a saturated fatty acid. Addition of subsequent double bonds has progressively smaller effects. Thus,
the molecular and spatial distribution of olefinic bonds is
probably of greater importance than the overall degree of unsaturation.

These problems are highlighted by the studies of Cremel
et al. (1979), who found large seasonal variations in the
fatty acid composition of liver mitochondria from the
European hamster, but which were not related to changes in
body temperature during hibernation. Wilkinson (1976)
attempted to alter lipid composition of hamster kidney by
feeding diets of high and low fatty acid saturation and to
see whether a "hibernation-like" increase in Na-K-ATPase
activity occurred with unsaturation or whether thermal sensitivity was made "rat-like" by increased saturation of fatty
fluids. He found changes in saturation only of specific
fatty acids resulting in small if any net changes in overall
saturation. The specific compositional changes he observed
were different from those he observed with onset of hibernation and there was no marked alteration in activity or

thermal sensitivity of (Na^++K^+)-ATPase.

Thus, there is no impression from the compositional literature that a bulk change in lipid composition occurs, although in the case of membrane enzymes with specific microenvironments within the membrane there might be substantial changes in lipid composition which would not affect the bulk matrix but would have important functional effects. Current understanding of the microheterogeneity of cellular membranes is rudimentary.

Direct evidence of a structural adaptation has been presented by Keith *et al.* (1975) in liver mitochondria of hibernating ground squirrels. The Arrhenius discontinuity of spin label found for the mitochondria of active ground squirrels was removed on hibernation. The description of thermotropic transitions and the assignment of transition temperatures using empirical motional parameters has recently been criticised (Cannon and Polnaszek, 1976; Schreier *et al.*, 1978) and such demonstrations must be treated with some caution. In any case, a comparison of the rotational parameters measured showed that the fluidity of hibernating individuals was considerably lower than that of active individuals, in complete contrast to what one might expect.

Charnock *et al.* (1980) compared the fluidity of liposomes prepared from myocardial membranes of hibernating and active ground squirrels. They found that the extracts from hibernating individuals were somewhat more fluid than corresponding extracts from active individuals. Phase transitions were observed by differential scanning calorimetry to be at 18-25°C in active individuals and at 5-16°C in hibernating squirrels. The authors maintain that only 2 percent of the lipids were involved in the transition, and the change in transition temperature during hibernation had no effect upon Arrhenius discontinuity of the (Na^++K^+)-ATPase.

Goldman and Albers (1979), however, have shown that the average fluidity of brain membranes from hibernating hamsters was greater than for the corresponding preparations from active hamsters. Unfortunately, they did not correct for light scattering artefacts which at 5 percent of total fluorescence might have marked effects upon the measured polarization and its temperature dependence (Cossins, 1981). In any case, the efficacy of the reported response on isolated membranes was low (10 to 12 percent; Cossins, this volume) and would seem insufficient to remove entirely a phase transition from the critical range of temperatures. On the other hand contamination of the changing membrane type by non-responding types (*e.g.* endoplasmic reticulum diluting plasma membranes) could lead to an underestimate of the efficacy, since different types of membranes do show quite different degrees of change with acclimation (Cossins, this volume).

In a similar fluorescence polarization study of synaptosomal and kidney microsomal membranes of hibernating and active hamsters and ground squirrels and of rat (Cossins and Wilkinson, 1982) it was not possible to demonstrate any sizeable or consistent difference in fluidity, certainly not on the scale, for example of that observed between cold- and warm acclimated goldfish (Cossins, 1977).

Using simple fluorescence of the dye ANSA (4-anilinonapthalene-1-sulphonic acid) as a measure of partitioning into liver membrane fraction (and, therefore, indirectly, phase separation) Houslay and Palmer (1978) found discontinuities in the Arrhenius plot at 25 and 13°C in preparations from awake hamster and 25 and 4°C in hibernating hamster which correlated with the enzyme results described above.

Finally, Aloia (1980) has discussed several studies with differential scanning calorimetry in membranes of non-hibernators indicating that bulk phase transitions occur below 0°C.

Thus, as with compositional studies and the transport studies described in the previous section, one is led to the conclusion that bulk membrane fluidity changes are unlikely to play a role in adaptation to low temperature in hibernators.

CONCLUSION

In this review we have examined evidence from four lines of study (membrane transport, membrane enzymes, lipid composition and physical measurements) on membrane involvement in cellular cold resistance of hibernating mammals both in terms of evolutionary adaptation (*i.e.* interspecies difference) and short term acclimation (*i.e.* differences related to physiological state). This evidence suggests that specific and selective modifications of membrane structure and activity contribute to the ability of hibernating mammals to survive low body temperature. Reduced temperature of phase separations for the various transport mechanisms is a possible explanation of such selective and specific changes which cannot be ruled out by the available evidence. More general alterations involving bulk lipid fluidity and parallel thermal behaviour of diverse transport systems such as have been found in prokaryotes and some poikilothermic eukaryotes do not appear to be present or important in hibernating mammals.

SCIENTIFIC NAMES OF SPECIES MENTIONED IN THE TEXT

Hedgehog	*Erinaceus europaens*
Mole	*Talpa europaea*
Squirrel monkey	*Saimiri sciurea*
Sheep	*Ovis aries*
Syrian hamster	*Mesocritus auratus*
European hamster	*Cricetus cricetus*
Rat	*Rattus rattus*
Russet suslik	*Spermophilus (= citellus) erythrogenus*
Golden-mantled ground squirrel	*S. lateralis*
Richardsons ground squirrel	*S. richardsonii*
Thirteen-lined ground squirrel	*S. tridecemlineatus*
Guinea pig	*Caria caria*
Rabbit	*Oryctolagus curriculus*

INJURY AT LOW TEMPERATURES

CHILLING INJURY IN PLANTS

J.M. Wilson and A.C. McMurdo

School of Plant Biology, University College of North Wales, Wales.

INTRODUCTION

Chilling injury is best defined as the injury which occurs to many tropical and semi-tropical plants when exposed to temperatures between 0 and 15°C. The speed of development of injury depends on the species, the previous temperature history of the plant, the chilling temperature, relative humidity and also other environmental conditions during chilling. Injury can occur within a few hours at 5°C, as in the case of the extremely chill-sensitive species *Episcia reptans* or symptoms may take many days to develop, as in the case of cucumber fruits. The first symptoms of chilling injury to the leaves of many agriculturally important tropical and sub-tropical species are rapid leaf wilting followed by the development of sunken necrotic patches within 24 hours of the start of chilling, *e.g.* *Phaseolus vulgaris* (french bean), *Cucumis sativus* (cucumber) and *Gossypium hirsutum* (cotton). On return of the plants to the warmth the leaf margins and sunken pits usually dry out giving the leaf a very mottled appearance with brown necrotic patches surrounded by green undamaged tissue. Chilling injury is of considerable economic importance in the transport of tropical fruits to the temperature regions since market life cannot be prolonged by lowering the temperature into the chilling range. In addition, chilling injury to seedling plants of cotton and sorghum can lead to complete failure and make re-sowing necessary.

CHILL HARDENING

In some crop plants, such as *P. vulgaris*, leaf wilting and injury at 5°C can be prevented for up to 9 days by chill hardening the plants at 12°C, 85% relative humidity(RH) for 4 days before chilling (Wilson and Crawford, 1974a). Injury to these species can also be prevented by drought hardening the plants at 25°C, 40% RH, by withholding water from the roots so that the leaves wilt over a 4-day period. Drought hardening has been shown to be as effective as chill hardening in preventing chilling injury. Chilling injury at 5°C can also be prevented in many crop species simply by maintaining a saturated (100% RH) atmosphere around the leaf by enclosing the plant inside a polythene bag before transfer from 25 to 5°C (Wilson, 1976).

Broadly speaking, chill-sensitive plants can be divided into two categories (Wilson, 1976) based on:

1) The sensitivity of the species to chilling injury.
2) The ability to harden against chilling injury.
3) Whether chilling injury can be prevented on direct transfer from 25 to 5°C by maintaining a saturated atmosphere around the leaf.

Species such as *P. vulgaris*, *G. hirsutum* and *C. sativus* are placed in category 2 (Table 1), as these plants are less chill-sensitive and usually only incur 50% leaf injury after 24 hrs at 5°C, 85% RH. These plants can also be chill and drought hardened against chilling injury. Injury can be prevented in *P. vulgaris* by maintaining a saturated atmosphere around the leaf on transfer from 25 to 5°C. This suggests that chilling injury to these species is primarily caused by water loss. However, when chilling is prolonged for several days by maintaining 100% RH, metabolic changes must eventually lead to cell death.

In the extremely chill-sensitive category 1 species, *E. reptans*, water loss is less important in the development of chilling injury since the rate of injury cannot be significantly reduced by maintaining a saturated atmosphere around the leaves. In addition it is not possible either to chill or drought harden the leaves of this species to withstand chilling injury. Even a prolonged period of acclimation in a cool, well ventilated greenhouse at 15°C resulted in little increase in chill tolerance. This suggests that chilling injury to the leaves of category 1 species is primarily metabolic. Tropical fruits also possess little ability to harden against chilling injury. Attempts at hardening sweet potatoes have not been success-

Table 1

The division of chill-sensitive species into two categories
based on their sensitivity to chilling, their ability to
chill and drought harden against chilling injury at 5°C,
85% relative humidity (RH), and on whether chilling injury
can be delayed on direct transfer from 25 to 5°C by main-
taining a saturated (100% RH) atmosphere.

Category 1.   *e.g. Episcia reptans, Episcia cupreata,*
Nautilocalyx lynchii.

1) Extremely chill-sensitive species which show injured
   spots after only 2 h at 5°C
2) These plants cannot be chill hardened at 12°C, 85% RH
   or drought hardened at 25°C, 40% RH to withstand
   chilling injury at 5°C. Even prolonged periods of
   acclimatization at 15°C result in little increase in
   chill tolerance.
3) Maintaining a saturated atmosphere at 5°C does not
   delay the onset of injury.

Category 2.   *e.g. Phaseolus vulgaris, Cucumis sativus,*
Gossypium hirsutum.

1) Less chill-sensitive species usually incurring severe
   leaf injury after 24 h at 5°C, 85% RH.
2) Chill hardening and drought hardening can protect the
   leaves against chilling injury at 5°C, 85% RH for up
   to 9 days in *P. vulgaris.*
3) Maintaining a saturated atmosphere at 5°C can prevent
   chilling injury for up to 9 days on direct transfer
   of *P. vulgaris* leaves from 25 to 5°C. Chilling injury
   can also be prevented for up to 3 days in *C. sativus*
   leaves and 2 days in *G. hirsutum* leaves by maintaining
   100% RH at 5°C.

---

ful in reducing chilling injury (Wheaton and Morris, 1967)
and hardening of cucumbers is only effective against slight
chilling (Apeland, 1966).

THE PHASE CHANGE HYPOTHESIS OF CHILLING INJURY

Recently the Lyons-Raison phase change hypothesis of
chilling injury (Lyons, 1973) has received a great deal of

attention. In its simplest form the hypothesis states that
at a certain critical temperature within the chilling injury
range the membrane lipids of chill-sensitive plants undergo
a transition from a liquid-crystalline to a solid gel state.
It is thought that the two main consequences of this
transition which eventually result in injury and death are
an increase in membrane permeability and an increase in the
activation energy of membrane bound enzymes (Fig. 1).
Whilst an increase in activation energy in itself may not be
damaging to a plant, it is thought that lethal increases in
ethanol and acetaldehyde content may occur due to a
metabolic imbalance between non-membrane bound glycolytic
reactions and membrane bound enzymes of the tricarboxylic
acid cycle. Stimulus for greater research on the belief
that "solidification of the membrane lipids" could account
for chilling injury was derived from two sources. Arrhenius
plots of oxygen uptake by mitochondria isolated from chill-
sensitive plant tissues were non-linear with a break at the
temperature at which it was believed that the plant became
chill injured (Lyons *et al*., 1964). Secondly, electron spin
resonance (ESR) techniques indicated that the lipids of
these mitochondrial membranes underwent a change in molecular
order at the same temperature as that of the "break" in the
Arrhenius plot of oxygen uptake (Raison *et al*., 1971a).
Similar changes in activation energy of membrane bound
enzymes or changes in molecular order as determined by ESR
were not thought to occur in chill-resistant plants until
well below 0°C because of the higher degree of unsaturation
of the membrane fatty acids.

The role of lipids in the prevention of low temperature
injury is suggested by increases observed in the degree of
unsaturation and often weight of lipid per cell during the
acclimatization of plants as well as ectothermic and endothermic animals to low temperatures (Wilson and Crawford
1974b; Clarke; Cossins; Herbert, this volume). It was
thought that an increase in the degree of unsaturation of
the membrane fatty acids of 5 to 12% prevented chilling
injury by lowering the phase transition temperature to below
5°C. Plant cell membranes usually contain at least 70%
unsaturated fatty acyl chains. In artificial mixtures of
unsaturated and saturated fatty acids at this concentration,
a 10% increase in unsaturation could lower the solidification temperature by as much as 20°C (Lyons and Asmundson,
1965). In agreement with this hypothesis Wilson and
Crawford (1974b) reported increases of 5 to 12% in the
degree of unsaturation of the fatty acids associated with
the phospholipids of *P. vulgaris* and *G. hirsutum* leaves

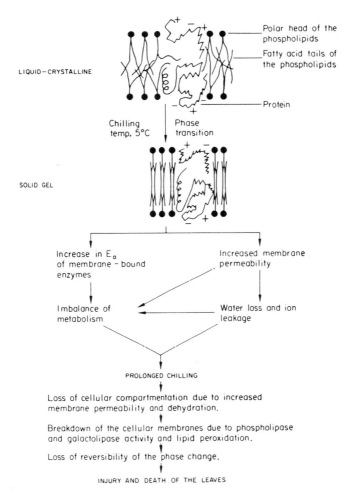

Figure 1. *A schematic pathway of the events leading to chilling injury in leaves according to the phase change theory. (The + and - signs indicate the ionic residues of the proteins.)*

during chill hardening at $12^{\circ}C$ (Table 2). These increases in total percentage unsaturated fatty acid were mainly due to an increase in the percentage of α linolenic acid (18:3ω3) No increase in the degree of unsaturation of the glycolipids was detected. Table 2 also shows that no significant increase in the degree of unsaturation of the phospholipids occurred during the ineffective hardening of *E. reptans* at $15^{\circ}C$ except for a slight increase in phosphatidylcholine.

Table 2

*Changes in unsaturated fatty acid associated with the
phospholipids during the chill hardening of Phaseolus
vulgaris and Gossypium hirsutum leaves at 12°C, 85% RH.
Changes in unsaturation during the ineffective hardening
of Episcia reptans at 15°C are included for comparison.*

Unsaturated fatty acid (% total wt)

Lipid	Phaseolus vulgaris		Gossypium hirsutum		Episcia reptans	
	25°C	12°C	25°C	12°C	25°C	15°C
Phosphatidylcholine	69.3	80.8	70.6	81.3	73.8	74.8
Phosphatidylethanolamine	63.2	66.8	60.9	67.7	72.8	71.1
Phosphatidylinositol	69.4	64.7	52.6	60.6	58.3	57.5
Phosphatidic acid	66.9	80.8	62.6	72.4	60.7	52.0
Phosphatidylglycerol	54.4	59.7	45.2	55.3	66.8	61.8

The phase change theory has been criticised because the
description of the phase transition as being a change from
a liquid crystalline to a solid gel state is a gross over-
simplification of the events occuring in the complex mixture
of lipids present in biological membranes. Indeed, it is
now clear that the events observed by ESR do not reflect this
type of change but instead some change in the molecular
ordering of discrete domains of lipid within the membrane
(Dalziel and Breidenbach, 1979). Of even greater concern are
reports of breaks in Arrhenius plots of oxygen uptake by
mitochondria of chill-resistant plants at temperatures within the chilling range (Marx and Brinkmann, 1979; Miller *et
al*., 1974; Pomeroy and Andrews, 1975) and of breaks in
Arrhenius plots of Hill activity in chloroplasts of chill-
sensitive and chill-resistant plants within the same
temperature range (Nolan and Smillie, 1976). In addition
investigations of changes in the molecular ordering of
membrane lipids at low temperatures using ESR have shown
that changes can occur at chilling temperatures in both chill-
sensitive and chill-resistant plants (Bishop *et al*., 1979).
    The validity of using ESR techniques for investigating
biological membranes has been criticised by Schrier *et al*.
(1978), since the transition temperature for a given sample
can vary by several degrees depending on the type of spin
label probe used. Some authors have been unable to detect

differences in Arrhenius plots of spin label motion or
fluorescence depolarisation which would enable them to
determine whether the plant is chill-sensitive or chill-
resistant (Bishop *et al*., 1979). In addition analyses of
either total cell lipid and of isolated membranes have not
always been able to demonstrate differences in the unsaturation of fatty acids between chill-sensitive and chill-
resistant or hardened chill-sensitive plants (Pomeroy and
Andrews, 1975; Wilson, 1976). In fact the membranes of both
chill-sensitive and chill-resistant plants have such a high
degree of unsaturation that a phase transition in the bulk
lipid would not be expected until well below $0^{\circ}C$. It is
possible that spin label probes are detecting transitions in
lipids which contain two saturated fatty acids. Experiments
on pure lipids of different fatty acid composition indicate
that only lipids with two saturated fatty acids show phase
transitions in the chilling injury range; the substitution
of one unsaturated fatty acid for a saturated one can lower
the transition temperature to below $-20^{\circ}C$ depending on the
lipid (Bishop *et al*., 1979). However, lipid molecules which
contain two saturated fatty acids are scarce in plant tissues
being most frequently found in the sulpholipid fraction. It
has been suggested by Canon *et al*. (1975) that apparent
breaks in Arrhenius plots of spin label data may simply be
artefacts due to an overestimation of the order parameters
and rotational correlation time of the spin label at low
temperatures. It has also been suggested that breaks in
Arrhenius plots of oxygen uptake by mitochondria may be biochemical artefacts because the Km value of the enzymes is
always assumed to be constant throughout the temperature
range and yet the Km can decrease with temperature so that
breaks may represent the onset of substrate limitation
(Silvius *et al*., 1978). It is possible that breaks in
Arrhenius plots of enzyme catalysed reactions represent a
direct effect of temperature on the protein independent of
any influence on either the bulk or boundary lipids.
Graham *et al*. (1979) consider that the increase in activation energy of the lipid free enzyme phosphoenol pyruvate
carboxylase at low temperatures may be due to the weakening
of its hydrophobic bonds or the strengthening of hydrogen
bonding and electrostatic interactions. Whatever the cause
of these breaks in Arrhenius plots of physiological and
biophysical measurements it is clearly necessary to have
a statistical procedure to determine whether the breaks are
real, if the breaks in ESR and oxygen uptake measurements are
coincident and whether or not breaks occur only in chill-
sensitive plants.

STATISTICAL METHODS FOR DETERMINING BREAK POINTS IN
ARRHENIUS PLOTS

Arrhenius (1889) reported that in certain ionic dissociation
reactions a plot of the natural log of the reaction rate
against the reciprocal of absolute temperature yielded a
straight line with a slope:

$$m = \frac{-Ea}{kB}$$

where E_a is the activation energy per molecule of the
activated complex, B is Boltzmanns constant. This has
become known as the Arrhenius Law though it only holds true
when there is only one rate limiting step (Fig. 2a; Arrhenius,
1889). If the rate limiting step changes at a certain point
the linear relationship will be preserved but the result will
be two straight lines joining at a break point coincident
with the change in rate limiting step (Fig. 2b). Figure 2c
shows the first problem encountered when using Arrhenius
plots. An experimentalist does not know whether the points
are linearly related or not (as there may be more than one
rate limiting step and these may vary independently of each
other) and so he may draw regression lines through unrelated
points. The more complex biological systems become the more
variance there will be in any observational data (Fig. 3).
For the majority of plant processes there are many rate
limiting steps; intuitively therefore we might not expect
the Arrhenius Law to hold true for such complex processes as
photosynthesis and respiration, processes which have been
widely used in studies on chilling injury. Another problem
is that the ommission of points from the end of an Arrhenius
plot will shift any break point simply as a statistical
artefact (Bagnall and Wolfe, 1978). This is of particular
importance when one considers that the most chill-sensitive
plants die rapidly at low temperatures and their metabolic
rates are so slow that they are difficult to measure; as a
result few experimental points are obtained at the low
temperature end of an Arrhenius plot and any apparent break
is likely to be shifted to a higher temperature. Furthermore, Arrhenius plots cannot be used to calculate activation
energies at temperatures above $28^{\circ}C$ because at these
temperatures the rates of many plant physiological processes
are decreasing from their optima. The Arrhenius plot will
thus have a positive slope at these temperatures and one
could therefore calculate a negative activation energy which
would clearly be misleading (Fig. 2d).

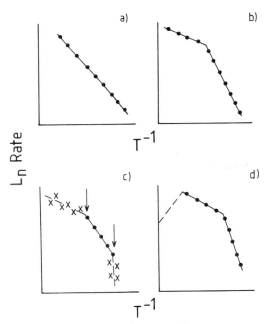

Figure 2. *Diagrams illustrating various features of Arrhenius plots. a) correct Arrhenius plot, all control parameters remain constant throughout the temperature range so that the Arrhenius Law holds true, b) an Arrhenius plot with a true breakpoint where the rate limiting step has changed but the Arrhenius Law still holds true either side of the break, c) an Arrhenius plot showing 2 false break points (arrowed). The points marked with crosses are not linearly related so Arrhenius' Law does not apply, d) an Arrhenius plot showing a possible negative activation energy (broken lines) at temperatures above the optimum.*

Until 1976 no statistical procedures had been used to fit the best lines to chilling injury data presented as Arrhenius plots. Lines were simply fitted to points by independent assistants who supposedly knew nothing about where the break points were supposed to occur. Consequently many contentious breaks were reported (Downton and Hawker, 1975; Kane *et al.*, 1978). In 1976 Raison and Chapman applied statistics to their data on growth rates and biophysical measurements of *Vigna radiata*. The method they employed was to work along each point on the graph and calculate the residual sums of squares for the best fitting pair of regression lines from each point. They considered a break to occur at minima in the residual sums of squares. This procedure can now be

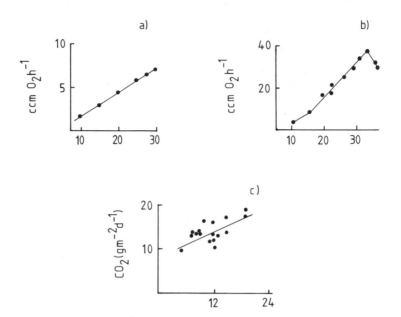

Figure 3. *The effect of temperature on the rate of respiration of a) Phycomyces on bread, b) Zea mays roots, and c) a forest in summer (taken from Wassink, 1972).*

performed by computer and is the basis of the maximum like-lihood programme that has been developed by Potter and Ross (1979). Figure 4 illustrates some growth results that have been analysed using the maximum likelihood programme. One of the main problems with this form of analysis is that one has to have smooth data with little variability or natural variation will cause small minima. In addition one has to assume *a priori* that straight lines are the best fit and, as one has to give an estimation of the break point, one is presenting the computer with a *fait accompli* in effect saying "there is a break here, isn't there?" rather than "is there a break here?". Bagnall and Wolfe (1978) suggested that a better fit can be obtained by using polynomial curves rather than straight lines, but the reduction of the residuals of their data is only a matter of 0.00028 when one uses a 3rd order polynomial rather than a straight line, so it is doubtful if this is of any significance. Other forms of statistical analysis have recently been developed (Willcox and Patterson, 1979; Wolfe and Bagnall, 1979) but these seem overelaborate considering that the methods of least squares and maximum likelihood are well established techniques.

The shortcoming shared by all these methods is that they

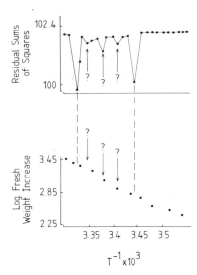

Figure 4. *An Arrhenius plot of the increase in fresh weight of cucumber seedlings at different temperatures tested for breaks using the maximum likelihood programme.*

simply attempt to produce a good fit for the data rather than a correct fit. As the existence and position of these breaks is of fundamental importance to the phase change theory it is essential to produce the correct model that gives an acceptable (though not necessarily the best) fit to all the available data. At Bangor we have developed a programme that, fits (by maximum likelihood) the best 2 line, 3 line and curved model (McMurdo and Wilson, 1980). It then analyses the distribution of the residuals around the line; if these are not random then the fit is incorrect even though it may offer quite low residuals. This programme was tested using some results from experiments on the growth and respiration of seedlings at different temperatures. The 2 straight line model offers the slightly better fit in both cases and the distribution of the residuals is more normal and therefore the straight lines are a more correct fit (Fig. 5). However, there is a break in the plot for rye (*Secale cereale*) which is chilling-resistant and should not be expected to show a break. Breaks in Arrhenius plots must therefore be used with extreme caution as evidence to support the phase change theory. This is supported by Bagnall (1979) and Smillie (1979) who reported that the so called critical temperature for injury as determined by ESR was not always concomitant with the onset of visible injury as one can hold

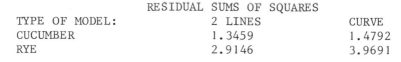

RESIDUAL SUMS OF SQUARES

TYPE OF MODEL:	2 LINES	CURVE
CUCUMBER	1.3459	1.4792
RYE	2.9146	3.9691

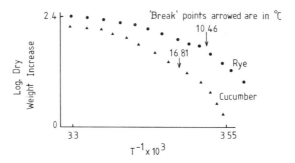

DISTRIBUTION OF THE RESIDUALS ABOUT THE MODEL. (THE MORE
NORMAL THE DISTRIBUTION THE MORE CORRECT THE FIT)

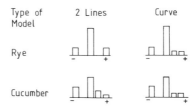

Figure 5. *Arrhenius plot of the dry weight increase in Rye
(chill-resistant) and Cucumber (chill-sensitive) seedlings.*

plants at or slightly below this temperature without injury
for relatively long periods of time.

In spite of the problems associated with the use of
Arrhenius plots Raison *et al.* (1979) have found that a very
good correlation exists between the transition temperature
of leaf lipids as determined by ESR and habitat temperature.
Temperate plants had transition temperatures in the range 0
to 2°C whilst tropical plants were nearly always much higher,
within the range $10\text{-}17^{\circ}$C. However, phase transitions can be
expected to occur at different temperatures within different
membranes from the same cell due to large differences in
fatty acid composition. It is possible that break temperatures in individual membranes are less important than the
fact that increases in activation energies are usually
greater at chilling temperatures in chill-sensitive plants
such as *C. sativus* and *Vigna radiata* than in chill-resistant
plants such as *Secale cereale* (Fig. 5). Whether or not the
cell is injured at 5°C may depend on whether the cell can

compensate for those changes resulting from any phase
transition such as alterations in membrane permeability or
the effects on metabolism of increases in the activation
energy of membrane bound enzymes.

CHANGES IN MEMBRANE PERMEABILITY AT CHILLING TEMPERATURES

From investigations of lipid bilayers Träuble and Haynes
(1971) suggested that an increase in membrane permeability
would be expected to accompany the phase transition due to:

1) A decrease in membrane thickness.
2) Changes in structure of the hydrocarbon chains
   important for diffusion across the membrane.
3) Changes in the arrangement of the polar head groups
   important for the entry of permeants into the membrane.

It has also been speculated that 'cracks' or 'channels' may
appear in the membrane at low temperatures caused by the
solidification of the lipid, thereby increasing membrane
permeability. In agreement with these hypotheses the
majority of studies on membrane permeability at chilling
temperatures have shown an increase in the rate of electrolyte leakage from chill-sensitive tissues. Liebermann *et al*.
(1958) were the first to show that the rate of leakage of
ions, mainly potassium, from sweet potato discs was increased
at 7.5°C. In addition Christiansen *et al*., (1970) and Guinn
(1971) detected an accelerated rate of leakage of electrolytes, proteins and carbohydrates from chilled cotton roots
and cotyledons. Enhanced leakage of electrolytes from
chilled leaf tissue has been reported by Wright and Simon
(1973), Creencia and Bramlage (1971), and Patterson *et al*.,
(1976). However, in all these studies the rate of leakage
only became rapid after many hours or days at the chilling
temperature and this argues against any rapid rise in
permeability which can be attributed to lipid phase transitions. In spite of this, increased rates of electrolyte
leakage at 5°C have been used to support the hypothesis that
a phase transition is responsible for the increased membrane
permeability and rapid leaf wilting that is characteristic of
the early stages of chilling injury to *P. vulgaris* and
G. hirsutum.
     To determine whether the rapid leaf wilting of *P. vulgaris*
leaves at 5°C could be due to a rapid increase in membrane
permeability, Wilson (1976) compared the rates of ion
leakage from *P. vulgaris* leaves after chilling in air at 5°C,
85% RH, with the rate of leakage on direct transfer from air
at 25°C to water at 5°C. Leaves of *P. vulgaris* leak

electrolytes if they are chilled in air for 24 h at 5^{0}C, 85%
RH, so that the leaves are badly injured and wilted before
they are transferred to distilled water at either 25 or 5^{0}C
(Fig. 6). The rate of leakage from the air chilled leaves
in water at 25^{0}C is almost twice as fast as the leaves
placed in water at 5^{0}C; this may be due to heat shock on
transfer from 5 to 25^{0}C. In contrast, the control leaves
transferred directly from 25^{0}C to water at either 25 or 5^{0}C
show a very slow rate of electrolyte leakage. The behaviour
of the control leaves appears to be contrary to the phase
transition hypothesis. If the phase transition in the membrane lipids causes an increase in membrane permeability
then we would expect the leaves to leak electrolytes when
transferred directly from 25^{0}C to water at 5^{0}C (*i.e.* without
chilling in air). However, the leakage of electrolytes from
unchilled leaves of *P. vulgaris* transferred directly to
water at 5^{0}C is very slow and that leakage is faster from
the control leaves transferred directly to water at 25^{0}C.
This result suggests that initially a phase transition in the
membrane lipids of the plasmelemma leads to a decrease in its
permeability to ions, such as K^{+}, at low temperature. In
support of this argument Blok *et al.* (1976) have shown that
the water permeability of liposomes prepared from synthetic
lecithin decreases drastically below the transition temperature. Furthermore, vesicles of *Escherichia coli* preloaded

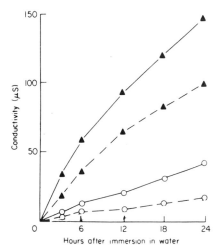

Figure 6. *Leakage of electrolytes from leaves of Phaseolus
vulgaris chilled at 5^{0}C, 85% RH for 24 h (▲) in comparison
to unchilled leaves (o), placed either in water at 25^{0}C (———)
or 5^{0}C (- - -).*

with labelled proline did not lose their radioactivity when incubated at temperatures below the transition (Esfahan *et al.*, 1971). Increased rates of ion leakage from leaf tissue may therefore occur only after prolonged chilling has resulted in cell dehydration and damage to cell membranes. Considerable water loss and cell damage must occur during chilling in air at $5^{\circ}C$, 85% RH, for the leaves to lose their electrolytes rapidly when transferred to distilled water (Fig. 6).

The leaves of some chill-sensitive species leak electrolytes more rapidly than *P. vulgaris* leaves on transfer from air at $25^{\circ}C$ to water at $5^{\circ}C$. Leaves of the extremely chill-sensitive category 1 species, *E. reptans*, lose approximately 25% of their total electrolytes within 5 h of transfer from $25^{\circ}C$ to water at $5^{\circ}C$ (Fig. 7). In this species visible signs of cell damage accompany the rise in the conductivity of the water so that it is not possible to distinguish if leakage is an immediate effect of a phase transition, or, a secondary effect due to cell injury. To resolve this problem, efflux experiments using leaf slices of *E. reptans* loaded with ^{86}Rb at $25^{\circ}C$ were performed. The results show that there is a far greater loss of ^{86}Rb 10 to 15 min after transfer to $5^{\circ}C$ than at $25^{\circ}C$ (Murphy, 1981). In contrast,

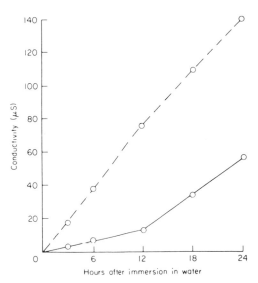

Figure 7. *Leakage of electrolytes from leaves of Episcia reptans transferred directly from $25^{\circ}C$, 85% RH to water at $25^{\circ}C$ (——), or $5^{\circ}C$ (– – –). Total quantity of electrolytes in leaves = 175 u S.*

the rate of efflux of ^{86}Rb from leaf slices of *Vicia faba*
(chill-resistant) was the same at 5 and 25°C. Whether this
increased rate of leakage of ^{86}Rb from *E. reptans* after only
ten min at 5°C can be attributed to a phase transition in
the membrane lipids or to a change in membrane protein is
still unsolved, but these experiments do indicate that
chilling can cause rapid changes in cell permeability in
some species. Changes in the permeability of leaf cells of
P. vulgaris during prolonged chilling at 5°C, 100% RH, are
currently being investigated.

THE CAUSE OF LEAF WILTING IN *PHASEOLUS VULGARIS* AT 5°C

If the Lyons-Raison hypothesis of chilling injury is correct
then we might expect that the increases in the degree of
unsaturation of membrane phospholipids during chill hardening
(Table 2) are essential for the prevention of leaf wilting
on transfer to 5°C, 85% RH. However, although drought
hardening was as effective as chill hardening in preventing
chilling injury at 5°C, 85% RH, no increase in the degree
of unsaturation of the phospholipids or glycolipids was
detected (Table 3). In general the degree of unsaturation
of the fatty acids associated with the phospholipids
decreased during drought hardening. For example, there is
a decrease in the proportion of unsaturated fatty acid
associated with phosphatidylcholine during drought hardening
so that, according to the phase transition hypothesis, we
would expect the drought hardened plants to be more chill-
sensitive and not chill-resistant. Chill hardening of
P. vulgaris leaves at 12°C is not effective if the plants
are maintained at 100% RH (Wilson, 1976). Although the
degree of unsaturation of the phospholipids increased during
an ineffective hardening treatment of 4 days at 12°C, 100%
RH, there was no increase in chill tolerance indicating that
chill hardening is not dependent on a high content of
unsaturated fatty acids. The phase transition hypothesis is
also unable to account for the prevention of chilling injury
to leaves of *P. vulgaris* by enclosing the plant inside a
polythene bag before transfer to 5°C. Plants maintained in
a saturated atmosphere for 7 days do not wilt at 5°C. If
lipid phase transitions resulted in an increase in the
permeability of the plasmalemma of the leaf cells at 5°C
then we would expect the leaves to wilt on transfer to 5°C,
100% RH, as the turgor pressure of the cell would facilitate
the loss of water and electrolytes.

The primary cause of chilling injury to *P. vulgaris*
leaves at 5°C, 85% RH, is water loss due to the opening of

Table 3

Fatty acid composition of phosphatidylcholine from leaves of Phaseolus vulgaris during chill hardening at 12°C, 85% RH and drought hardening at 25°C, 40% RH.

Fatty acid composition of phosphatidylcholine (%)

Fatty acid	Control	Chill hardened at 12°C 85% RH for 4 days	Drought hardened at 25°C 40% RH for 4 days
14:0	3.8	2.1	5.3
16:0	20.4	12.8	24.3
16:1	0.9	1.0	0.7
16:2	0.9	1.1	0.4
18:0	6.5	4.3	6.2
18:1	4.0	3.5	2.0
18:2	27.5	40.0	24.1
18:3	36.0	35.2	37.0
Total % unsaturated fatty acid	69.3	80.8	64.2

the stomata at a time when the permeability of the roots to water is low (Wilson, 1976). The opening of the stomata after 2 h at 5°C, 85% RH, (Fig. 8), is surprising as the leaf is wilted and in most plants the stomata close in the early stages of water stress before visible wilting occurs. The replacement of the water lost by evapotranspiration from the leaf is prevented by the low permeability of the roots to water at 5°C (Fig. 9), resulting in rapid leaf dehydration and injury. Hence the severity of chilling injury depends on a synergistic effect between stomatal opening and reduced permeability of the roots to water at 5°C. When only the leaves of *P. vulgaris* were chilled at 5°C, 85% RH, there was no detectable fresh weight loss after 24 h, except for a slight wilting after 3 to 4 h chilling (Fig. 10). Chilling the leaves alone for 24 h resulted in only 6% injury to the leaf after 2 days recovery at 25°C, 85% RH. In the reverse experiment, when roots alone were chilled and the leaves held at 25°C, 85% RH, there was a 30% decrease in fresh weight after 24 h but this resulted in

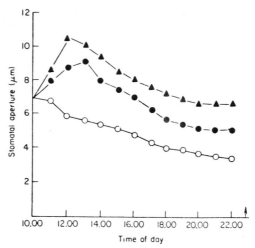

Figure 8. *Changes in stomatal aperture on transferring
entire plants of Phaseolus vulgaris directly from 25°C, 85%
RH to (a) 5°C, 85% RH (▲), (b) 12°C, 85% RH (●), compared
to the controls maintained at 25°C, 85% RH (O).   Arrow
shows start of night period.*

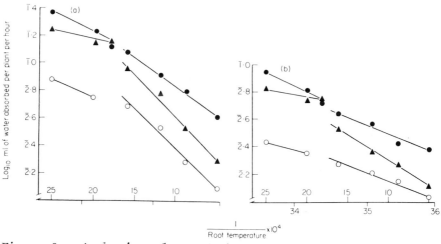

Figure 9. *Arrhenius plots of the effect of root temperature
on the rate of water absorption by Phaseolus vulgaris plants
grown at 25°C, 85% RH.   (▲), Chill hardened at 12°C, 85% RH
(●), and drought hardened at 25°C, 40% RH (O).   (a) shows
the rate of water uptake plus exudation under 50 cm Hg
vacuum and (b) the rate of exudation alone.   Each point
represents the average value from at least five plants.*

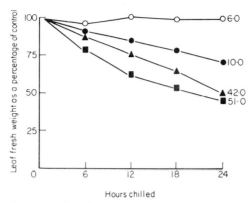

Hours chilled

Figure 10. *Changes in leaf fresh weight on chilling either the leaves alone (○), roots alone (●) or the whole plant of Phaseolus vulgaris (▲), for 24 h at 5°C. ■, denotes plants grown at 12°C, 100% RH, for 4 days before whole plant chilled at 5°C, 85% RH. Figures beside the points show the percentage of the leaf which became necrotic after 24 h chilling and 2 days recovery at 25°C, 85% RH.*

only 10% injury. Chilling the entire plant of *P. vulgaris* at 5°C, 85% RH, produced a more rapid fresh weight loss (50%) than chilling either the roots or leaves alone. The cause of stomatal opening in chill-sensitive species at 5°C is unknown.

THE MECHANISM OF CHILL AND DROUGHT HARDENING

Chill hardening at 12°C, 85% RH, prevents leaf dehydration by conditioning the stomata so that they close on transfer to 5°C, 85% RH, (Fig. 11). Similarly, drought hardening causes stomatal closure and the stomata remain closed on transfer to 5°C, 85% RH. Although chill hardening resulted in an increase in the permeability of the roots to water at low temperature, drought hardening produced a large decrease in root permeability (Fig. 9) and yet drought hardening was as effective as chill hardening in preventing leaf injury. This suggests that the most important factor in the prevention of chilling injury to *P. vulgaris* during chill and drought hardening is the closure of the stomata. This can be demonstrated by spraying the leaves of plants grown at 25°C, 85% RH, with 100 µM abscisic acid (ABA) which causes stomatal closure within 24 h. On transfer to 5°C, 85% RH, the sprayed leaves do not wilt as the stomata remain closed and injury is prevented for approximately 2 days by

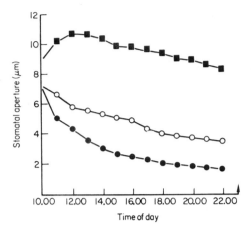

Figure 11. *Changes in stomatal aperture of Phaseolus vulgaris plants hardened at 12°C, 85% RH (●), and ineffectively hardened at 12°C, 100% RH. (by enclosure in a polythene bag) (■), on chilling at 5°C, 85% RH compared to the controls maintained at 25°C, 85% RH. (○). Arrows show start of night period.*

which time the effectiveness of the ABA has decreased. This decrease in the effectiveness of applied ABA is surprising as one might expect the plant to synthesize ABA during this period. Perhaps 5°C is too low a temperature for the synthesis of ABA in chill-sensitive plants.

During chill hardening at 12°C, 85% RH, the plant experiences a water stress (as shown by the temporary wilting of the leaves) due to the opening of the stomata (Fig. 8) and the decrease in the permeability of the roots to water (Fig. 9). However, at the intermediate temperature of 12°C the stress is not severe enough to result in damage and the wilting vanishes after 12 h. Similarly, during drought hardening the water stress is imposed simply by withholding water from the roots under conditions of high evapotranspiration so that the leaves wilt. Plants maintained at 5°C or 12°C, 100% RH, do not harden because they experience no water stress. Even though the stomata are open under these conditions no water can be lost from the leaf so that the stomata remain fully open on transfer to 5°C, 85% RH, (Fig. 11). Plants held at 12°C, 100% RH, for 4 days and then chilled for 24 h at 5°C, 85% RH, suffer a 55% decrease in fresh weight and approximately 50% leaf injury, after 2 days recovery at 25°C (Fig. 10). Therefore, enclosure in a polythene bag is not a method which can be used to lower

the hardening below 12°C. Although the above experiments
suggest that ABA is not synthesized in *P. vulgaris* leaves at
12°C, 100% RH, it is not known whether low temperature alone,
in the absence of water stress, can induce ABA synthesis.
However, the correlation between chill and drought hardening
has shown that an intermediate temperature of 12°C is not
essential for hardening. Therefore, water stress and not low
temperature *per se* is the primary factor inducing hardening
against chilling injury in *P. vulgaris* leaves.

CAUSES OF CHILLING INJURY TO *PHASEOLUS VULGARIS* LEAVES AT
5°C, 100% RH.

Although changes in stomatal aperture and root premeability
to water are able to explain the rapid wilting, dehydration
and ultimately injury to the leaves during chilling at 5°C,
85% RH, an alternative explanation must be sought for the
death of the leaves after 9 days at 5°C, 100% RH. The loss
of turgor, bleaching and necrosis of *P. vulgaris* leaves
under these conditions suggests that injury is probably due
to a combination of factors which may develop from a phase
transition in the membrane lipids.

Lipid and Fatty Acid Changes

In order to maintain membrane structure and function, cells
must be able to synthesize, transfer and incorporate fatty
acids into the numerous membrane systems throughout the cell.
(Clarke; Cossins, this volume). The effect of chilling on
these processes in chill-sensitive plants has received
surprisingly little attention in recent years. However,
using ^{14}C acetate as a radio-labelled precursor of fatty
acid synthesis McMurdo (1981) has shown that the half lives
of ^{14}C labelled fatty acids of *E. reptans* and *P. vulgaris*
leaves are very much shorter at chilling temperatures in
comparison to those of chill-resistant *Hordeum sativum* at
5°C. The free fatty acid and phosphatidic acid content of
E. reptans and *P. vulgaris* cells also increases at 5°C
reflecting an increased rate of lipid breakdown that did not
occur in chill-resistant *H. sativum* at 5°C. These changes
in turnover rates of membrane fatty acids may be initiated
by a phase transition in the membrane lipids of chill-
sensitive plants at 5°C and on prolonged chilling lead to
further changes in membrane permeability and function and
severely affect the reversibility of the phase change on
return of the plants to the warmth.

Photo-oxidation of Plant Pigments

The bleaching of some *P. vulgaris* leaves after 7 days
chilling at 100% RH indicates photo-oxidative degradation of
the leaf-pigments. Severe photo-oxidation of the leaf pigments and membrane lipids occurs in cucumber leaves after
2 to 3 days chilling at $1^{o}C$ (van Hasselt and Strikwerda,
1976). At temperatures higher than $1^{o}C$ this type of damage
develops more slowly.

Photosynthesis and Translocation

It is well known that photosynthesis is more sensitive to
low temperature than respiration so that starvation of plant
tissue may occur during prolonged chilling. Translocation
is inhibited in chill-sensitive species at $5^{o}C$ which may
lead to the starvation of non photosynthetic parts of the
plant, and accumulation of starch in the chloroplast may
further inhibit photosynthesis. Giaquinta and Geiger (1973)
suggested that the cessation of translocation in chill-
sensitive species at $5^{o}C$ is due to a phase change in the
membrane lipids of the plasmalemma of the sieve tube resulting in the collapse of the material lining the cell and the
blockage of the sieve plate by cytoplasm, organelles, P-
protein and membranes.

ATP Supply

The ATP and ADP levels in *P. vulgaris* leaves chilled at $5^{o}C$,
100% RH, increased over the first 24 h and remained high
during the following seven days (Fig. 12b). Even after 8
or 9 days chilling in a saturated atmosphere there was only
a slight fall in the ATP and ADP levels and this coincided
with a development of visible signs of chilling injury to
the leaf. A decrease in ATP supply below that necessary to
maintain the metabolic integrity of the cytoplasm cannot
therefore be considered to be the cause of chilling injury
to leaves at $5^{o}C$, 100% RH. In agreement with this result
Jones (1970) has also reported an increase in the ATP level
of *P. vulgaris* leaves at low temperatures. The increase in
ATP level of *P. vulgaris* leaves at $5^{o}C$ may be due to the
cold sensitivity of ATPase which is readily inactivated at
low temperatures (Penefsky and Warner, 1965). In contrast,
chilling leaves of *P. vulgaris* at $5^{o}C$, 85% RH, resulted in
rapid leaf wilting and injury and the ATP and ADP levels
decreased after 12 h chilling (Fig. 12a). Although the
leaves chilled at $5^{o}C$, 85% RH, were approximately 50%

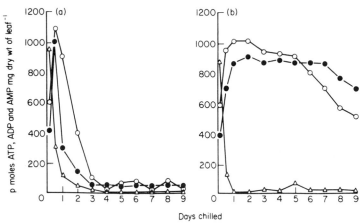

Figure 12. *Changes in the levels of ATP (●), ADP (○), and AMP (△) in the leaves of Phaseolus vulgaris, (a) during the development of chilling injury and water loss on direct transfer from 25°C to 5°C, 85% RH. and (b) during the prevention of chilling injury and water loss by maintaining a saturated (100% RH.) atmosphere around the leaves on transfer from 25°C to 5°C. Each point is the mean of three replicates.*

injured after 24 h, the level of ATP had decreased by less than 33%, suggesting that leaf dehydration and not a fall in ATP supply is the cause of cell death. The impaired phosphorylation of cotton leaves at 5°C reported by Stewart and Guinn (1969) can be attributed to the effects of water stress and not low temperature *per se*.

CAUSES OF CHILLING INJURY TO *EPISCIA REPTANS* LEAVES

ATP Supply

Although chilling leaves of *E. reptans* for 5 h at 5°C produced severe leaf injury there was no rapid decrease in ATP level during the first 5 h of chilling. The ATP and ADP levels fell by only 25% after 5 h chilling. Prolonged chilling between 5 and 24 h resulted in a further gradual decline in ATP level and a stabilization of ADP and AMP levels (Fig. 13). A reduced ATP supply is therefore unlikely to be the cause of chilling injury to *E. reptans* (Wilson, 1978).

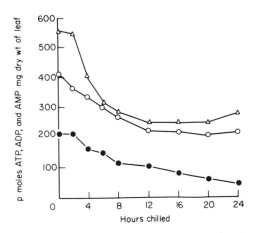

Figure 13. *Changes in the levels of ATP* (●), *ADP* (○) *and AMP* (△) *in the leaves of Episcia reptans during chilling in the dark at* 5°C. *Each point is the mean of three replicates.*

Leaf Respiration

Although changes in respiratory behaviour have been widely investigated in relation to chilling injury in fruits there have been comparatively few studies of changes in the respiration rate of chill-sensitive leaves at 5°C, especially in very sensitive leaves such as *E. reptans*. In cucumber fruits Eaks and Morris (1956) detected a doubling of the respiration rate after 8 days at 5°C and this increase coincided with the onset and development of chilling injury. This was followed by a decline in the respiration rate at the time of the general death of the tissue. Although Eaks and Morris were unable to explain the cause of this respiratory increase Creencia and Bramlage (1971) have shown, using the uncoupling agent 2,4-dinitrophenol (DNP), that the respiratory increase in *Zea mays* leaves after prolonged chilling for 24 to 48 h at 0.3°C is partly due to the uncoupling of oxidative phosphorylation at chilling temperatures.

In *E. reptans* leaves held at 5°C for 80 min the oxygen uptake rate is three times higher than in controls maintained at 25°C (Fig. 14a). The rate of carbon dioxide production also increased to a maximum after 80 min at 5°C but only to a maximum value of one-third of the rate of oxygen uptake (Fig. 14b). The respiratory peak after 80 min at 5°C coincided with the development of dark water-soaked

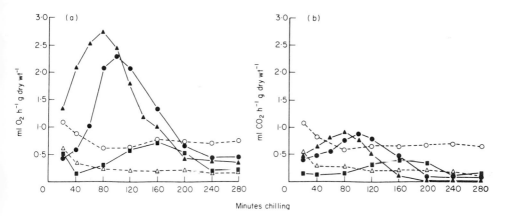

Figure 14. *(a) Oxygen uptake and (b) carbon dioxide evolution in the dark by leaf discs of Episcia reptans chilled (——) at 5°C (▲), 10°C (●) and 11°C (▪) in comparison to unchilled (- - - -) leaf discs at 12.5°C (△) and 25°C (○).*

patches on the leaf and the loss of leaf turgor. Plants chilled for 80 min at 5°C showed few signs of injury on return to 25°C. However, more prolonged chilling resulted in a gradual decline in respiration rate and an increase in leaf necrosis until after 5 h at 5°C none of the leaves recovered on subsequent transfer to 25°C.

Chilling at temperatures higher than 5°C delayed the onset of the respiratory rise and reduced its extent (Fig. 14a). In leaves chilled at 10°C the respiration rate was maximal after 100 min and was reduced by one-fifth in comparison to leaves chilled at 5°C. The first visible signs of chilling injury at 10°C also occurred at approximately the same time as maximum respiration rate. The extreme chill-sensitivity of *E. reptans* is demonstrated by the 90% injury incurred after 5 h at the relatively high chilling temperature of 10°C followed by 2 days at 25°C (Wilson, 1976). At 11°C there was an initial decrease in respiration rate followed by a gradual increase to the level of the control leaves maintained at 25°C after 160 min (Fig. 14a). The development of chilling injury was also slow at 11°C, the leaves incurring only 20% injury after 5 h chilling and 2 days recovery at 25°C, which indicates that 11°C is near the upper temperature limit for chilling injury in this species. An upper temperature limit of 11 to 12°C for chilling injury in *E. reptans* is supported

by the absence of any increase in the respiration rate of
the leaves held for 5 h at 12.5°C (Fig. 14a) and the
development of only 5% injury on transfer to 25°C (Wilson,
1976). At 12.5°C the respiration rate decreased to approximately one-third of the rate of the control leaves at 25°C
and remained at this level over the 5 h period. However,
prolonged chilling over several days at 12 and 15°C can cause
chilling injury to *E. reptans* (Wilson and Crawford, 1974b).
In contrast to *E. reptans* there was no increase in the
respiration rate of *P. vulgaris* leaves during the first 12 h
of chilling at 5°C (Wilson, 1978).

Experiments on the cause of this respiratory increase in
E. reptans leaves at 5°C, have been made using DNP and
potassium cyanide. Treating leaf discs of *E. reptans* with
1 mM DNP doubled the oxygen uptake rate at 25°C (Fig. 15).
However, transferring the DNP-treated discs to 5°C did not
lessen the effect of DNP, as would be expected if chilling
caused the uncoupling of oxidative phosphorylation.
Chilling the leaf discs treated with DNP resulted in an
extremely rapid three-fold increase in oxygen uptake within

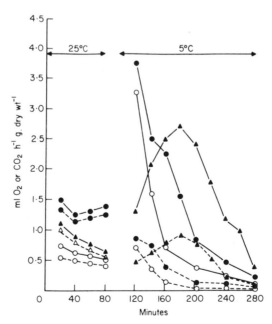

Figure 15. *The effects of 1 nM DNP (●) and 10 mM KCN (○)*
on the oxygen uptake (——) and carbon dioxide evolution
(- - -) rates of Episcia reptans leaf discs at 5°C in
comparison to untreated, control discs at 5°C (▲).

40 min of the start of chilling, whilst in the untreated controls at $5^{\circ}C$, the oxygen uptake did not attain its maximum value until 80 min after the start of chilling. The rapid acceleration of respiration by the addition of DNP showed that this method could not be used to determine whether any part of the respiratory increase was due to the uncoupling of oxidative phosphorylation. The respiration rate of leaf discs treated with 10 nM KCN at $5^{\circ}C$ was accelerated in the same manner as the DNP-treated discs. The rapid increase in the rate of oxygen uptake by the DNP and KCN treated leaf discs at $5^{\circ}C$ is not accompanied by a higher rate of carbon dioxide production.

It is suggested that the cause of this rapid wound-induced respiration in *E. reptans* leaves at $5^{\circ}C$ is an increase in the permeability of the cell membranes at low temperature caused by either lipid phase transitions, protein denaturation (Brandts, 1967), changes in lipid-protein interaction (Yamaki and Uritani, 1974) or a combination of these events. The magnitude of the phase change in the membranes of *E. reptans* leaves may be far greater than in *P. vulgaris* and this may account for the more rapid development of chilling injury in *E. reptans*. A rapid change in the permeability of the plasmalemma and tonoplast would account for the speed of injury to *E. reptans* and the rapid rise in respiration rate due to the loss of cell organisation which would allow enzymes increased access to substrates. It is speculated that the oxidation of phenols released from the vacuole as a result of a change in tonoplast permeability may account for the rapid rise in wound-induced respiration in this species.

Finally, it is perhaps rather naive to believe that all chilling injury can be related to a phase change event in the membrane lipids. An alternative explanation for chilling injury may reside in changes in cytoplasmic structure during chilling and re-warming. Patterson *et al.* (1979) have reported that within seconds of cooling living trichomes of tomato and other chill-sensitive plants that the thin strands of cytoplasm which normally span the vacuole disappear and are replaced by spherical vesicles. In contrast plants which can tolerate chilling, such as the onion, maintain normal cell structure down to $0^{\circ}C$. These changes in chill-sensitive plants at $5^{\circ}C$ may be due to the dissociation of microtubular elements making up the cytoskeleton of eukaryotic cells. On re-warming, polymerization of the tubulin sub-units is favoured and the cytoskeleton reformed so that the proto-plasmic strands are stabilized, provided chilling has not been of sufficient length to cause cell damage. Changes in

the structure and dynamic properties of cytoplasm during
chilling are now being investigated using spin labels (Keith
et al., 1979). Re-warming plants of *E. reptans* produces
unusual changes in cytoplasmic structure as seen by electron
miscroscopy. Upon re-warming *E. reptans* to 25°C after
chilling for 5 h at 5°C an extensive protein-like deposit
forms in the cytoplasm which has a very regular, almost
crystalline appearance in some sections (Murphy and Wilson,
1981). The precise nature of this deposit is at present
under investigation. Changes in chill-sensitive plants
during re-warming have received very little attention,
which is surprising as chilling injury to many species
develops more rapidly on return of the plant to the warmth
than if the plant is kept at the chilling temperature.

ACKNOWLEDGEMENT

Figures 2, 4 and 5 are reproduced with permission from
McMurdo, A.C. and Wilson, J.M. (1980). *Cryo-Letters* <u>1</u>,
231-238.

THE LOW TEMPERATURE LIMIT FOR GROWTH AND GERMINATION

E.W. Simon

The Queen's University of Belfast, Department of Botany, Northern Ireland.

INTRODUCTION

In the course of evolution organisms have become adapted to a wide range of temperatures. The aim of this paper is to present a comparative view of the range of temperatures at which bacteria, fungi and plants can grow; particular attention is paid to the minimum temperature for growth where this lies above $0^{\circ}C$. This survey of limiting temperatures leads to the view that similar biochemical mechanisms might be sensitive to temperature in each group. A number of mechanisms that have been proposed in recent years are next summarised. Finally the role of these mechanisms is considered in relation to questions raised by the effects of temperature on seed germination.

THE MINIMAL TEMPERATURE FOR GROWTH

As growth is the outcome of a multitude of biochemical reactions each of which should in theory continue, albeit at diminished rate, at reduced temperatures, one might expect growth itself to do the same. Many plants and mircoorganisms do show such a gradual decline in growth rate as temperatures approach freezing point. For other organisms growth rate declines gradually as temperature is lowered until a minimum is reached. Generally this minimum is so sharply defined that little difficulty is encountered in experiments of reasonable duration in distinguishing between temperatures that allow very slow growth and those at which there is no growth at all. The minimum temperature for the growth of microorganisms may be influenced by extreme

conditions of nutrient status, water activity or pH
(Michener and Elliott, 1964). Assessment of the maximum
temperature for growth also involves a time factor as
organisms may survive short exposures to temperatures that
would prove lethal after a longer period.

A selection of the data available for bacteria, fungi
and flowering plants is presented in Table 1. Because of
differences in experimental protocol it is not possible to
make detailed comparisons, but some general points can be
established.

Perhaps the most striking feature of the table is the
remarkably high temperature at which some microorganisms can
grow, fungi being found at 60°C, while bacteria extend the
range up to 80°C. The ability of these organisms to grow at
such high temperatures has stimulated research into their
biochemistry. Many of the participants in a recent
symposium (Friedman, 1978) address themselves to the protein
and lipid biochemistry of thermophilic microorganisms: how
is it that their proteins do not become denatured, and what
lipids are present in their membranes? Preoccupation with
such problems has drawn attention away from another aspect
of the life of thermophiles, namely their behaviour in the
cold. *Sulfolobus* is unable to grow below 55°C; for *Thermus
aquaticus* and *Thermoplasma* the minimum is around 40°C. The
same tendency can be seen in studies of mesophiles like
Escherichia coli, the relatively low maximum growth temperature tending to overshadow the problems of their lower
limit; why in fact are they unable to continue growth right
down to the freezing point?

Microorganisms are commonly divided into three broad
categories according to the temperature range in which they
grow, thermophiles growing at high temperatures, mesophiles
at more moderate temperatures and psychrophiles in the cold.
(Herbert, this volume). If these categories are quite
distinct it might be expected that the low temperature limit
for growth in each was determined by a different mechanism.
If, on the other hand, the division into these three
categories is only a convenience, the organisms themselves
forming a continuous unbroken series from those which will
grow at 0°C to those with a limit of 40°C or more, then a
single mechanism attuned to the appropriate environmental
temperature for each organism would be sufficient. The
figures for fungi in Table 1 taken in conjunction with others
to be found in the original references from which the Table
was compiled, leave no doubt that if there is a gap between
the categories it must be a very small one indeed.

This conclusion is reinforced by the report (Cochrane,

1958) that the optimum temperature for the growth of plant pathogenic fungi ranges continuously from 12°C to 37°C. As the range between minimum and maximum temperatures for growth is usually the order of 30 to 40°C (Table 1) this suggests that the minimum temperature for these pathogenic fungi must vary from 0 to about 20°C.

There are some gaps in the data for bacteria in Table 1 but arguably not so large as to rule out the possibility that comparable mechanisms operate to set the minimum in both thermophiles and mesophiles.

A temperature which is optimal or even maximal for one organism may yet prove too low for the growth of another. Thus *Rhizopus nigricans* will not grow above 30°C and yet this is the minimum temperature for *Humicola lanuginosa*; among bacteria the same applies for instance, to *Pseudomonas* and *Thermoplasma*. Clearly what constitutes a low temperature below which the organism will not grow varies from one organism to another. Low temperature in the sense used in the title of this volume should not be construed as meaning temperatures in the range immediately above freezing point. The human perception of cold is irrelevant; for some of the mircoorganisms listed in Table 1 low temperature may span the whole range from 0°C up to 40 or 50°C.

Little precise information on minimum and maximum temperatures for the growth of flowering plants is available in the literature, perhaps because of the complication that plants generally grow better with alternating day and night temperatures than under constant conditions. Tomatoes for instance make no growth at a constant temperature of 5°C but will continue to grow with night temperature of 5°C if the day temperature rises to at least 17°C (Went, 1944). In addition the minimum temperature for growth may depend on the degree to which the plants have been hardened by previous exposure to cool conditions. The figures in Table 1 can nevertheless be taken as an indication of range of minimum temperatures encountered among plants. A number of species including the temperate cereals and grasses will continue to make slow growth at temperatures near freezing, but others such as maple cease growth in early autumn, only starting again when dormancy is broken and temperatures have risen in spring. By contrast, many plants of tropical and subtropical origin such as maize and squash fail to grow at temperatures below 10 or 14°C respectively, and indeed suffer injury if exposed to such temperatures. This chilling injury limits the geographical range over which these plants can be grown and is an obstacle to refrigeration of their produce.

Table 1

The temperature range for growth

	Minimum	Maximum	Reference
	($^{\circ}$C)		
Bacteria			
Sulfolobus	55	80	Brock, 1978
Thermus aquaticus YT-1	40	79	"
Thermoplasma	37	65	"
Lactobacillus thermophilus	30	65	Gaughran, 1974
Streptococcus thermophilus	20	53	Deibel and Seeley, 1974
Clostridium perfringens	15	–	Michener and Elliott, 1964
Escherichia coli	8	46	Ingraham and Stokes, 1959
Pseudomonas	0	34	"
Fungi			
Humicola lanuginosa	30	60	Cooney and Emerson, 1964
Mucor miehei	25	57	"

Table 1 (Continued)

	Minimum	Maximum	References
		(°C)	
Mucor pusillus	20	55	Cooney and Emerson, 1964
Aspergillus fumigatus	12	52	"
Rhizopus nigricans	8	30	"
Fusarium coeruleum	5	35	Wolf and Wolf, 1947
Ceratostomella coerulea	3	33	"
Polyporus versicolor	0	40	"
Angiosperms			
Cucurbita pepo	14	46	Stiles and Cocking, 1969
Zea mays	9.5	46	"
Acer platanoides	7	26	"
Triticum vulgare	0	42	"

The Effect of Sub-Minimal Temperatures

The symptoms of chilling injury in plants depend on the
tissue in question and the conditions to which it is subject
(Lyons *et al.*, 1979; Wilson and McMurdo, this volume).
Rather little information is available on the critical
temperature below which tissues become injured. Wade (1979)
gives the following figures for fruits: $12-13^\circ$C for banana,
$10-13^\circ$C for mango, $7-10^\circ$C for lime and pineapple, 7°C for
cucumber, aubergine and peach, and about 5°C for apple. The
development of symptoms is complicated by interaction with
a time factor, for although the lower the temperature below
the critical value, the more severe the symptoms will
ultimately become, the longer they take to develop to their
full extent.

Many microorganisms including mesophiles can be stored
under refrigeration. This suggests that they can withstand
temperatures below the minimum for growth, but in fact
there is some evidence of damage under these conditions.
When bacterial cells in the exponential phase are transferred
suddenly to a low temperature they suffer a cold shock
losing viability within a matter of minutes (Herbert, this
volume). With stationary phase cells (or growing cells that
are cooled slowly), viability is lost in an exponential
manner over a period of days. Thus the viable count of cells
of *Clostridium perfringens* fell by several order of magnitude
in 10 days at temperatures just below the minimum for growth
(20°C) and more slowly at 10 or 1°C (Ingram and Mackey, 1976).

This brief exercise in the comparative physiology of
microorganisms and plants shows that the minimum temperature
for growth may be anywhere between 0 and 40 or 50°C. The
absence of any clear-cut distinction between thermophiles
and other bacteria supports the view that the same mechanism
might govern the minimum temperature for growth of one
organism at 30°C and another at 5°C; what has been learned
about the minimum temperature for the growth of thermophiles
may therefore be relevant to the behaviour of plants and
their seeds in the cold.

Seen in this wider context, the flowering plants differ
from microorganisms in two respects: first, none of them
can survive for long at temperatures much above 50°C; and
second, they develop particularly severe symptoms at sub-
minimal temperatures, possibly a reflection of their more
complex structure.

Hypotheses that have been developed about the minimum
temperature for growth of organisms may be relevant to the
problems raised by the germination of seeds at low temp-

erature. The minimum temperature for the growth of microorganisms is generally so sharply defined as to suggest that
it is due to a co-operative molecular phenomenon such as a
lipid phase change or a change in protein conformation.

Membrane Lipid Phase Change

Phospholipids undergo a thermotropic gel to liquid-
crystalline phase transition as a result of the co-operative
melting of the hydrocarbon chains (Pringle and Chapman, this
volume). As membranes contain a variety of lipids the phase
transition is spread over a broad temperature range in which
gel and liquid-crystalline phases coexist. The membrane
composition of *Acholeplasma* can be modified by exogenous
fatty acids, and it is possible to raise the minimum temperature for growth from $8^{\circ}C$ up to $28^{\circ}C$ (McElhaney, 1976).
In each instance growth stopped at the temperature at which
about 90% of the membrane lipid was in the gel state,
possibly because the membrane is then less permeable and new
lipid molecules are less readily inserted into the membrane
(Amulunxen and Murdock, 1978). The minimum growth temperature could not be reduced below $8^{\circ}C$ even in the presence
of linoleic ($18:2\omega6$) and oleic ($18:1\omega9$) acids, which have
melting points of -5 and $+10.5^{\circ}C$ respectively; evidently
some other factor comes into play below $8^{\circ}C$.

   Although the main evidence for this type of cold sensitive
mechanism comes from a prokaryote it is likely that something
comparable occurs in eukaryote cell-types despite the frequent
presence in the membranes of steroids which broaden or abolish
the highly co-operative nature of the gel to liquid-
crystalline transition (Herbert; Pringle and Chapman, this
volume).

   In addition to its effects on permeability and membrane
synthesis, such a phase change may have further consequences,
the activity of membrane-bound enzymes becoming affected by
the physical state of neighbouring phospholipids (Lyons *et al.*,
1979). As the membrane is cooled a temperature is reached at
which the particular phospholipids in the vicinity of a
membrane-bound enzyme enter the gel-state; this change of
state may affect the conformation of the enzyme, raising its
activation energy and giving rise to the more or less abrupt
break in Arrhenius plots for such enzymes (Wolfe, 1978). One
possible consequence of such a depression in the activity of
membrane bound enzymes is an imbalance in metabolism. The
precise relationship between the temperature of the break
(or breaks) in Arrhenius plots and chilling injury is unclear
(Wilson and McMurdo, this volume).

Protein Conformational Changes

Studies conducted with cold-sensitive microorganisms have
demonstrated several mechanisms by which alterations to
proteins restrict growth at low temperatures. Ingraham and
his colleagues have worked with cold-sensitive mutants of
E. coli with a similar growth rate to the wild-type at 37°C
but which ceased to grow at 20°C although the wild-type
continued to grow down to 8°C (Ingraham, 1973). In this
mutant the first enzyme of the histidine pathway was blocked
at 20°C, although in cell-free preparations the enzyme
remained fully active at this temperature. This enzyme is
subject to feedback inhibition by histidine in both wild-
type and mutant, but in the latter sensitivity to feedback
inhibition at 20°C became so great that synthesis of
histidine became blocked at a concentration too low to
support protein synthesis. If the medium was enriched with
histidine the mutant grew as well as the parent at all
temperatures. A number of other allosteric enzymes are now
known to become more sensitive to their effectors at low
temperature (Inniss and Ingraham, 1978). Such enhanced
inhibition of enzyme activity could possibly account for the
inability of organisms to grow in the cold, but so far there
is little direct evidence.

A second class of cold-sensitive mutants includes those
in which there is a more or less complete failure in the
assembly of proteins to form cell organelles. The assembly
of ribosomal proteins to form functional ribosomes has been
examined in detail. In a number of mutants of *E. coli* the
assembly of the 50S and 30S ribosomal subunits fails at 20°C,
smaller particles accumulating in their place (Nomura, 1970);
such mutants are therefore incapable of protein synthesis
and so cannot grow below this temperature. Ribosome
production is also defective in some *Aspergillus* mutants
at low temperatures (Waldron and Roberts, 1974). Another
low temperature lesion of protein synthesis occurs in *E. coli*
at 8°C. When 70S ribosomes have completed translation they
normally dissociate into 30S and 50S units which combine
with new mRNA in the formation of fresh polysomes. If
E. coli cells are incubated at 8°C or less the 30S and 50S
subunits accumulate suggesting a failure to initiate protein
synthesis (Friedman *et al.*, 1969).

A few enzymes are now known to be cold-sensitive, many
of them dissociating into inactive subunits at low temperature (Beyer, 1972; Bock and Friedman, 1978). Arrhenius
plots of some of these enzymes show quite sharp discontinuities, the enzymes being relatively inefficient in the

cold, possibly leading to an imbalance between metabolic pathways. Cessation of growth can only be explained meaningfully by such cold-sensitive enzymes when the increase in activation energy occurs at a temperature close to, or above the limiting temperature. Ljungdahl *et al.* (1978) present a convincing case for *Clostridium thermoaceticum* which has only low activity below 40°C and contains enzymes with discontinuities at 55, 43, 40 and 35°C respectively. Graham *et al.*, (1979) have compared the soluble enzyme phosphoenolpyruvate carboxylase in two tropical and chill-sensitive plants (*Lycopersicon* and *Passiflora*) with the enzyme in some temperate and alpine species. The Arrhenius plot for the enzyme in *Lycoperison* was non-linear with much reduced activity below about 10°C. Cooling from 20°C down to 1.5°C reduced the activity of the enzyme from the tropical species much more than that from the alpine or temperate species. It is thus possible that the metabolism of the tropical, chilling-sensitive species is disrupted by a direct effect of cold on the enzyme.

Growth might therefore become blocked in the cold because protein subunits were unable to assemble normally into polymeric enzymes, proteins or other subcellular entities; or because allosteric proteins become especially sensitive to their effectors. These processes all depend on the precise conformation of the protein molecules in question which must therefore differ in some respect from that in related, cold-tolerant organisms. The conformational changes in temperature-sensitive proteins is possibly related to a reduction in hydrophobic bonding at low temperatures (Brandts, 1967).

Evidence can thus be found in the literature to support either of two hypotheses:

1) That growth is restricted in the cold because of an effect on membrane lipids, which may in turn influence the growth and permeability of the membrane, or may increase the activation energy of membrane-bound enzymes.
2) That growth is restricted in the cold because the conformation of enzymes is changed in a critical manner below a certain temperature; the enzyme in question may be membrane bound or soluble.

THE MINIMAL TEMPERATURE FOR GERMINATION OF NON-DORMANT SEEDS

Like growth, the germination of seeds is only possible over a restricted range of temperatures, becoming slower as the extreme limits of the range for a particular species are approached. It is now accepted practice, following Thompson (1970a), to characterise the temperature range for

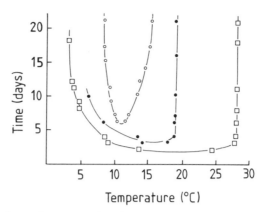

Figure 1. *The time required for 50% germination of seeds of Silene concoidea at various temperatures. Seeds tested at the time of harvest (o), three months (●), and one year (□), after harvest. Redrawn from Thompson (1970a).*

seed germination by the minimum and maximum temperatures at which 50% of the seeds germinate (Fig. 1, Table 2). Within any one species there may be a small range of variation in the limiting temperatures according to the source of the seeds or the particular cultivar in question (Thompson, 1970a; 1973). Some species like lettuce and mustard (Simon *et al.*, 1976) will germinate down to near freezing point while others cease to germinate at a temperature well above 0°C. The cut-off is very abrupt.

The minimum temperature for some species of water-logged or disturbed habitats is remarkably high, 30°C for *Polygonum persicaria*, for instance (Table 2). However it should be pointed out that the data in Table 2 refer to germination at constant temperature, very different from the diurnal variation found under natural conditions. Although it is known that seeds may germinate at lower temperatures under fluctuating rather than constant conditions (Thompson, 1970b; Thompson *et al.*, 1977) it is still appropriate to enquire what mechanism prevents the germination of *P. persicaria* below a constant temperature of 30°C. It seems likely that the seeds of such a species require higher (constant) temperatures for germination than does the growth of the mature plants. There is some evidence for this in cucumber, the minimum temperature for germination being a little higher than the minimum for the growth of seedling roots (Simon *et al.*, 1976).

Table 2

The temperature range for germination

	Minimum	Maximum	References
		(°C)	
Lactuca sativa	2	27–32	Thompson, 1973
Triticum sativum	3–5	30–43	Noggle and Fites, 1974
Primula palinuri	5	19	Thompson, 1970a
Zea mays	8–10	40–44	Noggle and Fites, 1974
Primula farinosa	10	28	Thompson, 1970a
Cucumis sativus	12–13	42	McMenamin, 1978
			Simon et al., 1976
Ajuga reptans	16	36	Thompson, 1970a
Carex nigra	23	37	Mason, 1976
Polygonum persicaria	30	39	Mason, 1976

The Effect of Sub-Minimal Temperatures

Chilling-sensitive plants are by definition damaged by
exposure to sub-minimal temperatures, but the seeds of such
species may be much less sensitive to cold. Thus cucumber
seeds will not germinate below 12-13°C, but the seeds remain
viable even after 14 days at 5°C (Simon *et al.*, 1976); likewise sorghum seeds remained viable after 20 days inbibition
at 8°C (McWilliam *et al.*, 1979).

It should be mentioned in parenthesis that the seeds of
one group of tropical plants behave differently, suffering
injury in the cold. These seeds are said to be recalcitrant,
not conforming to the usual relationships between temperature,
moisture content and longevity (King and Roberts, 1979).
Whereas most seeds remain viable for extended periods at low
moisture contents, the seeds of cacao, mango and oil palm for
example have a high water content (>30%) at maturity and are
killed by attempts to reduce their water content. In addition
they are damaged by exposure to chilling temperatures; the
critical limiting temperature still remains to be established
for each species.

Another type of chilling damage is peculiar to seeds,
with no counterpart in mature plants. The seeds of cotton,
soybean and some bean species are injured if they are
imbibed in cold water. Cotton seeds for instance may be
killed by exposure to water at 5 or 10°C for a period,
before transfer to the warm; seeds that survive the cold
have an abnormal root system, the radicle tip aborting with
the development of laterals (Christiansen 1963, 1967, 1968).
This type of injury is characteristic of the early hours of
imbibition by dry seeds and so could not account for the
symptoms observed in recalcitrant seeds which have a high
water content when mature. It is indeed an imbibitional
chilling injury and has been attributed to a lesion in the
orderly reestablishment of bilayer membranes as the seeds
imbibe (Bramlage *et al.*, 1978); it is not yet clear why the
seeds of these few species should be especially sensitive to
cold during imbibition.

Mechanisms Which May Limit Germination at Reduced Temperatures

Evidence suggests that the low temperature limit for
germination differs in several respects from the lower limit
for growth. The minimum for the germination of some seeds
is as high as 30°C, substantially more than the minimum for
growth (Table 2). Secondly, plants are more liable to be
injured at sub-minimal temperatures than seeds. And finally,

it seems that among the species that require high temperatures
for seed germination are some in which the adult phase is
also chilling sensitive (maize, cucumber) and others in which
it is not (*Ajuga, Polygonum*). These differences suggest that
the underlying mechanisms may not be the same in each case.

Seeds like mustard which can germinate, albeit slowly,
down to temperatures of 5°C or less yield linear Arrhenius
plots for germination rate (measured as the reciprocal of the
time required for 50% germination). On the other hand there
is a sharp break in the plot for germination of *Silene nutans*
at about 6°C, for cucumber and mung bean at around 14°C and
for *Ajuga* at 16°C (Simon, 1979; Simon *et al*., 1976). Above
the temperature of the discontinuity germination proceeds
with a temperature coefficient (Q_{10}) of around 2-3, but at
lower temperatures the coefficient rises to 10 to 1,000. The
discontinuities are clearly related to the limiting temperatures for germination, for in each case at a temperature
$1-2^{\circ}$C below that of the discontinuity germination fails
completely, (that is to say that less than 50% of the seeds
germinate even after a prolonged period). One possibility
to account for these results would be a lipid phase change
affecting membrane-bound enzymes such as those of the
mitochrondia, leading to an imbalance of metabolism and the
accumulation of toxic substances. This hypothesis requires
a discontinuity in Arrhenius plots of membrane-bound enzyme
activity, but no such discontinuity was observed in the
respiration of imbibed cucumber of mung bean seeds, or in the
respiration or growth of 1 cm roots from young seedlings.
Moreover when cucumber seeds were held at sub-minimal temperatures for several days and then transferred to 15 or 20°C
they began to germinate promptly, without any suggestion
that a toxic product, accumulated in the cold, had first to
be metabolised (Simon *et al*., 1976).

An alternative view is that there is a change in protein
conformation at the temperature of the discontinuity in the
Arrhenius plot. Germination is still possible, though slow,
at temperatures $1-2^{\circ}$C lower, but it fails completely under
yet colder conditions as protein conformation departs
further from the native condition (Simon *et al*., 1976). As
it is generally supposed that the high temperature limit for
biological activity is set by the denaturation of some
protein it would then follow that both high and low temperature limits for seed germination were determined by
changes in protein conformation, as has indeed been suggested
for the growth of *Bacillus stearothermophilus* and *Clostridium
thermoaceticum* (Babel *et al*., 1972; Ljungdahl *et al*., 1978).

THE MINIMAL TEMPERATURE FOR GERMINATION OF DORMANT SEEDS

The minimum and maximum temperatures for seed germination
can be regarded as characteristic features of a particular
species. Nevertheless quite marked shifts in the minimum
and maximum limits are commonly found as seeds emerge from
dormancy.

Fully dormant seeds are unable to germinate at any
temperature. One might expect to find that as dormancy was
lost an increasing proportion of the seeds would be able to
germinate over the whole range of temperature characteristic
of the species, but the experimental data indicate otherwise. Thus seeds of the cereals and *Festuca* will only
germinate at temperatures around 10°C when they are freshly
harvested and still somewhat dormant, but they acquire the
ability to germinate also at higher temperatures if they
are first stored dry for a month or more when they lose their
dormancy (Kearns and Toole, 1939; Stokes, 1965). On the
other hand *Amaranthus arvense* can at first only germinate at
temperatures around 40°C, but the minimum temperature for
germination continually falls as dormancy gradually dis-
appears in dry storage (Vegis, 1964). Much the same is true
of seeds released from dormancy by stratification (storage
of seeds under cold, moist conditions). Seeds of apple first
become able to germinate at 5°C, the temperature of stratification, but after 6 or 7 weeks they also germinate rapidly
at temperatures up to 30°C (Thevenot and Come, 1973). Birch
shows the opposite change; at first the seeds only germinate
around 30°C, but they will germinate at 15°C after 1 month
stratification, and at 0°C after 5 to 6 months, a remarkably
wide drop in the low temperature limit. Finally it must be
said that the temperature range for some species widens at
both ends of the range (Fig. 1).

Vegis (1964) and Stokes (1965) summarise many experiments
of this sort and make the generalisation that seeds first
become able to germinate over a relatively narrow and
restricted temperature range which then widens as the seeds
are stored for longer periods and gradually emerge from
dormancy. The minimum temperature may fall in this way by
as much as 20 or 30°C. The fall seems to occur smoothly
and gradually rather than stepwise. The opposite change (a
narrowing of the temperature range for germination) occurs
as the seeds mature and enter dormancy.

The basic observations were established 15 years ago, but
little attention has been paid to them in the intervening
period. They present an intriguing problem as regards
control mechanisms, for the upper and lower temperature

limits for some seeds can evidently change as the seeds emerge from dormancy under quite mild storage conditions – either dry at room temperature, or imbibed in the cold. If we could understand the mechanisms governing these shifting temperature limits we might come closer to understanding the nature of dormancy itself.

If the low temperature limit were determined by some type of protein denaturation, one could suggest a mechanism for such an expansion of the temperature range. Suppose that one or two essential protein(s) become denatured in dormant seeds, perhaps as a result of the stresses set up as the seed dried out on maturation. Dry after-ripening and stratification would then be periods of restoration of protein conformation, enabling the protein(s) to withstand a wider temperature range without losing activity, and so permitting germination over a wider temperature span. The protein(s) in question would clearly play some role of central importance in the life of the seed, possibly as enzyme(s), controlling the ability of the seed to germinate.

Increases in peroxidase and catalase activities were already reported 50 years ago in stratified seeds (Stokes, 1965) and more recent accounts indicate enhancement of other enzymes including chromatin-bound RNA polymerase (Tao and Khan, 1976) and a gibberellin-producing system (Arias *et al.*, 1976). However it should be emphasised that protein synthesis can occur in *in vitro* systems from stratified seeds (Barnett *et al.*, 1974) and, more important, work with inhibitors indicates that some of the enzymes that appear during stratification are in fact newly synthesised (Tao and Khan, 1974, 1976). These enzymes only appear during stratification at low temperature, and not in control seeds held at 20°C (Davies and Pinfield, 1980; Tao and Khan, 1974; Yankelevich and Nikolaeva, 1975). This seems to indicate that some other, overriding change occurs in the cold allowing both germination and enzyme synthesis to proceed. If this master control is itself a change in enzyme conformation, one can only conclude that it still remains to be detected.

There are also reports of a rise in enzyme activity in seeds losing dormancy during dry storage. Thus a transaminase becomes more active in wild oat seed as well as systems producing gibberellic acid and sucrose (Chen and Varner, 1969; Kovacs and Simpson, 1976; Simpson, 1965). It is possible that here again the enhancement of enzyme activity should be attributed to enzyme synthesis which can apparently continue at very slow rates even in relatively dry seeds (Edwards, 1976). There is in short, no evidence

at present that pre-existing proteins become activated by
some change in conformation during dry storage or stratification.

   An alternative hypothesis sees lipid molecules as the
key to control of the temperature range for germination.
One possibly mechanism is suggested by the work of Raison
et al. (1978) on the tubers of Jerusalem artichoke and
potato. At the initiation of dormancy and again at its
termination, mitochondrial membrane lipids were fluid
between 22 and 3°C, but at mid-dormancy the range fell to
9 to -5°C. A change in membrane lipid structure evidently
occurs during the winter months enabling the tissue to
withstand low temperatures. If such a change in membrane
lipids occurred in seeds it might be thought to govern the
temperature range for germination; however it is open to
debate (or experiment!) whether such a change in membrane
lipids would occur for instance as cereal grains emerge from
dormancy in dry storage. Nor is it easy to envisage a change
in lipids that could allow germination to occur at both
higher and lower temperatures (Fig. 1).

   The nature of the mechanism controlling the temperature
span over which seeds can germinate still remains to be
established; the widening of this span as seeds emerge from
dormancy is especially puzzling. The speculations above
are offered in the hope that they will stimulate experiments
designed to elucidate the mechanisms at work.

THE EFFECTS OF COLD SHOCK ON SPERM CELL MEMBRANES

P.F. Watson

*Department of Physiology, The Royal Veterinary College,
London, England.*

INTRODUCTION

When the semen of certain mammalian species is cooled rapidly
to a few degrees above $0^{\circ}C$ a high proportion of the spermatozoa become irreversibly immotile and infertile.   Milovanov
(1934) coined the term 'temperature shock' to describe this
phenomenon, and although a considerable research effort has
been invested to elucidate it in the intervening 45 years, a
complete explanation is still wanting.
     While much is now known of the cellular changes caused by
temperature shock (or cold shock as it is more commonly
called today), it is only recently that clues to the nature
of the lesion have begun to emerge.   The impetus for the early
research was a desire to obtain fertile stored semen for use
in stock improvement by artificial insemination.   The general
tenor of the research was thus concerned with devising ways
of avoiding cold shock.   Hence, it was soon realised that
ram or bull spermatozoa could be revived upon rewarming if
the original rate of cooling was no greater than about
$10^{\circ}C$ h^{-1} (Birillo and Pulhaljskii, 1936; Gladcinova, 1937).
During the next 25 years many aspects of cellular damage
were described.   Several groups of workers investigated the
susceptibility of spermatozoa from a variety of different
sources and after various treatments.   In the last 15 years
research has been directed towards finding an explanation
of cold shock at the membrane level.

THE PHENOMENON OF COLD SHOCK

If a tube containing a small sample of ram or bull semen at

30°C is transferred to an ice-water bath for a few minutes, a large proportion of the spermatozoa are found to be immotile on rewarming. The loss of motility is accompanied by an increase in the proportion of cells stainable by a vital dye (Table 1). This technique has frequently been used following cold shock to estimate the percentage of cells alive (unstained) at the time of staining (Hancock, 1951; Kampschmidt et al., 1953; Lasley et al., 1942b; Lasley and Mayer, 1944; Mayer and Lasley, 1945; Quinn et al., 1968a; Wales and White, 1959). It is doubtful whether the staining reaction of spermatozoa is related simply to cell death but rather reflects significant changes in surface cell membranes (Bangham and Hancock, 1955; Dott and Foster, 1972) which may presage, or are consequent upon, losses of function necessary to the sustaining of the cell's integrity. Rapid warming from 0 to 37°C causes no comparable cell damage (Chang and Walton, 1940).

Species Differences

The susceptibility of spermatozoa to cold shock varies with species. Boar spermatozoa are extremely sensitive immediately after ejaculation, almost no spermatozoa survive a cold shock stress lasting a few minutes, although they gain resistance on incubation (Benson et al., 1967; Lasley and Bogart, 1944a; Pursel et al., 1972a; Pursel et al., 1972b, 1973). Bull, ram and stallion spermatozoa are highly susceptible (Blackshaw, 1954; Blackshaw and Salisbury, 1957; Bogart and Mayer, 1950; Lasley et al., 1942a; Mayer and Lasley, 1945) but dog (Wales and White, 1959) and cat (T.E. Glover, pers. comm.) spermatozoa are less sensitive. Rabbit, human and fowl spermatozoa are the most resistant (Wales and White, 1959); there is thus a continuum of sensitivity (Table 2). The effects of cold shock have been most intensively studied in ram, bull and boar semen.

Morphological Studies

Morphological studies have revealed that the acrosome in particular is affected by cold shock (Boender, 1968; Hancock 1952; Iype et al., 1963; Onuma, 1963; Pursel et al., 1972a; Quinn et al., 1969). Light micrographs reveal that after cold shock the acrosome shows swelling or bubbling in a high percentage of cells (Plate 1). Acrosomal membrane defects are apparent and in some spermatozoa the anterior portion of the acrosome becomes detached leaving the equatorial segment intact. Conventional thin-section electron microscopy (Jones,

Table 1

Motility and staining reaction of cold shocked and slowly cooled ram spermatozoa. Motility scored on a scale of 0 to 4

Incubation time at 37°C after treatment (h)	Control			Cold-shocked[a]			Slowly cooled[b]		
	% sperm motile	motility score	% sperm unstained	% sperm motile	motility score	% sperm unstained	% sperm motile	motility score	% sperm unstained
0	70	3.7	73	7	1.3	1	73	3.7	77
2	50	3.0	77	7	0.5	6	47	2.7	67
4	47	2.5	72	0	0	1	33	2.0	63
Mean	55	3.1	74	4	0.6	3	51	2.8	69

(a) *Spermatozoa diluted in an egg yolk/glucose/phosphate diluent and subjected to sudden cooling to 0°C for 5 min.*

(b) *Spermatozoa diluted in an egg yolk/phosphate diluent and cooled to 5°C over 2 h.*

1973) showed that the plasma membrane was often absent over
the acrosomal region (Plates 1 and 2; W.V. Holt pers. comm.).
The outer acrosomal membrane was convoluted except in the
equatorial segment which appeared little altered. Frequently
the acrosomal substance was less electron dense probably indicating a loss of contents. The plasma membrane overlying
the post acrosomal region was present but was often damaged
or broken and the material underlying this region was disorganised (Plate 2). The nucleus and nuclear membranes were
apparently unaltered. Mitochondrial changes were variable,
with some spermatozoa showing some reduction in the density
of the ground substance as previously described (Quinn *et al.*,
1969), while others appeared little affected (Plates 1 and 2).
The membrane over the midpiece was usually present but often
damaged (Plate 2). No changes were noted in the longitudinal
fibres. These observations are in accord with those of Quinn
et al., (1969) apart from the fact that the alterations observed were less severe both for control and treated spermatozoa suggesting an improved fixation technique. No reliable
conclusions can be drawn from other published micrographs
(Hammerstedt *et al.*, 1976), since the plane of section is
variable.

The light micrographs (Plate 1) suggest a much greater
degree of disruption of the acrosome than is apparent in the
electron micrographs (Plate 2). These observations are
difficult to reconcile unless it is assumed that as well as
staining membranes Giemsa also stains acrosomal contents.
Even so, the appearance of stained membranous material sloughing from some of the sperm heads is unlikely to be solely
plasma membrane. Thus, it may be that the method of pre-
paring Giemsa slides (Watson, 1975) may induce further
changes in spermatozoa rendered unstable by cold-shock.

Plate 1 (opposite). *Light micrographs of Giemsa-stained ram
spermatozoa. (A) Control, acrosome (a). (B) Cold-shocked.
Acrosome swollen with irregular margin, and evidence of
membrane tearing (b) and sloughing (c). The equatorial
segment (d) is intact. The acrosome stained less densely
after cold shock. (C) Low power electron micrograph of
cold-shocked ram spermatozoa. In the head sections the
plasma membrane is absent and the outer acrosomal membrane
(e) is irregular. Some midpiece sections are apparently
undamaged (f) while others show loss of density and disruption
of internal structure (g). Free membrane fragments are
numerous (h). Electron micrograph by courtesy of Dr W.V. Holt.*

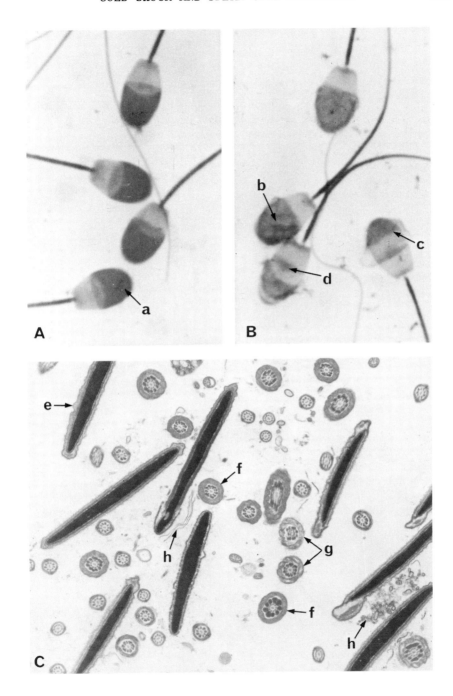

Table 2

*Mean motility and staining reaction of spermatozoa subjected
to cold shock at 0°C. Motility scored on a scale of 0 to 4*

	Temp. °C	Ram	Bull	Dog	Rabbit	Man	Fowl
Motility score	30	3.8	3.3	3.8	3.5	3.8	4.0
	0	0.3	2.0	3.0	3.0	3.5	3.5
% Motile	30	69	67	84	72	54	91
	0	3	15	50	50	45	80
% Unstained	30	78	78	91	85	74	93
	0	1	11	43	45	52	92
No of replicates		4	7	4	4	4	5

*Modified from Wales and White (1959). Reproduced with
permission of J. Endocr and the authors.*

Plate 2 (opposite). *Electron micrographs of ram spermatozoa.
(A) and (B). Saggital sections of heads of control and cold-
shocked spermatozoa, respectively. The plasma membrane (a)
is absent over the acrosome of the cold-shocked spermatozoon.
After cold-shock the smooth contours of the outer acrosomal
membrane are lost in the anterior region (b) with undul-
ations and a tendency to form vesicles. The equatorial
segment (c) is not altered. The acrosomal substance is less
dense. The post-acrosomal plasma membrane is present but
broken (d), and the sub-membranal substance in this region is
disorganised.
(C) and (D). Transverse sections of midpieces of control and
cold-shocked spermatozoa, respectively. In these sections
cold-shock has caused no significant changes and the plasma
membranes are present.
(E). Longitudinal section of midpiece of cold-shocked
spermatozoon. The plasma membrane (e) has been considerably
disrupted, and within a single spermatozoon the mitochondrial
appearance ranges from seemingly unaltered (f) to severely
damaged (g). Electron micrographs by courtesy of Dr W.V. Holt.*

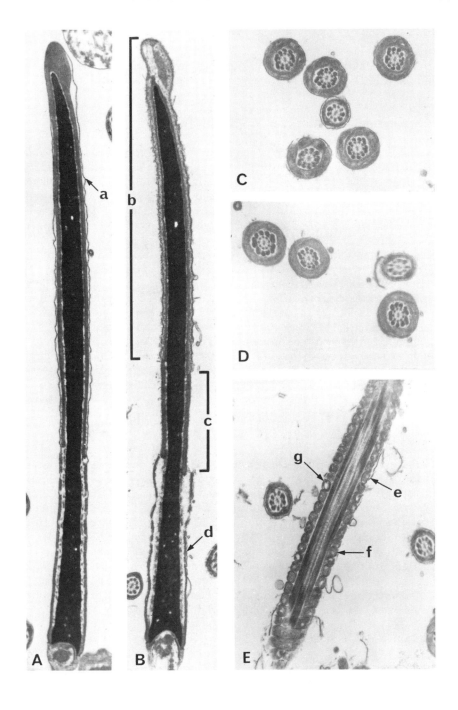

Biochemical Changes

Coinciding with the morphological changes the biochemistry
of susceptible spermatozoa is considerably altered by cold
shock.

1) Carbohydrate Metabolism

Anaerobic glycolysis and respiratory activity were found to
be diminished in ram and bull spermatozoa (Blackshaw, 1958;
Blackshaw and Salisbury, 1957; Chang and Walton, 1940; Mayer,
1955). Fructolysis in ram semen was reduced almost to zero
both in spermatozoa motile before cold shock and in spermatozoa rendered immotile by fluoride (Mann and Lutwak-Mann,
1955). Mayer (1955) claimed that respiratory activity and
aerobic glycolysis were more sensitive to cold shock than
anaerobic glycolysis. This finding is of interest since
intracellular proteins, particularly cytochrome c, are
released from cold-shocked spermatozoa, although this was
considered to be an after-effect rather than an immediate
response to cold shock (Mann and Lutwak-Mann, 1955).
Evidence of the loss of protein from midpiece regions has
also been detected by means of histochemical techniques and
this correlates with the mitochondrial changes observed by
electronmicroscopy (Quinn *et al.*, 1969). ATP levels were
found to diminish promptly on cold shock and no resynthesis
was possible (Mann and Lutwak-Mann, 1955).

2) Release of Intracellular Enzymes

Release of intracellular enzymes has been demonstrated as a
result of cold shock. These enzymes include lactate dehydrogenase, glutamic-oxaloacetic transaminase, alkaline and acid
phosphatase, glucose-6-phosphate dehydrogenase and glucose
phosphate isomerase (Harrison and White, 1972; Moore *et al.*,
1976; Murdoch and White, 1968; Pursel *et al.*, 1968, 1970). Some
of the enzyme activity released into seminal plasma
originates from cytoplasmic droplets and not from spermatozoa
(Harrison and White, 1972) which urges caution in interpreting
the above results; nevertheless, loss of enzymes is undoubt-
edly linked to the decline in metabolic activity.

3) Release of Lipids

Lipids are lost from ram, bull and boar spermatozoa as a
result of cold shock (Blackshaw and Salisbury, 1957;
Hartree and Mann, 1959; Pickett and Komarek, 1967). These

are predominantly phospholipids and the particulate nature
of the material released from ram spermatozoa suggests membrane fragmentation which correlates with the morphological
data (Darin-Bennett *et al.*, 1973).

Cold-shocked bull spermatozoa released most phospholipid
and boar spermatozoa least, but the methods of processing
semen prior to cold shock might have contributed to these
differences. Choline phosphoglycerides were preferentially
released from ram and boar spermatozoa while phosphatidyl
ethanolamine was also released from boar spermatozoa. The
more severe treatment of freezing, which causes a much more
extensive membrane disruption, was required to release this
latter phospholipid from ram and bull spermatozoa, suggesting
differences in location within the cell (Darin-Bennett *et al.*,
1973).

In contrast to the release of specific phospholipids from
ram and boar spermatozoa, the more general loss of phospholipid from bull spermatozoa (including cardiolipin from
the inner mitochondrial membrane) implies a more complete
breakdown of some cells while others were little affected.
This agrees with the motility observations of Choong and Wales
(1962).

In view of the evidence that the acrosome is particularly
damaged by cold shock Darin-Bennett *et al.* (1973) speculated
that the specific phospholipids released by ram and boar
spermatozoa originate from acrosomal membranes. However,
the plasma membrane cannot be excluded as a source of membrane
fragments. It is probable that differences in the release
of specific phospholipids reflect differences in membrane
composition between species.

4) Alterations in Cation Distribution

Cold shock induces profound cation disturbances in ram, bull
and boar spermatozoa. Na^+ and Ca^{2+} are gained by cells
whereas K^+ and Mg^{2+} are lost (Blackshaw and Salisbury, 1957;
Hood *et al.*, 1970; Pursel *et al.*, 1969; Quinn and White,
1966). Interestingly, Quinn and White (1966) found that although the movements of Na^+, K^+ and Mg^{2+} were by passive
diffusion, the accumulation of Ca^{2+} in undiluted or diluted
ram spermatozoa or diluted bull spermatozoa as a result of
cold shock was against a concentration gradient. This suggests
that following membrane permeability changes induced by cold
shock, calcium is taken up into spermatozoa in a bound form.
Calcium binding sites have been demonstrated on the outer acrosomal membrane of human spermatozoa (Roomans, 1975) and are
presumably involved in the calcium-dependent acrosome reaction

(Talbot *et al.*, 1976; Yanagimachi and Usui, 1974). Species less susceptible to cold shock showed little or no ion flux, (Quinn and White, 1966).

The implication of Ca^{2+} in the mechanism of cold shock is supported by the fact that the presence of EDTA lessened both the accumulation of Ca^{2+} and the loss of motility and increase in staining reaction of cold-shocked ram and bull spermatozoa; other ion movements were not affected (Beljkevic *et al.*, 1959; Quinn and White, 1968). Moreover, of the ions studied, Mg^{2+} alone was found to prevent the decline in motility and increase in staining reaction during incubation after cold shock (Quinn and White, 1968).

5) *Miscellaneous Biochemical Changes*

Other biochemical changes which have been noted following cold-shock in spermatozoa are an acquired ability to metabolise succinate, an increase in ammonia formation indicating breakdown and deamination of intracellular protein (Mann, 1964), a loss of membrane-bound protein (Mann and Lutwak-Mann, 1955) and a progressive breakdown of DNA (Quinn *et al.*, 1969). Release of polysaccharide has also bben detected, probably from the acrosome (Quinn *et al.*, 1969).

In all of this work there is a problem in distinguishing primary from secondary changes. Some biochemical events, such as breakdown of DNA, become manifest only on incubation (Quinn *et al.*, 1969) while others, *e.g.* incapacity to synthesis ATP, are evident immediately (Mann and Lutwak-Mann, 1955). The oxygen consumption of bull spermatozoa was found to decline if, following cold-shock, they were held at a low temperature for between 5-30 min (Mayer, 1955). It is not clear whether this is a progressive response to low temperature *per se* or, more likely, whether such changes would also occur at higher temperatures following a brief cold shock.

Cold-shocked bull spermatozoa incubated at $37^{\circ}C$ were unable to maintain their metabolic functions whereas at $21^{\circ}C$ no differences were noted between shocked and control cells (Blackshaw and Salisbury, 1957). This could imply that losses of vital cell constituents proceed at a faster rate at higher temperatures. Moreover, the motility of ram and bull spermatozoa fell during incubation after cold shock (Blackshaw and Salisbury, 1957; Quinn and White, 1968).

These observations suggest that cold shock induces a sudden irreversible alteration in the cell membranes, the effects of which take time to be fully expressed, although

they occur more rapidly at higher temperatures. The primary
event, then, is considered to be loss of membrane integrity
with a general increase in permeability. Direct evidence of
increased membrane permeability resulting from cold shock in
spermatozoa of susceptible species, but not in those of
resistant species, has been obtained by electron spin resonance studies (Hammerstedt *et al.*, 1976, 1978).

LIPID COMPONENTS OF SPERM MEMBRANES

In view of the importance of membranes to an understanding
of cold shock some consideration of studies of membrane lipid
composition is given below.

Phospholipids

Phospholipids are present at about 2 μmol per 10^9 ejaculated
spermatozoa (Darin-Bennett *et al.*, 1973a, 1974; Poulos and
White, 1973; Quinn and White, 1967) and constitute 60-70%
of total lipids (Komarek *et al.*, 1964; Miller *et al.*, 1965;
Pursel and Graham, 1967; Selivonchick *et al.*, 1980). In the
bull and the ram, some 45% are plasmalogens while in other
species the plasmalogen fraction is 20% or less (Darin-
Bennett *et al.*, 1973; Hartree and Mann, 1959; Poulos *et al.*,
1972; Selivonchick *et al.*, 1980). The phospholipid distribution of several species is shown in Table 3. Although there
are some minor variations in the literature, these figures
are generally in broad agreement with those of other studies
(bull - Komarek *et al.*, 1964; Neill and Masters, 1972;
Pursel and Graham, 1967; ram - Hartree and Mann, 1959;
Neill and Masters, 1973; Quinn and White, 1967; Scott *et al.*,
1967; boar - Johnson *et al.*, 1969, 1972).
    While ram and bull spermatozoa are distinguishable from
other species by their higher proportion of choline phosphoglycerides, boar spermatozoa (which are highly susceptible to
cold shock) are apparently similar in phospholipid distribution to the spermatozoa of the more resistant species.
However these studies relate only to the total phospholipid
distribution and do not distinguish species differences in
individual membrane composition. They do suggest, however,
that the membrane composition of boar spermatozoa is different from that of bull and ram spermatozoa, and this view
is strengthened by the several unique responses of boar
spermatozoa to cooling and cold shock which may be related
to membrane phenomena.

Table 3

Phospholipid composition of spermatozoa. Data are means of 3 to 6 analyses for each species, expressed as a percentage of the total phospholipid extracted.

	Ram	Bull	Boar	Rabbit	Dog	Human	Monkey	Fowl
Sphingomyelin	11.4	10.5	12.6	23.4	18.3	21.4	8.1	13.1
Phosphatidylserine	1.0	1.8	2.4	2.5	3.6	2.7	1.6	8.9
Phosphatidylinositol	nd	nd	nd	3.3	2.7	1.9	nd	1.6
Choline plasmalogen	40.8	44.8	9.8	14.0	3.6	2.7	6.9	7.8
Phosphatidylcholine*	17.3	15.5	37.4	28.7	27.5	28.8	33.0	39.6
Ethanolamine plasmalogen	5.9	5.5	10.8	8.3	15.3	9.4	16.1	5.2
Phosphatidyl-ethanolamine*	5.6	3.5	13.5	8.2	20.1	21.6	25.0	8.5
Cardiolipin	5.6	3.4	3.0	2.7	3.0	1.6	4.5	1.7
Other lipids	12.4	15.0	10.5	8.9	5.8	9.9	4.8	13.6

nd – Not detected or present in amounts below 1%. * – These fractions may contain corresponding alkyl phospholipids. Modified from White and Darin-Bennett (1976) and Darin-Bennett et al., (1977). Reproduced with permission of J. Reprod. Fert. and the authors.

Phospholipid-Bound Fatty Acids and Aldehydes

The phospholipid observations give no indication of the constituent fatty acids which influence membrane characteristics (Chapman and Wallach, 1968). Docosahexaenoic (22:6) acid was found to be the principal phospholipid-bound fatty acid in all mammalian species so far examined except the rabbit, the dog and the rhesus monkey in which docosapentaenoic (22:5) acid and/or palmitic (16:0) acid predominated (Table 4). These results agree with other published data (Johnson *et al.*, 1969, 1972; Neill and Masters, 1972, 1973; Pursel and Graham, 1967) although in some earlier studies the presence of C_{22} fatty acids was often not detected (Dott and Dingle, 1968; Gray, 1960; Scott *et al.*, 1967). The distribution of docosahexaenoic (22:6) and docosapentaenoic (22:5) acids differs with species (Table 4) and there is some disagreement whether 22:5 exists at all in bull spermatozoa (Ahluwalia and Holman, 1969; Poulos *et al.*, 1973a; Selivonchick *et al.*, 1980).

In bull, boar and ram spermatozoa (species susceptible to cold shock) over 60% phospholipid-bound fatty acids are C_{22} polyenoics (Table 4) whereas more resistant mammalian species have only 30-40% (Darin-Bennett *et al.*, 1974, 1977; Poulos *et al.*, 1973a). Also, the spermatozoa of susceptible species contain lower proportions of palmitic (16:0), stearic (18:0) and oleic (18:1) acids. Fowl spermatozoa differ in having docosatetraenoic (22:4) acid as the predominant polyunsaturated acid and higher proportions of oleic and stearic acids than palmitic acid (Darin-Bennett *et al.*, 1974). As a result, the ratios of polyunsaturated to saturated fatty acids in ejaculated ram, bull and boar spermatozoa are 2.5 or greater, whereas those of the resistant species (human, fowl, rabbit, dog, monkey) approximate to unity (Darin-Bennett *et al.*, 1974, 1977; Poulos *et al.*, 1973a).

Studies of the distribution of phospholipid-bound fatty aldehydes have shown that palmitaldehyde (16:0) predominates in spermatozoa of all species studied (Darin-Bennett *et al.*, 1974, 1977; Gray, 1960; Johnson *et al.*, 1969, 1972; Poulos *et al.*, 1973a; Pursel and Graham, 1967). Again, it is interesting to note a distinction between the species susceptible to cold shock, which have 80% or more of their sperm fatty aldehydes as palmitaldehyde (16:0), and the resistant species, which while having a predominance of 16:0 also have substantial amounts of other aldehydes (15:0, 16:1, 18:0, 18:1 or 18:2). Aldehydes with other chain lengths were present in only minor proportions and no longer chain fatty aldehydes were detected (Darin-Bennett *et al.*,

Table 4

Phospholipid-bound fatty acids of mammalian spermatozoa. Expressed as a percentage of total fatty acids; molar corrections for detector response applied when known; mean ± standard error.

Fatty acid	Ram	Bull	Boar	Rabbit	Dog	Monkey	Man	Fowl
14:0	5.2±1.5	2.0±0.4	3.0±0.4	1.5±0.2	1.0±0.6	nd	nd	nd
16:0	14.7±0.5	14.1±0.4	13.0±0.9	24.2±0.6	27.2±0.7	29.2±1.0	24.8±0.3	13.6±0.9
18:0	5.1±0.4	5.3±0.2	5.1±0.6	18.7±0.2	13.2±1.4	11.2±0.9	11.3±1.0	20.3±1.1
18:1	3.8±0.3	5.7±0.5	2.6±0.4	9.1±0.7	11.8±0.9	14.8±1.5	8.1±0.4	19.2±0.7
18:2	1.7±0.1	3.9±0.0	2.1±0.1	4.8±0.3	3.2±1.3	3.8±0.2	4.0±0.2	1.6±0.3
18:3	nd	nd	nd	nd	1.3±0.3	nd	nd	nd
20:0	nd	nd	nd	1.6±0.6	nd	3.1±0.1	nd	2.9±0.1
20:4+20:3	4.5±0.1	3.5±0.1	3.2±0.2	nd	6.6±0.7	8.9±0.4	5.1±0.4	10.2±1.8
22:0	1.1±0.1	nd	1.5±0.1	1.2±0.1	nd	nd	2.9±0.8	nd
22:3	nd	nd	nd	nd	nd	nd	nd	3.8±0.4
22:4	nd	nd	1.6±0.1	nd	nd	nd	nd	24.6±2.6
22:5	nd	nd	27.9±0.8	39.0±1.4	28.4±3.0	2.8±0.8	nd	nd
22:6	61.4±0.9	61.3±0.9	37.7±1.7	nd	3.9±1.8	25.2±0.9	35.2±2.4	2.3±1.1
Others*	2.5	4.2	2.3	nd	0.5	1.2	8.6	0.3
No of replicates	3	3	3	6	5	3	3	4

nd – fatty acid either not detected or present in amounts less than 1%. * – Minor acids identified included 14:0,15:0,16:1,17:0,18:3,20:0,20:2 and 22:4. From Poulos et al. (1973), Darin-Bennett et al. (1974, 1977). Reproduced with permission of Comp. Biochem. Physiol. and J. Reprod. Fert. and the authors.

1974, 1977; Poulos *et al.*, 1973).

The distribution of fatty acids varies between classes of phospholipid. Thus, in bull and ram spermatozoa much of the docosahexaenoic (22:6) acid is located in the choline phosphoglycerides (Neill and Masters, 1972, 1973) and in the ram, 90 percent of the fatty acid of choline plasmalogen and 75 percent of phosphatidylcholine fatty acid is 22:6 (Darin-Bennett *et al.*, 1976). A similar distribution can be inferred from published data for the bull (Poulos *et al.*, 1973a; Selivonchick *et al.*, 1980) although precisely comparable data are not available.

In ram spermatozoa, the major fatty acid of ethanolamine plasmalogen is 22:6 (80%) while phosphatidylethanolamine contains only 13% 22:6 and the major components are 16:0 (33%), 20:4 (26%) and 18:0 (18.5%) (Darin-Bennett *et al.*, 1976). Arachidonic (20:4) acid is also concentrated in phosphatidylethanolamine of bull spermatozoa (Neill and Masters, 1972; Selivonchick *et al.*, 1980). Cardiolipin is predominated by myristic (14:0) acid (40%) together with substantial proportions of 16:0 (17%), 18:1 (14%) and 18:2 (11%) (Darin-Bennett *et al.*, 1976).

Of relevance to the consideration of cold shock is that human spermatozoal phospholipids possess a quite different distribution of fatty acids. While 22:6 is the major fatty acid present, no large disparity in proportion exists between the five major phospholipid classes (Darin-Bennett *et al.*, 1976). As previously mentioned the shorter chain C_{16} to C_{20} fatty acids are also present in greater proportion than in ram spermatozoa and are in a more even distribution between phospholipids.

This variation in phospholipid-bound fatty acyl chains between species is further emphasised in that in bull spermatozoa, the ether phospholipids consist almost exclusively of 16:0 chains at the *sn*-1 position and C_{22} chains at the *sn*-2 position. This contrasts with the diacyl phospholipids which contain a wider spectrum of molecular species (Table 5). The presence of polyunsaturated C_{22} fatty acids exclusively at the *sn*-2 position may influence the interactions with cholesterol (Selivonchick *et al.*, 1980). Comparative studies of detailed phospholipid structure in sperm membranes of other species are unfortunately not yet available.

Jones and Mann (1977) have published a phospholipid-bound fatty acid analysis of ram spermatozoa fragmented into an outer acrosomal + plasma membrane fraction, denuded heads, and midpieces + tails. While 22:6 predominated in all three fractions, higher proportions of shorter chain saturated fatty acids were present in the acrosomal + plasma membrane

Table 5

Fatty acyl chain at the sn-1 and sn-2 positions of alk-1-enylacyl (A), alkylacyl (B) and diacyl (C) choline and ethanolamine phosphoglycerides. Values are expressed as percentage by weight.

Fatty acid	Choline phosphoglycerides						Ethanolamine phosphoglycerides					
	sn-1			sn-2			sn-1			sn-2		
	A	B	C	A	B	C	A	B	C	A	B	C
14:0	0.5	3.3	2.6	–	–	–	3.6	8.6	–	–	–	–
16:0	97.8	95.5	70.1	–	–	–	91.1	80.8	34.2	–	–	1.6
16:1	–	–	–	–	–	–	–	–	3.4	–	–	–
18:0	0.8	0.4	15.9	–	–	7.5	3.4	0.2	50.9	–	–	2.1
18:1	0.5	–	1.3	–	–	10.9	1.4	9.5	9.7	–	–	5.5
18:2ω6	–	–	1.1	–	–	18.5	–	–	1.4	–	–	5.3
20:3ω6	–	–	0.6	–	0.4	5.0	–	–	–	0.4	0.4	–
20:4ω6	–	–	2.3	–	1.1	16.1	–	–	–	5.0	6.7	70.4
20:5ω3	–	–	–	–	–	–	–	–	–	1.0	2.5	–
22:4ω6	–	–	–	0.5	0.3	1.8	–	–	–	1.7	1.0	–
22:5ω6	–	–	–	23.4	19.8	7.2	–	–	–	22.6	20.2	3.2
22:5ω3	–	–	–	–	1.6	2.2	–	–	–	4.4	4.6	0.8
22:6ω3	–	–	–	71.9	74.7	30.6	–	–	–	59.9	62.2	10.5

From Selivonchick et al., (1980).
Reproduced with permission of Elsevier/North-Holland Biomedical Press and the authors.

fraction. The denuded heads fraction had the highest proportion of 22:6. It is clear that a lower ratio of polyunsaturated to saturated fatty acids exists in the membranes
most readily disrupted by cold shock, compared with the
remainder of the cell membranes. In view of the findings of
Darin-Bennett *et al*. (1976), these observations might be
interpreted as evidence for a differential distribution of
phospholipid classes between various membrane systems. Thus,
the outer acrosomal + plasma membranes may contain a relatively
high proportion of phosphatidylethanolamine and phosphatidylcholine. This distribution combined with a lower cholesterol
mole percentage might predispose to lateral phase separations.
Further studies of this nature, using refined techniques for
separating individual membranes and regions of membranes
are required to provide a detailed knowledge of lipid
distribution within the various areas of the spermatozoon.

Cholesterol

The cholesterol to phospholipid ratio is an important
determinant of membrane fluidity (Shinitsky and Henkart, 1979;
Pringle and Chapman, this volume). Darin-Bennett and White
(1977) compared the phospholipid and cholesterol contents of
the spermatozoa of two species resistant to cold shock
(rabbit, human) with those of two susceptible species (ram,
bull). The cholesterol to phospholipid molar ratio was 0.88
and 0.99 in the resistant species, and 0.38 and 0.45 in the
susceptible species (Table 6). Similar values for cholesterol
content (Pursel and Graham, 1967; Quinn and White, 1967;
Scott *et al*., 1967) and cholesterol to phospholipid molar
ratio (Selivonchick *et al*., 1980) have also been obtained for
ram and bull spermatozoa. Cholesterol is believed to modulate
the fluidity of membranes by interacting with the fatty acyl
chains of the phospholipids.

Thus, it seems from the lipid data that appropriate membrane fluidity is achieved in spermatozoa of different species
by either of two distinct mechanisms. Fluidity may be maintained by a high proportion of polyunsaturated fatty acyl
chains in the phospholipids, which influence the partial
specific volume of the molecule (Shinitsky and Henkart, 1979;
Pringle and Chapman, this volume). Alternatively, fluidity
may be maintained by an increase in the cholesterol to
phospholipid ratio.

It was initially thought that spermatozoa with a high
ratio of polyunsaturated to saturated fatty acids would be
more resistant to cold shock, since polyunsaturated fatty
acids, having lower transition temperatures, would confer

Table 6

The cholesterol and phospholipid content of mammalian spermatozoa.

Species	Cholesterol (μmol per 10^9 sperm)	Phospholipid P (μmol per 10^9 sperm)	Cholesterol to phospholipid molar ratio	Cholesterol mole % *
Ram	0.722	1.920	0.38	27
Bull	0.893	1.991	0.45	31
Rabbit	1.411	1.607	0.88	62
Human	1.438	1.447	0.99	50

* mole % = 100 x cholesterol/cholesterol + phospholipid

Modified from Darin-Bennett and White (1977).

Reproduced with permission of Cryobiology and the authors.

greater fluidity on membranes at low temperature. This
belief was based on evidence from temperature-insensitive
plants and poikilothermic organisms, and from acclimation
studies of organisms to low temperatures. These organisms
respond by increasing the proportion of unsaturated fatty
acids in response to low temperature apparently in order
to maintain lipid fluidity (Quinn and Chapman, 1980).

However, the comparison with spermatozoa is not valid
since spermatozoa are adapted to the normal range of body
temperatures of endothermic animals. Differences between
species merely reflect different solutions to a common
problem of maintaining membrane fluidity at body temperature,
and need not relate to responses at other temperatures. In
fact, it appears that species having sperm membranes with
lower ratios of polyunsaturated to saturated fatty acids and
higher cholesterol to phospholipid ratios are incidentally
the more resistant to temperature changes.

THE ROLE OF LIPIDS IN SUSCEPTIBILITY TO COLD SHOCK

The evidence presented suggests that cold shock susceptibility
may be determined by the membrane cholesterol content and the
proportion of polyunsaturated fatty acids, both of which
influence lipid fluidity. Darin-Bennett and White (1977)
suggested that the presence of cholesterol at less than
approximately 30 mole percent might allow the formation of
cholesterol-free phospholipid crystalline domains below
their transition temperature, as has been found for artificial
membranes (Oldfield and Chapman, 1972; Phillips and Finer,
1974). With equimolar proportions of cholesterol and phospholipid no lateral phase separation would take place and
membrane fluidity would be maintained. The wide divergence
of distribution of fatty acids between different phospholipid
classes in susceptible species (Darin-Bennett *et al.*, 1976)
might exacerbate this phenomenon since the transition temperature is influenced by fatty acid composition. It is not
known to what extent different membrane systems vary in their
phospholipid composition, fatty acid distribution or cholesterol content. It is, perhaps, significant that freeze-
fracture studies of filipin-treated ram and bull spermatozoa
revealed a differential distribution of cholesterol in the
plasma membrane, the tail regions possessing more cholesterol
than head regions. Also negatively-charged phospholipids
were confined predominantly to the anterior head region of the
plasma membrane, with a low concentration in the posterior head
region and none in the tail (Bradley *et al.*, 1979,1980). Thus
the plasma membrane in the anterior head region of spermatozoa

of cold-shock susceptible species may be particularly prone
to lateral phase separations at low temperatures.

Exposure of spermatozoa to 2,6-di-*tert*-butyl-*p*-cresol
(butylated hydroxytoluene, BHT) prior to cold shock prevented
the increased membrane permeability in ram and bull spermatozoa, but boar sperm membranes were still apparently
rendered permeable (Hammerstedt *et al.*, 1976, 1978). However, boar spermatozoa were actually protected by BHT in
that fertility was retained (Pursel, 1979). This action of
BHT was attributed to its membrane lipid perturbing ability
(Hammerstedt *et al.*, 1976, 1978), which may broaden phase
transitions and prevent lateral phase separations in lipids.

Experiments with models and model membrane systems have
suggested that the consequences of lateral phase separations
are the exclusion of intrinsic proteins into fluid lipid
regions and the development of crystalline lipid domains
intersected by packing faults (Chapman *et al.*, 1979).
Aggregation of proteins may interfere with their metabolic
functions, *e.g.* transport mechanisms, and membrane permeability may be increased (Gingell, 1976).

VARIATION IN SUSCEPTIBILITY TO COLD SHOCK

The Influence of Sperm Cell Maturity

Testicular and proximal epididymal spermatozoa are known to
be more resistant to cold shock than spermatozoa from the
cauda epididymidis and ejaculated spermatozoa (Lasley and
Bogart, 1944a; Lasley and Mayer, 1944; Quinn and White, 1967;
White and Wales, 1961). It is also generally agreed that
the lipid content of spermatozoa decreases markedly from the
testis to the cauda epididymidis. Phospholipids of all
classes are affected but choline plasmalogen, although the
major phospholipid at all stages in the ram and bull,
declines proportionately less than others (Poulos *et al.*,
1973b, 1975; Quinn and White, 1967; Scott *et al.*, 1967). In
boar spermatozoa there is a similar reduction in total phospholipid content (Johnson *et al.*, 1972; Pickett *et al.*,
1967), and although choline plasmalogen represents only a
small proportion it is also preferentially retained (Grogan
et al., 1966).

The relative stability of ether phospholipids (Gottfried
and Rapport, 1962; Lands and Hart, 1965; Selivonchick *et al.*,
1980) may explain the preferential retention of choline
plasmalogen in ejaculated spermatozoa. Further evidence is
provided by the fact that fatty aldehydes of bull spermatozoa,
in contrast to fatty acids, show little change in overall

composition during passage through the epididymis (Poulos *et al.*, 1973b).

Only slight differences in lipid composition were found in spermatozoa after they had left the cauda epididymidis. Sensitivity to cold shock in ram spermatozoa develops between the caput and cauda epididymidis suggesting a relationship between changing lipid membrane components and susceptibility to cold shock (Quinn and White, 1967). The phospholipid-bound fatty acids also change during maturation with a relatively greater loss of saturated fatty acids, particularly 16:0, in ram and bull (Poulos *et al.*, 1973b, 1975), but not in boar spermatozoa (Johnson *et al.*, 1972). Consequently, testicular spermatozoa of the bull and ram have ratios of polyunsaturated to saturated fatty acids of 0.9 and 1.0 respectively (Darin-Bennett *et al.*, 1974; Poulos *et al.*, 1973b) correlating with their resistance to cold shock. Cholesterol levels were also found to decline during passage through the epididymis (Quinn and White, 1967; Scott *et al.*, 1967).

The relatively greater resistance of immature spermatozoa to cold shock probably accounts for the observation that samples of ram cauda epididymal spermatozoa with a high proportion of attached cytoplasmic droplets (considered an index of immaturity) were more resistant than samples with a low percentage of attached droplets (Wales and White, 1959; White and Wales, 1961). A similar explanation may be advanced to account for the slightly greater resistance of second ejaculates in the bull since a greater proportion of spermatozoa would come from more proximal regions. The effect was slight, however, and in series of successive ejaculates from the ram an opposite trend was observed, due perhaps to the effects of seminal plasma or declining sperm concentration (Quinn *et al.*, 1968a).

Individual Variability

Apart from species differences in susceptibility of spermatozoa to cold shock and the changing sensitivity with cell maturity, considerable differences exist between individuals of the same species (Lasley and Mayer, 1944; Lasley *et al.*, 1942a; Pursel *et al.*, 1973) and, at least for bull spermatozoa, within ejaculates between spermatozoa (Choong and Wales, 1962). Some 25 percent of bull spermatozoa was found to be resistant to repeated cold shocks (Table 7). Thus, although the phenomenon is regarded as all-or-none for an individual cell, there are variations in susceptibility between cells in a population.

Table 7

Percentage of diluted bull spermatozoa motile after repeated sudden cooling to 0°C

Number of times cold-shocked	Mean number of motile spermatozoa (%)
0 (Control)	78.6
1	37.8
2	28.6
3	26.8
4	25.0

From Choong and Wales (1962). Reproduced with permission of Aust. J. biol. Sci. and the authors.

Variability at this level in response to cold shock implies at least minor variations in membrane composition. While there is no direct evidence to support this view for spermatozoa, there is considerable evidence from other cell types showing variability in membrane components. Moreover, continuous metabolic processing of membrane components is recognised, the turnover of lipids being more rapid than that of protein elements. The significance of such membrane variability for the functions of cells is not understood (Quinn and Chapman, 1980). In spermatozoa, both the presence of phospholipases (Scott and Dawson, 1968) and the synthesis of phospholipids (Selivonchick *et al.*, 1980) have been demonstrated. It seems probable, therefore, that sperm cell membranes are in a dynamic state which may explain variations in susceptibility to cold shock and differences in reported lipid composition. Whether the membrane composition of an individual cell remains constant or is continuously altered, resulting in greater or less resistance to cold shock, is unknown.

The Influence of Seminal Plasma

It is generally considered that resistance to cold shock is an inherent characteristic of the spermatozoa (Pursel *et al.*, 1973; Wales and White, 1959) but accessory secretions may

influence the susceptibility. Thus, bull and ram semen
diluted more than three times with additional seminal plasma
before cold shock showed a greater loss of sperm motility
than undiluted semen (Choong and Wales, 1962; Quinn *et al.*,
1968a). Ram semen collected by electroejaculation was found
to be more susceptible to cold shock than semen collected by
artificial vagina (Quinn *et al.*, 1968a) and this was
attributed to the fact that ram spermatozoa are more resistant
to cold shock at lower pH (Quinn and White, 1968). They
argue that because semen collected by artificial vagina has
both a lower proportion of accessory secretions (and thus a
lower buffering capacity), and also a higher sperm density,
it would be more likely to become acid by production of
lactate from seminal fructose by spermatozoa (Quinn and
White, 1968). That this was not attributable to the sperm
density *per se* was shown by exchanging seminal plasma between
semen samples collected by the two methods before cold shock
(Quinn *et al.*, 1968a).

Dilution of seminal plasma while maintaining the sperm
density constant was found to have little influence on the
responses of bull spermatozoa to cold shock (Wales and White,
1959), whereas boar spermatozoa from the sperm-rich fraction
of the ejaculate were more severely damaged by this treatment,
suggesting a protective effect of seminal plasma (Pursel *et
al.*, 1973). In a comparison of cold-shocked spermatozoa from
the sperm-rich fraction and the whole ejaculate the percentage
of normal acrosomes was lower in the whole ejaculate, while
the motility was better protected (Pursel *et al.*, 1972a). In
contrast the spermatozoa of the sperm-rich fraction of boar
semen were more resistant to cold shock than those of the
whole ejaculate (Polge, 1956). It is clear that complex
interactions exist between spermatozoa and seminal plasma in
the boar. The nature of harmful constituents of ram or bull
seminal plasma has not yet been investigated, but basic
proteins have been implicated in the case of boar spermatozoa
(Moore *et al.*, 1976).

The Special Case of Boar Spermatozoa

The contrast between boar semen and that of other susceptible
species in regard to cold shock is quite pronounced in
several respects. After ejaculation boar spermatozoa are
highly susceptible to cold shock but resistance develops
during subsequent incubation prior to cold shock. Thus,
4-7 h after ejaculation some 50-80 percent of spermatozoa
have acquired resistance as judged by motility and morphology
(Butler and Roberts, 1975; Pursel *et al.*, 1972a, 1972b, 1973).

A similar, but much less pronounced, effect has been observed
for ram spermatozoa following incubation for up to 2 h (Quinn
et al., 1968a). The development of resistance of boar
spermatozoa during incubation is influenced by pH, dilution
and extender composition (Butler and Roberts, 1975; Pursel
et al., 1972a, 1972b, 1973). Moreover, egg yolk, a protective substance for ram and bull spermatozoa provided no
protection for boar spermatozoa (Benson *et al.*, 1967) and
interfered with the development of resistance (Pursel *et al.*,
1972b).

Moore *et al.*, (1976) showed that cold shock sensitivity of
ejaculated boar spermatozoa was increased by the binding of
basic proteins from seminal vesicles to the sperm surface.
Since basic proteins increase the permeability of membranes
(Drew and McLaren, 1970; Hibbitt and Benians, 1971; Ryser
and Hancock, 1965) it is possible that vesicular basic
proteins may similarly affect boar spermatozoa rendering them
susceptible to damage during cooling (Moore *et al.*, 1976).

At present these observations cannot be correlated with
membrane composition. However, the development of resistance,
the effect of pH, the influence of seminal plasma constituents
and other additives all imply that membrane structure may be
modified after ejaculation.

THE CONDITIONS LEADING TO COLD SHOCK

Methods of achieving cold shock are:

1) Adding undiluted semen at 30°C to precooled diluent or
   staining mixture (Choong and Wales, 1962; Quinn *et al.*,
   1968b; Wales and White, 1959).
2) Allowing undiluted semen to run over a precooled glass
   surface (Hancock, 1952).
3) The most common method, transferring a small volume of
   undiluted or diluted semen in a glass tube at room
   temperature, 30° or 37°C to a water bath at 0° or 5°C
   and maintaining that temperature for up to 10 min to
   allow equilibration to occur.

In each case, the rate of cooling is of the order of $10-15^{\circ}$C
min^{-1} or greater. The rate of cooling is important in
determining cold-shock injury. Ram or bull spermatozoa
cooled from 37°C gradually over 3-4 h would survive temperatures of $0-5^{\circ}$C, whereas the motility after rewarming was
reduced if the rate of cooling was increased (Birillo and
Puhaljskii, 1936). In contrast, boar spermatozoa collected
in the complete ejaculate are apparently damaged simply by
cooling irrespective of the rate (Moore *et al.*, 1976; Polge,

1956) and only survive after an incubation period of several hours prior to cold shock. The sperm-rich fraction, however, will survive slow cooling (Polge, 1956). These findings have been utilised in the processing of semen for freeze-preservation.

Sudden cooling above 15-20°C has little detrimental effect even when the temperature interval is of the order of 20°C (Quinn et al., 1968b; Wales and White, 1959). Below 15°C, however, the effect is more pronounced as the final temperature decreases (Blackshaw, 1958; Chang and Walton, 1940; Gladcinova, 1973; Quinn et al., 1968b; Wales and White, 1959). These studies frequently confounded the temperature interval with the final temperature attained and, incidentally, the rate of cooling. In order to distinguish between these Quinn et al., (1968b) subjected ram spermatozoa to a constant reduction in temperature of 22°C after cooling slowly to different initial temperatures (Fig. 1). The absolute temperature range was found to be crucial. However, the temperature interval required to produce cold shock diminished as the initial temperature decreased. The importance of the final temperature in the causation of cold shock suggests that lipid phase transitions are involved: the lower the temperature, the greater the proportion of lipids which have undergone a phase transition.

The effect of the duration of cold shock, i.e. the holding period at low temperature following cold shock, has been found to vary with the parameter measured. For example, the amount of phospholipid released from spermatozoa was unaffected by the duration of cold shock from 5-120 min (Darin-Bennett et al., 1973). On the other hand, metabolic parameters showed increasingly severe changes with longer periods of treatment (Chang and Walton, 1940; Mayer, 1955). This is consistent with an assumption of a primary membrane change followed by subsequent metabolic derangement as a consequence of the primary event.

PROTECTIVE ADDITIVES

Egg yolk has long been known to provide protection to ram and bull spermatozoa undergoing cold shock and exposure to low temperature (Blackshaw, 1954; Kampschmidt et al., 1953; Lardy and Phillips, 1939; Lasley et al., 1942b; Phillips, 1939). Several other substances, e.g. milk and milk proteins, bovine albumin, egg albumin (Choong and Wales, 1962; Quinn and White, 1968) and testicular and sperm lipoprotein (Miller and Mayer, 1960) also provide a measure of protection against cold shock, but cholesterol has no effect (Gebauer

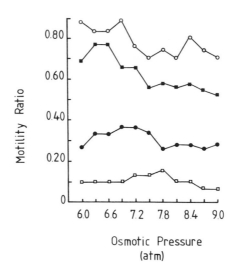

Figure 1. *Motility ratios of ram spermatozoa in semen diluted 1:3 with sodium citrate diluents of varying tonicity. Temperature gradients:* O *37-15°C,* ∎ *32-10°C,* ● *27-5°C,* □ *22-0°C. The effect of the osmotic pressure of the diluents was not significant. Motility ratio is the ratio of the motility score of the sample (over a range 0 to 5) to that of the control.*

From Quinn et al., (1968b). Reproduced with permission of Aust. J. biol. Sci. and the authors.

et al., 1970; Lovelock, 1955b). The active fraction of egg yolk is the low density lipoprotein (Foulkes, 1977; Kampschmidt *et al.*, 1953; Pace and Graham, 1974; Watson, 1976) of which phosphatidylcholine appears to be instrumental in providing protection (Blackshaw, 1954; Kampschmidt *et al.*, 1953; Quinn *et al.*, 1980). The possible superiority of egg lipoprotein over egg lecithin for cold storage (Kampschmidt *et al.*, 1953) or freezing (Gebauer *et al.*, 1970) may be due to either its greater solubility or binding of the protein to the sperm surface (Foulkes, 1977; Watson, 1981).

While egg yolk or lecithin provide excellent protection for bull spermatozoa undergoing cold shock (Blackshaw and Salisbury, 1957; Kampschmidt *et al.*, 1953) they are less effective for ram spermatozoa (Blackshaw, 1954, 1958) and are of little value for boar spermatozoa (Benson *et al.*, 1967; Pursel *et al.*, 1972b). Boar spermatozoa are also unique in that other additives beneficial in other species are not effective, although egg lipoprotein and casein were stated to be of some use (Pursel *et al.*, 1973). Butler and Roberts (1975) showed that phosphatidylserine but not phosphatidylcholine or phosphatidylethanolamine protected boar spermatozoa. This suggests differences in the sperm cell surface between species. Lecithin was more effective for ram spermatozoa at pH 6.5 than at neutral or alkaline pH, but this was not so for bull spermatozoa (Blackshaw, 1958); this effect may be related to the isoelectric point of lecithin (pH 6.4) but it could also be due to the increased resistance of ram spermatozoa to cold shock at low pH (Quinn and White, 1968). Lecithin was also found to reduce the accumulation of calcium ions in cold-shocked ram spermatozoa (Quinn and White, 1968). In view of earlier comments about the possible role of calcium ions in cold shock damage, one mechanism for the action of lecithin, casein and other proteins might be that they may act as calcium chelating agents. It is significant that the protective effect of lecithin was not additive with that of EDTA (Quinn and White, 1968).

The protection by large molecular weight proteins and the apparent absence of permanent modification of the cell membranes by lecithin suggests that these substances act at the cell surface, implying that the primary damage induced by cold shock occurs in the plasma membrane (Quinn *et al.*, 1980). It is still not certain how such substances could exert their effects, but three possibilities should be considered:

1) Chelation of Ca^{2+} (see above).
2) Modification of lipid fluidity and hence phase separation events.
3) Stabilisation of protein elements in the membrane lipid matrix.

The second possibility has its origins in the proposal by Lovelock (1954a, 1955b) that lecithin acts by modifying the ratio of cholesterol to phospholipid in the membrane. However, no evidence was found with ram spermatozoa either for enrichment of phospholipid or for depletion of cholesterol in the membranes (Quinn *et al.*, 1980). While these findings do not rule out the possibility of membrane lipid modification it is unlikely that any permanent alteration occurs, such modifications being the result of associations of lecithin liposomes with superficial membrane components.

THE EFFECTS OF SLOW COOLING

Spermatozoa that have been cooled slowly (which avoids cold shock: Table 1) are not unaffected by continued exposure to low temperature. Such treatment leads to changes in the

plasma membrane, the acrosome and mitochondria (Jones and
Martin, 1973), loss of fertility (Watson, 1979) and, if
protective substances are not included in the diluent, loss
of motility (Jones and Martin, 1973; Watson, 1976).
Carbohydrate metabolism is depressed and ionic movements
are similar in kind to those observed in cold shock
(O'Shea and Wales, 1966a, b).

The interpretation of these results is complicated by the
inability to separate the effects of low temperature *per se*
from senescent changes which, although reduced at 0-5°C,
continue to occur. This is difficult to investigate because
the changes produced are similar in both cases. Nevertheless,
it is well recognised that cold treatments increase the
dissociation of fertility and motility. For example, ram
spermatozoa cooled slowly in protective diluents and stored
for 24-48 h at 5°C had a very low fertility, while the
motility on rewarming was indistinguishable from fresh
diluted semen (Watson and Martin, 1976). Also, egg yolk has
little effect on the survival of ram spermatozoa incubated
at 37°C (Watson, unpublished results) whereas it profoundly
reduces the loss of motility during storage at 5°C (Jones
and Martin, 1973; Watson, 1976). These data suggest that
exposure to low temperature *per se* causes alterations to
spermatozoa.

The similarity of the nature of the changes in cold-
shocked and slowly cooled, stored spermatozoa raises the
intriguing question of whether there is a common underlying
cause. There are, however, considerable differences in the
degree of membrane alteration which require explanation.
Since calcium has been implicated in a wide range of metabolic functions within other cells (Randle *et al.*, 1979;
Rasmussen *et al.*, 1979) it is tempting to speculate that
access of calcium ions to the interior of the cell is the
common link between these phenomena. Inability to regulate
intracellular calcium levels may have serious metabolic and
structural consequences. In cold shock, it is conceivable
that lateral phase separations in membrane lipids with exclusion of proteins into remaining heterogeneous fluid domains
cause the increased permeability. These changes may be less
severe in slow cooling since remixing of membrane components
would tend to maintain membrane fluidity. However, calcium
pumps would be less active at low temperature leading to a
slow accumulation of intracellular calcium, and any tendency
towards protein clustering may adversely affect the functioning of these pumps, or may increase permeability (Gingell,
1976). These ideas remain conjectural at present; there is
a need for further experimentation to support or disprove

this thesis. It is still not clear why cold shock leads to such gross disruption of the plasma membrane overlying the anterior acrosome. The structural changes demonstrated by light and electron-microscopy indicate as yet unappreciated specialised membrane features within this region.

CONCLUSION

Considerable evidence points to the importance of membrane lipid characteristics in determining the susceptibility to cold shock, but there remains a need to demonstrate phase transitions or lateral phase separations in sperm cell membranes over the range of $0-20^{\circ}C$. Much progress could be made from studies of isolated pure membrane fractions of spermatozoa (Cillis *et al.*, 1978; Lunstra *et al.*, 1974; Silvestroni *et al.*, 1980; Zahler and Doak, 1975).

The role of calcium in cold shock damage is suggested by a number of studies. Further work is necessary to establish that an increase in intracellular calcium concentration is an essential stage in the disruption of cellular structure and function. It is worth noting that an influx of calcium ions into the cell is recognised as an essential step in the final modification of the spermatozoa prior to penetration of the egg cell mass (Talbot *et al.*, 1976; Yanagimachi and Usui, 1974). Premature elevation of intracellular calcium levels could well interfere with subsequent normal function.

The species differences in susceptibility to cold shock cannot be correlated with the survival of spermatozoa after freezing and thawing (Watson, 1979). Thus, bull spermatozoa can be frozen with success whereas other species present greater difficulties. It should be noted, however, that slow cooling above $0^{\circ}C$ is essential for the survival of ram and bull spermatozoa, while a mandatory stage in the freeze-processing of boar spermatozoa is a holding period above $15^{\circ}C$ for several hours before cooling. This discussion has demonstrated that the response of cell membranes to low temperatures *per se* is at least as important as a consideration of the phenomena associated with ice crystal formation (Watson, 1979) to the successful freeze-preservation of mammalian spermatozoa.

The mammalian spermatozoon undergoes a sequential series of membrane modifications in its passage from the testis to its final union with the ovum in the fallopian tube of the female. The practical problem of sperm preservation is that of arresting and reactivating this series of changes without altering the delicate interrelationships of the cell with its environment. The study of cold shock may contribute towards a better understanding of membrane bio-

chemistry and, hence, promote the development of more
satisfactory preservation techniques.

ACKNOWLEDGEMENTS

I am indebted to Dr W.V. Holt for the electron micrographs
and for his helpful comments on their interpretation. My
thanks also go to Mr W.J. Anderson who assisted with all
stages in the preparation of the manuscript.

COMPUTER ASSISTED ANALYSIS OF FREEZE-FRACTURED MEMBRANES FOLLOWING EXPOSURE TO DIFFERENT TEMPERATURES

A.W. Robards

Department of Biology, University of York, England

G.R. Bullock, M.A. Goodall and P.D. Sibbons

Ciba-Geigy Pharmaceuticals Division, Horsham, England

INTRODUCTION

Freeze-etching provides a unique view of cell membranes at a resolution unobtainable by any other contemporary method. For that reason, and despite the technical difficulties, freeze-etching techniques are indispensable among the various methods of studying normal or experimentally treated membranes.
  The particles seen on membrane faces from freeze-fracture replicas are generally considered to be protein or glyco-protein macromolecules intercalated within a fluid lipid matrix. It is thought that the relative frequency of particles is an indication of the metabolic activity of the membrane. Thus relatively inactive membranes, such as myelin, have very few particles, if any; plasma membranes of cells such as erythrocytes have particle frequencies in the range 2 to 3,000 μm^{-2}, whilst membranes across which there are very high rates of water transport such as the descending loop of Henle in the kidney can have frequencies in excess of 4,000 μm^{-2} (Verkleij and Ververgaert, 1978). It has been suggested that at certain temperatures membranes may undergo lipid phase transitions which affect their permeability and activity (Simon, 1974; many authors, this volume). In some cases it has been possible to show a correlation between intramembrane particle (IMP) aggregation associated with lipid phase separations and changes in fatty acid unsatur-ation during temperature acclimation (Martin *et al.*, 1976).
  In common with many other methods in cell biology, the sampling and statistical analysis of images obtained from

freeze-fracture replicas require scrupulous attention if
significant results are to be obtained. IMP frequencies are
often seen to vary considerably, both within different areas
of the same membrane and between membranes from different
cells. Similarly, different membranes may show a greater or
lesser degree of particle aggregation: this has been
correlated with phase transitions within the membranes of
some cells (Martin *et al.*, 1976).

   Our work concerns the study of two very different cell
types: muscle cells from rat hearts and cells from the roots
of maize (*Zea mays*). In both cases, but for different
reasons, the effects of cooling on the structure and properties of the cell membranes are of considerable interest.
Consequently, we have developed methods for freeze-etching
these tissues and subsequently subjecting the micrographs to
statistical analysis. In this way it will be possible to
compile reliable estimates of particle frequency and
aggregation in normal membranes so that these may be compared
with membranes subjected to the experimental effects of
chilling.

Effect of Temperature on Nutrient Uptake by Maize Roots

In much of Northern Europe, forage maize (*Zea mays*) is grown
in the early part of the season when the temperature of the
soil is only 10 to 12°C and may increase only slowly (Carr
and Hough, 1978). Different varieties of maize show clear
differences in tolerance to low temperatures with respect to
such features as root elongation and mineral nutrient uptake
(Clarkson and Gerloff, 1979). When the rate of root growth
or the rate of nutrient uptake is plotted against temperature,
the Q_{10} becomes progressively larger when moving from warm to
cool conditions. In particular, there appears to be a sharp
transition around 13°C. The organisation of cell membranes
may be affected by cooling and lipid phase transitions may be
responsible for some of the observed phenomena (McMurdo and
Wilson; Pringle and Chapman, this volume). Analysis of
freeze-fracture replicas from normal or chilled maize roots
should, therefore, provide evidence as to whether structural
changes, including phase transitions, do occur as the growth
temperature is reduced below about 13°C.

Calcium Paradox in Mammalian Hearts

When the heart of any mammal is perfused with calcium-free
medium above a critical temperature (32°C for the rat), the
sarcolemma separates from the basement membrane with the

concomitant formation of fluid filled vesicles. If calcium
is then restored, there may be a massive contracture of the
muscle fibres, loss of enzymes and development of the 'stone
heart', a phenomenon which has been termed by Zimmerman and
Hulsmann (1966) 'the calcium paradox'. If, however, the
temperature of the calcium-free period is kept at, or below,
the critical temperature the extensive damage is prevented
and few cells show signs of contracture (Hearse *et al.*, 1978).
To maintain hearts in a healthy state during coronary by-pass
surgery, both the constitution and temperature of the
perfusing medium are of vital importance and hypothermic
conditions are generally employed. It is probable that much
of the damage during the reperfusion period is due to massive,
uncontrolled uptake of calcium resulting from changes in the
sarcolemma during the calcium-free period when this is above
the critical temperature. For this reason, a careful
analysis of the structure of the heart sarcolemma before and
after such experimental procedures is clearly important.

While the damage to the muscle cells as seen by the
separation of the basement membrane and sarcolemma is very
clear cut and the ensuing damage to the cells when calcium
is restored is patent, little attention has been paid to the
membranes of the vascular endothelial cells. Although
membrane separation has not been observed, it seems likely
that some alteration to the membrane structure would follow
calcium-free perfusion and reperfusion. Consequently, we
have extended our studies on the sarcolemma to the plasmalemma of the vascular endothelial cells which have been
exposed to the same experimental conditions and evaluation.

TECHNIQUES

Freeze-etching is now a well established method of studying
cell membranes. If the frozen specimen is fractured and a
replica is prepared without an etching stage, then the process
is usually referred to as freeze-fracture. In studies such
as this, where the distribution of IMPs is concerned, there
is little point in etching. It is usually assumed that the
fracture plane through the frozen specimen proceeds along the
mid-line of the membrane: we have seen nothing in our micrographs to suggest that this is not the normal case in the
membranes currently under study.

A major difficulty in the evaluation of particle
distribution and frequency on freeze-fractured membranes is
the accumulation of data that can be shown to be statistically
significant. In many published papers the authors have been
content to illustrate differences using one or two

'representative' micrographs. Some attempt must be made to
improve on this. This involves: careful sampling procedures
from the original biological specimens, preparation of a
sufficient number of replicas from different specimens to
eliminate spurious data from random fluctuations, selection
of an adequate number of micrographs from different fields
and a sufficient number of areas on each micrograph to
provide a significant statistical analysis, and use of
appropriate statistical methods to calculate IMP frequencies,
coefficients of dispersion and other parameters. This paper
reports our attempts to move in this direction and, while by
no means claiming to have overcome all of the many problems
involved, does point to some of the ways that freeze-fracture
experiments can be made more useful than they have sometimes
appeared in the past.

Maize Roots

Maize seeds were germinated and grown on Hoagland's medium
for three weeks at approximately 20°C with full aeration.
6.0 mm long root segments were excised and frozen without
any cryoprotective treatment in subcooled liquid nitrogen.
These 6.0 mm segments were fractured in the freeze-etching
unit so that the fracture plane was half way along the
segment (Fig. 1). Replicas were obtained from approximately
3.0 mm, 40 to 50 mm and >100 mm from the apex. Some seedlings were transferred to growth media at 8, 6, 4 or 0°C for
different periods of time (0.5 to 4.0h) before freezing.

Rat Hearts

Isolated rat hearts were perfused with a balanced salt
solution for 5 min at 37°C (Hearse *et al.*, 1978). Calcium-
free perfusion was carried out at 37, 32, or 29°C. Calcium-
containing perfusate was reintroduced after 5 min and
continued for 20 min at 37°C after which these hearts were
perfuse-fixed.
    After fixation, the hearts were perfused with 10 ml of
25% glycerol as a cryoprotectant and left immersed for
several hours prior to excision of samples taken from the
left ventricle, which were frozen in the same way as for the
maize roots (Fig. 1). Samples were also taken from the
same hearts for light and transmission electron microscopy.
At least three hearts were sampled for each treatment.

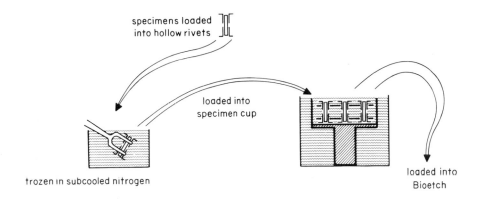

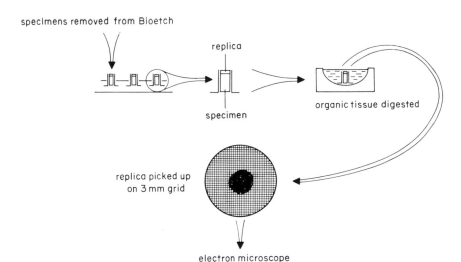

Figure 1. *Diagrammatic representation of the freeze—fracture process using a Leybold-Heraeus Bioetch 2005 automatic freeze-etching unit.*

Freeze-Fracturing

Specimens were transferred into a Leybold-Heraeus Bioetch 2005 automatic freeze-etching unit. Freeze-fracturing and replica formation were carried out under standard conditions (Sleytr and Umrath, 1976). An advantage of the Bioetch system is that three pairs of complementary replicas (six individual replicas in all) are produced during each run.

Furthermore, once the specimen cup has been loaded, the rest
of the run proceeds entirely automatically according to pre-
set operating conditions.

It was found useful to record images so that the final
magnification of some micrographs was 20,000X. This allowed
different cells to be identified and the relationships
between various parts of the replica to be interpreted.
Further micrographs were recorded at higher magnifications
so that IMPs could be studied and analysed at a final
magnification of 100,000X. In some cases, and especially so
when very close-packed particles were present, or when the
size of particles was being investigated, even higher final
magifications were necessary.

Analysis of Micrographs

A sample was defined as an arbitrary area of an electron
micrograph containing a suitably flat membrane surface; any
micrograph can thus provide several homogeneous samples.
Each sample is recorded during an interactive data-logging
session with the microcomputer and digitiser. Each sample
is identified to the computer using a sample code which
incorporates a subcode for the micrograph from which it
comes. This allows the grouping of homogenous sample results.

The operator digitises each observed particle (in this
study, any particle >8.0 nm) in the sample region, and, the
total area of the sample region is recorded by scribing
around its periphery. Care needs to be taken that the area
analysed is suitably planar. However, bias can also occur
by using too large an area so that a false level of
clustering is recorded. In order to overcome these problems,
templates of fixed area have been used to obtain regularised
areas on the micrographs.

Each sample is analysed completely independently of all
other areas. The microcomputer imposes over the whole
sample region a grid, scaled such that the average number of
observations per quadrat is two. The frequency per quadrat
is determined and the sample mean and variance computed.

The nearest neighbour index, D, is then calculated
using the first point nearest to each observation only.
This provides a coefficient of dispersion (C_d) which can
be used as a measure of randomness, clustering or regularity
in the observations. Randomly dispersed particles would
yield a coefficient of 1.0, completely clumped particles
would give 0, and regularly arranged particles would give a
value of 2.14. A second measure of randomness is in the
mean to variance ratio. If this is close to unity there is

a high probability that the particles have a random distribution. If the variance is smaller than the mean this implies a regular (binomial) distribution, if higher, clustering (negative binomial) is implied.

A Poisson distribution is fitted and also a binomial or negative binomial depending on the mean to variance ratio. Finally, the mean number of particles per quadrat and the mean number of particles per square micrometre are calculated. These analyses were all derived from Rogers (1974).

The system described above allows the accumulation of micrographs from large areas (up to about 2.0 mm diameter) of either maize roots or rat hearts. This is important, because it allows representative areas to be selected for quantitative analysis and, most importantly, it is possible to make an accurate identification of the particular cell types. The computer assisted system has provided a means of quantifying some of the parameters associated with particle distribution on membranes and, consequently, points to real differences or similarities between membranes that might be misinterpreted from casual observation of a few micrographs. These points are highlighted in the examples below.

MAIZE ROOTS

Large areas of replica from young maize roots have been obtained consistently (Plate 1). This allows positive identification of cells from which membranes were examined: a feature not always noted in other published freeze-fracture studies of roots. Superficial observation of different plasmalemma membrane faces showed considerable variation in both particle frequency and distribution (Plate 2). In common with most other membranes studied by freeze-fracture, P-faces had more particles than their corresponding E-faces. The situation is made more complex in plants by the presence of plasmodesmata through the cell walls. Although no quantitative data were obtained, there was some visual evidence that cooling resulted in increased particle aggregation around the mouths of plasmodesmata as seen on plasmalemma E-faces.

The main body of data arising from the maize root study is presented in Table 1. The majority of fractures were approximately 3 mm from the root tip. At this position, the various tissues had been formed from the meristem and cell expansion was almost complete. However, the cells were at an early stage of differentiation so that, for example, xylem vessel elements still retained living contents and cortical parenchyma cells had not completed vacuolation. It was

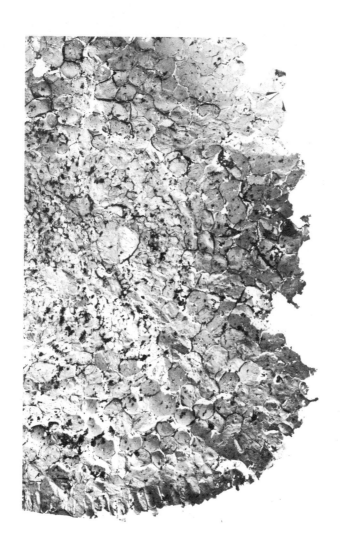

200 μm

Plate 1. *Low magnification view of replica from transverse fracture across a young maize root.*

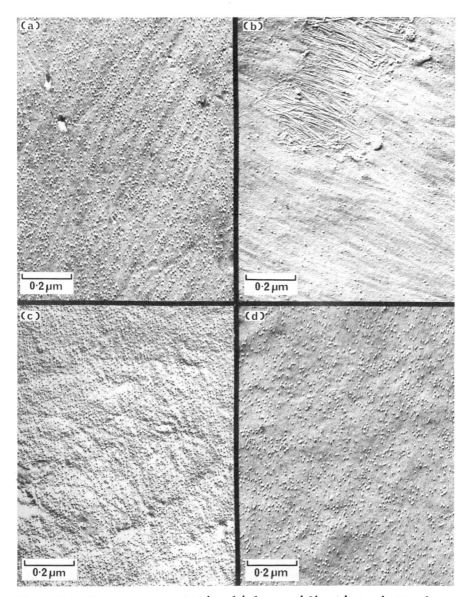

Plate 2. *Some representative high magnification views of different cell membranes from maize root cells. (A) Cortical cell P-face, precooled for 1h at 4°C. (B) Cortical cell E-face, precooled for 1h at 0°C. (C) Xylem parenchyma P-face, ambient temperature (approx. 20°C). (D) Endodermal E-face following polyvinylpyrrolidone treatment.*

Table 1

Intramembrane particle frequencies and coefficients of dispersion (C_d) in the plasmalemma of different cell types from maize roots

Cell Type	Source/Treatment	n	C_d	Frequency			
				Normal	Mature	Cryoprot.	Cooled
Epidermis-P	Ambient 3mm	10	1.08±0.02	647±56			
Epidermis-E	0°C,1h 3mm	15	1.12±0.05				106±26
Cortex-P	Ambient 3mm	36	1.18±0.01	1766±68			
	Ambient 3mm	–	1.18±0.01	1740±67			
	Ambient 3mm	–	–	1140±65			
	Ambient 4-5 cm	5/10	0.97±0.01		1241±202		
	Ambient Mature	3	1.25±0.04		1254±134		
	Ambient 3mm PVP	–	–			1690±142	
	Ambient 3mm Glycerol	–	–			1105±30	
	0°C,1h 3mm	42	1.21±0.01				1741±56
	0°C,1h 3mm	–	1.22±0.02				1738±57
	4°C,1h 3mm	21/22	1.29±0.04				2277±141
	4°C,1h 3mm	–	1.29±0.03				2275±130
	6°C,1h 3mm	15	1.19±0.01				1920±76

Table 1 (Continued)

Cell Type	Source/Treatment	n	c_d	Frequency			
				Normal	Mature	Cryoprot.	Cooled
Cortex-E	Ambient 3mm	–	–	350±20			
	Ambient 3mm	4	1.19±0.02	465±24			
	Ambient 3mm	9	1.11±0.03	447±42			
	Ambient 3mm	10	1.15±0.04	398±60			
	Ambient 3mm	23	1.14±0.02	429±31			
	Ambient Mature	5	1.06±0.05		170±23		
	Ambient 4–5cm	8	1.01±0.07		297±56		
	Ambient 3mm PVP	4	1.11±0.07			269±30	
	Ambient 3mm PVP	–	–			655±85	
	Ambient 3mm Glycerol	–	–			415±10	
	0°C,1h 3mm	71	1.06±0.02				414±32
	4°C,1h 3mm	65	1.05±0.01				845±114
	6°C,1h 3mm	19	1.08±0.02				561±34
Endodermis-P	Ambient 3mm	7	1.09±0.03	1577±80			
	Ambient 3mm	–	–	2500±110			
	Ambient 3mm PVP	16	1.11±0.03			1180±110	
	Ambient 3mm PVP	–	–			2985±145	
	Ambient 3mm Glycerol	–	–			2805±200	
	0°C,1h 3mm	15	1.14±0.02				1538±62

Table 1 (Continued)

Cell Type	Source/Treatment	n	C_d	Frequency			
				Normal	Mature	Cryoprot.	Cooled
Endodermis-E	Ambient 3mm	7/12	1.07±0.02	653±109			
	Ambient 3mm	–	–	1295±25			
	Ambient 3mm PVP	9	1.09±0.02			952±45	
	Ambient 3mm PVP	–	–			1450±30	
	Ambient 3mm Glycerol	–	–			1240±95	
	0°C,1h 3mm	5	1.01±0.09				252±43
	4°C,1h 3mm	5	1.12±0.02				1436±43
Pericycle-E	0°C,1h 3mm	14	1.02±0.04				583±43
Xylem Parenchyma-P	Ambient 3mm	8	1.16±0.02	2107±170			
Xylem Parenchyma-E	Ambient 3mm	15	1.04±0.03	356±26			
Metaxylem-E	Ambient 4–5 cm	6/11	1.04±0.03		538±42		

Data presented as mean ± standard error. n = number of samples evaluated (n/n indicates different numbers for C_d and frequency respectively). Frequency = number of particles > 8nm diameter μm^{-2}. Normal = 3mm fracture from ambient grown roots. Mature = replica from root > 10mm from apex. Cryoprotected = treated with glycerol or polyvinylpyrrolidone Cooled = roots cooled for 1 to 1.5 h at either 8, 6, 4 or 0°C.

relatively easy to prepare large, intact replicas from this
zone of the root. No attempt has been made to combine data
to give overall means, it has been thought preferable to
present the full data so that the variability can be clearly
seen. Data for epidermal cells is limited but suggests that
both P and E-faces may have low frequencies of IMPs.
'Normal' cortical cells gave P-face frequencies in the range
1100 to 1800 μm^{-2} while E-faces had much lower frequencies,
in the range 300 to 500 μm^{-2}. There was some indication
that particle frequencies fell during maturation of cells and
that chilling produced slightly higher frequencies. Cryoprotective treatment did not produce any clear differences
although polyvinylpyrrolidone (PVP) was sometimes correlated
with higher than normal frequencies. A feature from all of
these experiments was that the coefficient of dispersion
showed considerable uniformity: it was usually slightly
larger than 1.0, indicating that the particles were randomly
dispersed with a slight tendency towards regularity in
arrangement rather than towards clumping.

Endodermal cell membrane IMP frequencies overlapped with
the range encountered in the cortex but consistently yielded
some very high relative counts from both P- and E-faces so
supporting the earlier findings of Robards et $al.$, (1980).
Thus no normal cortical P-face mean gave frequencies greater
than 1800 μm^{-2} whereas the endodermal P-faces ranged up to
2500; similarly cortical E-faces went no higher than about
500 μm^{-2} while endodermal E-faces reached 1300. The partition coefficient between E and P-faces thus appeared to be
significantly larger in cortical cells (3 to 4) than in
endodermal cells (2 to 3). No significant effects of cooling
were noted on IMP distribution or frequency in endodermal
cells.

Some data were collected from pericycle, xylem parenchyma
and metaxylem cells but these are, as yet, inadequate for
detailed analysis. There is however some evidence that the
xylem parenchyma P-face are highly populated with particles
and has a very high (>5) partition coefficient.

Cooling maize roots for 1.0 to 1.5 h at temperatures down
to 0°C therefore has no consistent effect on the appearance
of the P-faces of cortical or endodermal cells although some
clumping of particles was noted on E-faces and especially so
around the mouths of plasmodesmata. Such structural changes
require different forms of evaluation from those described
here and will be reported separately. Most significantly,
no clear change in membrane structure could be perceived
when membranes were cooled below 13°C, the temperature below
which the Q_{10} for various root processes increases rapidly.

The hypothesis that a marked change in the Q_{10} of different
processes may be correlated with membrane lipid phase
transitions therefore does not receive support from this work
assuming that such transitions would have been manifested in
freeze-fracture replicas of cell membranes (as they have been
previously *e.g.* Martin *et al*., 1976; Fujikawa, this volume).
It might also be argued that such transitions would not take
place over such relatively short periods of time. This seems
improbable but longer term experiments are now in progress.

RAT HEARTS

Observation of heart samples by light microscopy indicated
that when compared with control tissue (Plate 3a), calcium-
free perfusion at $29^{\circ}C$ followed by reperfusion at $37^{\circ}C$
produced overall cellular shrinkage and some separation of
the intercalated discs, but no evidence of intracellular
damage (Plate 3b). In contrast, calcium-free perfusion at
$37^{\circ}C$ followed by reperfusion at the same temperature showed
the characteristic picture of the calcium paradox with all
cells affected (Plate 3c). Observations by conventional
transmission electron microscopy, demonstrated that at $29^{\circ}C$
calcium-free perfusion induced some minor alterations in the
cells but at $37^{\circ}C$ contracture of the cells and mitochondrial
extrusion were marked. The appearance of the same tissue
after freeze-fracture techniques can be seen in Plate 4. In
Plate 4a the appearance of a fracture running through the
sarcolemma of a single muscle cell can be seen, with the
striations equating to the Z-bands being represented by
ridges in the material. The fracture has exposed a very
considerable area of the P-face of the membrane, providing
a large area of replica sufficiently flat for analysis.

At higher magnification (Plate 4b)), the P-face of a
control heart sarcolemma shows large numbers of IMPs among
which are dispersed the caveolae which appear to be relatively
few in number and generally fracture across the neck. Plate
4c shows a P-face taken from the $29^{\circ}C$ hearts, the essential
difference being an increase in the particle size for,
although particles appear to be greater in frequency, analysis
has shown this not to be the case (Table 2). Plate 4d again
shows a comparable face from the $37^{\circ}C$ calcium-free perfusion,
the essential difference observed here being the increased
number of particles which are comparable with control
material in appearance.

Analysis of the frequency and coefficient of dispersion of
the particles from the three experimental conditions (Table 2)
has demonstrated that, while the coefficient of dispersion

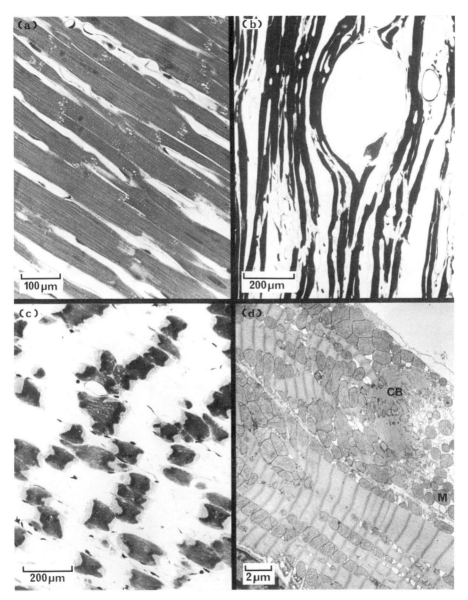

Plate 3. *Light micrographs of toluidine blue stained longitudinal sections. (A) control; (B) 29°C calcium-free perfusion; and (C) 37°C calcium-free perfusion hearts. (D) an electron micrograph of 37°C calcium-free perfusion illustrating typical contracture bands (CB) and extruded mictochondria (M). All calcium-free periods were followed by calcium reperfusion.*

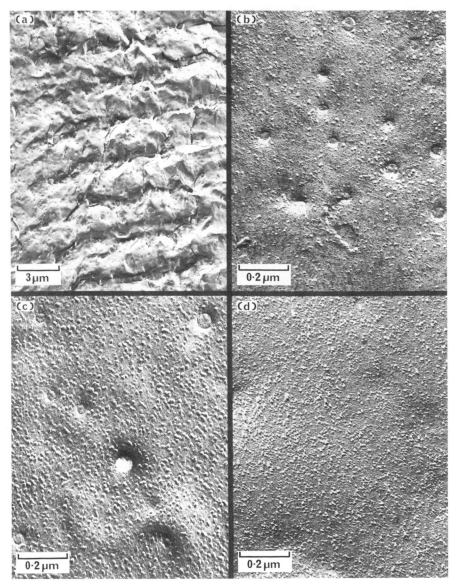

Plate 4.   *Freeze-fractured membranes from heart muscle cells.*
(A) Low magnification micrograph showing sarcolemma of a
single cell.   (B) High magnification micrograph of sarco-
lemma P-face from control heart to show IMP particles.   (C)
and (D) Sarcolemma P-faces from 29°C (C) and 37°C (D) calcium-
free perfused hearts.   Calcium-free period followed by
calcium reperfusion.

Table 2

Intramembrane particle frequencies and coefficients
of dispersion (C_d) on P-faces of rat heart sarcolemma

Treatment	n	C_d	Frequency
Control hearts	19	1.22±0.02	1492±28
Calcium deprivation at 29°C	20	1.19±0.02	1368±60
Calcium deprivation at 37°C	13	1.20±0.02	1732±65

for all three is the same, indicating a random distribution,
prevention of the calcium paradox by lowering the temperature
had no affect on the number of particles even though they
appeared larger. In contrast, sarcolemma analysed following
the full paradox shows an increase in the frequency of
particles.

Analysis of the plasmalemma from the vascular endothelium
proved considerably more difficult due to the high frequency
of pinocytotic vesicles present (Plate 5A-D). Sampling
therefore had to be restricted to vesicle-free areas and this
may have biassed the results. However, the particle frequency
again showed an increase for the P-face of vascular endothelium taken from tissue showing the full paradox compared
with the control (Table 3). In contrast, the membranes from
tissue perfused at 29°C showed a considerable fall in IMP
frequency and the particle size was similar in all cases.

Thus it appears that, when the temperature of calcium-
free perfusion is kept at 29°C, the size of the particle is
larger than in either the control or treated hearts from the
higher temperature (37°C). Analysis of the frequency of the
particles, however, shows there is no increase in number
compared with the control hearts whereas at the higher
temperature there is a significant increase in number. The
particles appeared not to differ greatly in patterns of
dispersion.

In view of the separation of the sarcolemma from the
basement membrane seen at higher temperatures (Hearse *et al.*,
1978), it may be possible that an alteration in the protein
content of the sarcolemma is associated with the uncontrolled
calcium uptake which occurs when calcium is restored.

While the increased number of particles seen in the
plasmalemma of the vascular endothelium from the 37°C

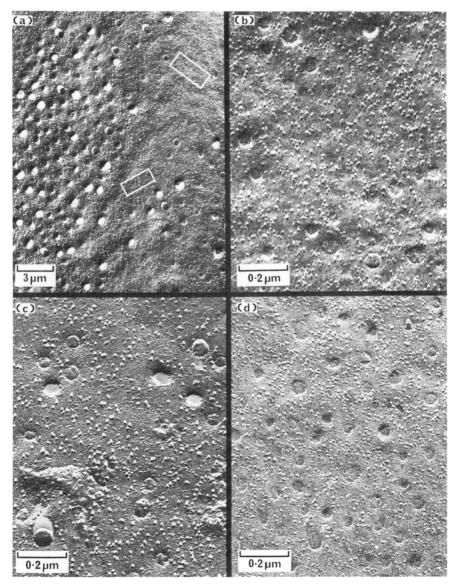

Plate 5. *Freeze-fractured membranes of vascular endothelial
cells. (A) Control heart membranes showing the irregular
distribution of caveolae and particles, the rectangles
enclose selected sampling areas. (B) High magnification
control heart. (C) High magnification 29°C heart and (D)
high magnfication 37°C heart.*

Table 3

Intramembrane particle frequencies and coefficients of dispersion (C_d) on P-faces of the plasmalemma of rat cardiac vascular endothelial cells

Treatment	n	C_d	Frequency
Control hearts	16	1.16±0.01	1237±56
Heart perfused with calcium-free medium at 29°C, 10 min then with calcium medium for 20 min at 37°C	12	1.14±0.02	884±29
Heart perfused with calcium-free medium at 37°C, 10 min then for 20 min at 37°C (*i.e.* full paradox)	24	1.20±0.01	1476±62

hearts is similar to that seen in the sarcolemma, the fall in number of particles in the 29°C hearts as compared with the controls is very interesting. With the extensive washing out of calcium during the calcium-free period, restoration of the calcium levels would be expected to take place from the vasculature. If the reduced number of particles could be equated with a lower rate of calcium transport, this would explain the protective effect of lowered temperature against the excessive uptake of calcium into the muscle cells. It has, therefore, become essential to study the same membranes during the calcium-free period only.

SUMMARY

Freeze-fracturing, in combination with suitable quantitative methods of analysis, can now be used to evaluate different experimental effects on membrane structure. In the case of plant roots, the major problems to be overcome in future experiments remain correct sampling procedures (both sampling from roots and sampling areas from micrographs), retaining large enough areas of replica to allow positive identification of different cell types (especially difficult with more mature parts of the roots), and ensuring either that essentially flat areas of membrane are studied or that some geometrical correction procedure is used to compensate for

curved or sloped surfaces. In both cases the experimental
work is at an early stage and further work is necessary before
the full benefit will be apparent. However, it should be
clear that such a system of working is indispensable if statistically significant results are to be obtained. Such results
should contribute to a better understanding of membrane
permeability and will doubtless emphasise the critical need to
study particular membranes from specific cells and tissues.

ACKNOWLEDGEMENTS

Many of our colleagues deserve our grateful thanks for help
during this project. In particular, we are indebted to
Hilary Quine and Meg. Schutz (York) for electron microscopy,
particle analysis and photography; to Terry Newman (York)
for much of the initial work in freeze-fracturing maize roots
and to John Baker (St Thomas' Hospital, London) for supplying
experimental heart material. We are also grateful to Dr David
Clarkson of the ARC Letcombe Laboratory for numerous helpful
discussions during the course of this work and to the ARC for
the provision of a grant under which the maize root work was
carried out.

FREEZING INJURY

LIPOSOMES AS A MODEL SYSTEM FOR INVESTIGATING FREEZING INJURY

G.J. Morris

Institute of Terrestrial Ecology, Culture Centre of Algae and Protozoa, Cambridge, England.

INTRODUCTION

Many investigations into the nature of cellular freezing injury have been reported, and it is now generally accepted that the primary sites of this damage are at the membrane level although the specific mechanism of this injury is not well understood. Until recently the strategies used to analyse freezing injury could be divided into two categories.

Firstly, cellular ultrastructure or general biochemical composition have been compared before and after freezing. This approach, however, does not allow a distinction to be made between the primary mechanisms of injury and secondary pathological events caused by the release of lytic enzymes. Secondly, many attempts have been made to correlate the response of a cell to freezing with the composition of its membrane lipids, particularly the phospholipid and fatty acid compositions. A major difficulty with this approach is that the molecular organisation of membranes is not homogeneous, and differences in composition exist both between the various membrane systems in a cell and also within individual membranes (Clarke; Pringle and Chapman, this volume). The cellular response to freezing could therefore be determined by a critical environment within a membrane and bulk lipid extraction and analysis would not be expected to provide insights into the local composition of these regions (Clarke, this volume). As these methods of analysis appear redundant, novel approaches examining the mechanical (Steponkus *et al.*, this volume) and thermodynamic properties of membranes (McGrath, this volume) have been instigated.

Liposomes (lipid bilayer vesicles) have been used extensively to model the structure and function of biological membranes (Bangham *et al.*, 1967, 1974; Tyrell *et al.*, 1976) and this system has many potential advantages for investigating the biochemistry of freezing injury. The composition of liposomes can be altered easily and reproducibly and the effects of these modifications on the biophysics of the bilayer are well understood. During formation liposomes entrap solutions and the integrity of the bilayer following freezing and thawing can then be assessed by measuring the leakage of entrapped solutes (Kinsky *et al.*, 1968). The characteristics of these solutes may be chosen so that specific features of the freeze-thaw lesion may be investigated. This principle of intracellular solute leakage is often used to determine the viability of cells upon thawing. Finally, multilamellar liposomes, within limits, behave as perfect osmometers (Bangham *et al.*, 1967). This enables the effects of shrinkage and rehydration, stresses important at low rates of cooling, to be investigated at a constant temperature. In constrast, unilamellar liposomes are non-ideal in their response to osmotic stress (Johnson and Buttress, 1977).

There are three assumptions implicit in using multilamellar liposomes to examine cellular freezing injury:-

1) Damage to the cellular membranes is the primary cause of injury rather than a secondary, pathological event.
2) Perturbations to the lipid moiety of the membrane are the critical events of freezing injury.
3) Concentric bilayers will respond to the stresses of freezing and thawing in a manner similar to a biological membrane.

This article discusses whether or not liposomes are a valid model system for investigating freezing injury in cells. Experiments to determine the nature of the freeze-induced lesion are then reviewed, and a hypothesis of injury at low rates of cooling examined.

Assessment of Liposomal Damage

Before any discussion of the effects of freezing and thawing on liposomes, the criteria used to determine whether the bilayer has been damaged must first be considered.

Above the phase transition temperature, phospholipids in aqueous solutions form structures that consist essentially of lipid bilayers completely surrounding aqueous compartments. Usually these structures are made up of many bilayers, but their precise shape, size and number depend on the method of preparation. A pure phosphatidylcholine liposome of many

lamellae and no significant central compartment contains
little osmotically active solution because the separation
between adjacent bilayers is so small. If, however, a highly
charged lipid such as phosphatidic acid, dicetylphosphate or
octadecylamine is incorporated into the liposome an electrostatic repulsion between the bilayers arises, whose magnitude
is inversely proportional to the square root of the electrolyte concentration (Bangham et al., 1967). Liposomes with
charged lipids prepared in low electrolyte concentrations
therefore contain large volumes of osmotically active solution.

Lipid bilayers in the liquid-crystalline state are only
slightly permeable to electrolytes, sugars and other highly
polar molecules (Pringle and Chapman, this volume). Thus if
liposomes are prepared in the presence of these compounds they
are entrapped in the aqueous compartments. Extraliposomal
molecules can then be removed, in the case of low molecular
weight markers by dialysis. The release of these compounds
following various treatments has been used to examine the
integrity of liposomal membranes. Glucose is a widely used
marker, which may be assayed either enzymatically (Kinsky et
al., 1968) or as ^{14}C-glucose. The former method is preferable
as it measures only free glucose and can be used to determine
glucose flux in the presence of intact liposomes. Release of
^{14}C-glucose is a more sensitive index, but does not distinguish between glucose in solution and that entrapped in non-
sedimentable microvesicles.

Energy transfer between molecules within a bilayer has
also been used to determine integrity of the bilayer. The
energy transfer quantum yield ($\emptyset_{DA}$) decreases with the sixth
power of the intermolecular distance (Strauss and Ingenito,
1980). Any membrane fragmentation or a selective loss of
probe molecules from the bilayer would be observed as a
reduction in $\emptyset_{DA}$.

Absorbance studies on liposomal suspensions are useful in
determining any average size changes within a population. A
decrease in optical absorbance at 450 nm is evidence of either
aggregation into larger units or of swelling, whilst an increase signifies a reduction in average size as a result of
fragmentation or osmotic shrinkage.

THE RESPONSE OF LIPOSOMES TO FREEZING AND THAWING

Many variables determine the survival of biological systems
following freezing and thawing. These variables include the
final temperature attained, rates of cooling and warming and
the type and concentration of cryoprotective additive. With
many cell-types an optimum rate of cooling has been observed.

It has been suggested that two distinct mechanisms of injury
occur, and this has been formulated into the 'two-factor'
hypothesis of freezing injury (Mazur, 1970; Mazur *et al.*,
1972). Specifically at low rates of cooling all ice formation
is extracellular and injury is the direct result of osmotic
stress. At high rates of cooling cells are damaged by the
nucleation of intracellular ice and its subsequent recrystal-
lization during warming; an optimal recovery is then observed
at an intermediate rate of cooling. It is generally assumed
that the optimum rate of cooling for any cell-type is determined by its permeability to water and its surface area to
volume ratio. A comparison of the optimum rate of cooling
with the known water transport properties of several cell-
types gives qualitative support to this relationship (Leibo,
1977). The incorporation of cryoprotective additives modifies
the response to the rate of cooling, and in general as the
initial concentration of a permeating compound is increased
the optimal cooling rate moves to lower values (Mazur *et al.*,
1972).

Leakage of Intraliposomal Markers

At a low rate of cooling the loss of glucose from liposomes
of dipalmitoyl phosphatidylcholine was less than 20 percent
upon thawing from $-25^{\circ}C$ (Morris and McGrath, 1981a). Following the incorporation of cholesterol and dicetylphosphate
liposomes become more sensitive to freezing injury. This
increased susceptability was not caused by the anionic nature
of the liposomes for similar results were obtained with
cationic liposomes (octadecylamine supplying the positive
charge) as well as liposomes with no added highly charged
lipid. The response of liposomes to low rates of cooling is
correlated directly with their cholesterol content. This
relationship is pronounced at high sub-zero temperatures but
becomes less obvious at lower temperatures (Fig. 1), and upon
thawing from $-196^{\circ}C$ complete loss of glucose occurs irrespective of cholesterol content.
   The addition of cholesterol to dipalmitoyl phosphatidylcholine reduces the temperature of the phase transition. With
no cholesterol present the phase transition is at $41.5^{\circ}C$ wherea
at concentrations greater than 20 mole percent there is no
phase change detectable by differential scanning calorimetry
(Pringle and Chapman, this volume).
   The phase transition temperature of phosphatidylcholine
liposomes can be reduced at a constant cholesterol level by
the incorporation of successive double bonds into the fatty
acid chains. This also reduces the glucose yield upon thawing

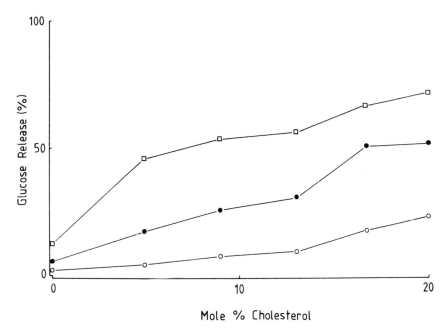

Figure 1. *Glucose release (% total intraliposomal glucose) from dipalmitoyl phosphatidylcholine:cholesterol:dicetylphosphate liposomes containing varying molar proportions of cholesterol. Glucose release was monitored following freezing to and thawing from −5 (O), −15 (●) and −25°C (□).*

(Morris and McGrath, 1981a). The increased susceptibility of liposomes to freezing damage associated with the incorporation of cholesterol therefore cannot be due simply to the reduction in the phase transition temperature, but must be due to some other factor.

Cryomicroscopy

Direct observation of liposomes during freezing and thawing is possible with a specialised microscope system (Diller and Cravalho, 1970; McGrath *et al.*, 1975). At low rates of cooling liposomes dehydrate and distort extensively, demonstrating that at high sub-zero temperatures liposomes are osmotically active (Plate 1). Similar observations have been made by Siminovitch and Chapman (1971). A crinkling of some liposome surfaces was observed during cooling, and upon thawing and rehydration dark bodies were associated with those membranes which previously exhibited crinkling. No such structure was observed with those liposomes in which the

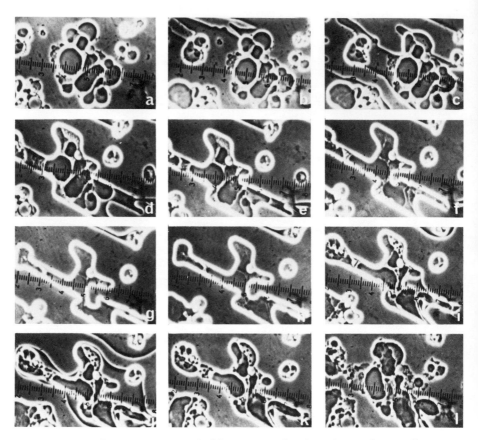

Plate 1. *The response of liposomes during freezing and
thawing (magnification x 562.5). (a) Control, (b)-(g) during
cooling at 0.25°C min⁻¹ (b) -1, (c) -5, (d) -7.5, (e) -10,
(f) -15, (g) -25°C. (h)-(i) during warming at 2.5°C min⁻
(h) -15, (i) -7.5, (j) -5, (k) -1, (l) +5°C.*

surfaces remained smooth during cooling. Under these conditions approximately 75 percent of entrapped glucose was released (Fig. 1). Whilst there are obvious morphological
changes, visible in the cryomicroscope, particularly fragmentation and a general swelling, a suspension of liposomes
nonetheless remains upon thawing. Freezing and thawing thus
causes the release of glucose, but does not destroy the liposomal structure extensively enough to form a colloidal suspension of phospholipid.

    As the rate of cooling is increased the time available for
osmotic shrinkage is reduced and at rapid rates of cooling

(10 to $400^{\circ}C$ min^{-1}) liposomes reached low temperatures without significant shrinkage, but intraliposomal ice was not apparent by 'black flashing'. These results differ from those of Siminovitch and Chapman (1971) who clearly demonstrated such flashing at a cooling rate of $10^{\circ}C$ min^{-1}. Until recently, black flashing has been considered to be diagnostic of intracellular ice nucleation and crystal growth as it was assumed that optical darkening was caused by the diffraction of light as the result of the formation of many small ice crystals (McGrath et al., 1975). More recently it has been suggested that this darkening is an optical interference effect at the surface of bubbles of gas forced out of solution by the formation of intracellular ice (Morris and McGrath, 1981b; Steponkus and Dowgert, 1981). Whilst optical flashing is indicative of intracellular ice its absence cannot be used as evidence that intracellular ice has not formed. Some cell-types cooled at rates at which the probability of intracellular ice formation is very high do not exhibit black flashing (Steponkus and Dowgert, 1981). Optical diffraction studies do reveal intraliposomal ice in rapidly cooled liposomes but the crystal size is very small (Costello and Gulik-Kryzwicki, 1976). Thus at rapid rates of cooling intraliposomal ice formation can be assumed to occur, but is not visible by cryomicroscopy. The observations of flashing in rapidly cooled liposomes (Siminovitch and Chapman, 1971) may be due to the use of either large unilamellar liposomes or multi-lamellar liposomes with a large aqueous core.

Absorbance Changes

Immediately upon thawing from $25^{\circ}C$ the optical absorbance at 450 nm of liposome suspensions is greater than that of unfrozen controls, indicating liposome shrinkage. Warming at the rate used in this study ($90^{\circ}C$ min^{-1}) does not allow osmotic equilibrium to be established during thawing. With increasing time of incubation at $+25^{\circ}C$ the absorbance falls below that of the control (Fig. 2). The extent of this long-term decrease in absorbance is dependent upon the final temperature attained, at lower final temperatures there is a greater degree of swelling (Fig. 3).

With unilamellar liposomes freezing and thawing induces an aggregation into larger units (Gerritsen et al., 1978; Hui et al., 1981; Yu and Branton, 1976) although this could be partially reversed by sonication (Strauss and Ingenito, 1980).

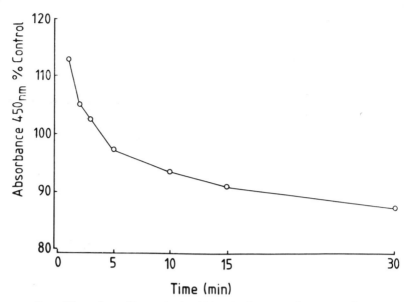

Figure 2. *The absorbance at 450 nm (control = 100%) of a suspension of liposomes frozen to and thawed from −25°C and maintained at +25°C for different periods of time.*

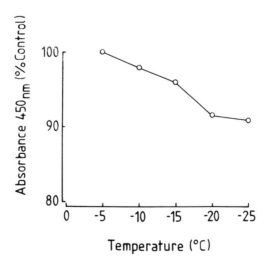

Figure 3. *The absorbance at 450 nm (control = 100%) of liposomes frozen to different final temperatures.*

Cryoprotective Additives

The effects of three commonly employed cryoprotective additives on glucose release from liposomes maintained at $20^{\circ}C$
have been examined (Morris and McGrath, 1981a)

Glycerol induced a loss of glucose at concentrations
normally used for cellular cryoprotection (>0.25M). At low
concentrations (50 mM) glycerol is freely permeable to liposomes (de Gier *et al.*, 1968) but at higher levels it has been
demonstrated to induce membrane fusion (Ahkong *et al.*, 1975;
Chandler and Heuser, 1979). The observed release of glucose
may be related to this fusion process. On a molar basis
dimethylsulphoxide (Me_2SO) was less damaging, and methanol
induced no significant loss of glucose at all. All these
compounds induced a decrease in the absorbance of liposomes
in the range 0.1 to 1.0 M, suggesting that they are freely
permeable. In studies of monolayers no evidence of glycerol
or Me_2SO distributing into the phospholipid was obtained
(Williams and Harris, 1977). This is in contrast to the
observation that Me_2SO is membrane active, increasing the
phase transition temperature of liposomes (Lyman *et al.*,
1976) and decreasing probe mobility (Goodwin and Farrant,
personal communication). Methanol has the opposite effect
increasing the mobility of membrane incorporated electron
spin probes (Grisham and Barnett, 1972; Hubbel *et al.*, 1970).
At concentrations greater than 1.5 M both glycerol and Me_2SO
destroy membrane integrity as demonstrated by a reduction
of energy transfer within the bilayer (Strauss and Ingenito,
1980).

Following freezing to and thawing from $-196^{\circ}C$ there was a
complete release of glucose from unprotected liposomes,
independent of the rate of cooling in the range 0.25 to
$1,000^{\circ}C\ min^{-1}$. The addition of glycerol (0.25 M) or methanol
(1.0 and 2.0 M) did not protect liposomes. Dimethylsulphoxide
was cryoprotective, the extent of this protection being
dependent on the rate of cooling and the concentration of
Me_2SO (Fig. 4). At a concentration of 0.75 M Me_2SO minimal
glucose release was observed at a rate of cooling of $1.7^{\circ}C$
min^{-1}; at both faster and slower rates of cooling there was
an increase in the amount of glucose released upon thawing.
With a lower concentration (0.50 M) the optimum rate was
$10^{\circ}C\ min^{-1}$. This response to a spectrum of cooling rates
and the shift of the optimum rate to lower values following
an increase in the level of cryoprotectant, is analogous to
that observed with a variety of cell-types (Mazur *et al.*,
1972).

At any given concentration of Me_2SO the phospholipid to

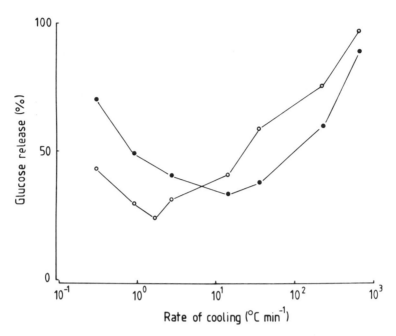

Figure 4. *Glucose release (%) from liposomes following cooling at different rates to $-196^{\circ}C$ in the presence of 0.75 M (O) or 0.50 M (●) dimethylsulphoxide.*

cholesterol ratio of liposomes affects the value of the optimal cooling rate. As the cholesterol content of the liposomes is decreased the optimal rate of cooling shifts to lower rates (Morris, in press a). The effect of cholesterol on the water permeability of phospholipid bilayers is complex. With dipalmitoyl phosphatidylcholine at temperatures below the phase transition, conditions relevant to this study, the addition of cholesterol decreases the activation energy of water permeability (Blok *et al.*, 1977). The incorporation of cholesterol has no significant effect on the size distribution of multilamellar liposomes. Thus in liposomes with a low cholesterol content and consequently a reduced water permeability, intraliposomal ice would be predicted to form at lower rates of cooling. The observed shift in the optimum cooling rate to slower values as the cholesterol level is decreased is thus consistent with the optimum rate of cooling being determined largely by the permeability to water. This relationship is only qualitative as it is difficult to model water transport in a system of concentric compartments (Fettiplace and Haydon, 1980).

Liposomes as Models for Freezing Injury

The response of liposomes to freezing and thawing simulates much of the behaviour of intact cells, as follows:-

1) At low rates of cooling dehydration occurs whilst at rapid rates intraliposomal ice is nucleated.
2) Following freezing and thawing a leakage of entrapped solutes occurs, and this leakage is proportional to the final temperature attained.
3) The leakage of entrapped marker is reduced by the addition of dimethylsulphoxide and the extent of this protection is dependent on the rate of cooling.
4) Alterations in the composition of the liposome which are known to reduce the water permeability shift the optimal rate of cooling to lower values, an observation which is consistent with the 'two-factor' hypothesis of freezing injury.

Liposomes thus offer a valid model system for investigating the mechanism of cellular freezing injury. However, the analogy between liposome bilayers and biological membranes must not be overstated. The lipid composition of biological membranes is heterogeneous, whilst liposomes are homogeneous. In addition, until recently it has not been possible to incorporate proteins into liposomal bilayers which then demonstrate enzyme or transport activity *in situ* (Hoffman *et al.*, 1980; Pringle and Chapman, this volume). The addition of a glycoprotein fraction to lipid bilayers reduces freeze-thaw damage (Strauss and Ingenito, 1980). Some consequences of these structural differences are highlighted with liposomes of erythrocyte lipid extracts, which are more resistant to freezing injury than are erythrocytes (Fig. 5).

NATURE OF THE FREEZE-THAW INDUCED LESION

For any complete understanding of the mechanism of freezing injury it is necessary to characterise precisely the membrane lesion. The nature of this lesion was examined using liposomes containing dipalmitoyl phosphatidylcholine, cholesterol and dicetyl phosphate at a mole ratio of 7:2:1 respectively. This phospholipid to cholesterol ratio approximates to that of many mammalian cell-types. In these experiments a low rate of cooling was utilised, and damage will thus be due primarily to 'solution effects' (Mazur, 1970; Mazur *et al.*, 1972).

The purity of the dipalmitoyl phosphatidylcholine was examined following freezing and thawing by thin layer chromatography on silica gel. The only spot visible under ultra-

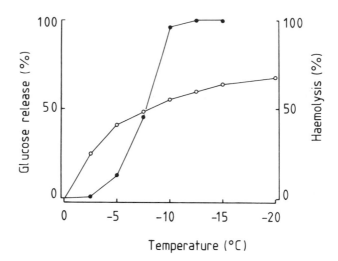

Figure 5. *Glucose release (%) from liposomes of total erythrocyte lipid extract (O) and haemolysis of erythrocytes (●) following cooling at a rate of 0.25°C min⁻¹ to different final temperatures.*

violet light or by acid spray was unaltered dipalmitoyl phosphatidylcholine; there was no evidence of phosphatidic acid, lysophosphatidylcholine or of any degraded material at the origin (Clarke, unpublished results). Freezing and thawing thus does not directly alter the phospholipid struct-ure, and damage must be associated with an alteration of the bilayer arrangement rather than direct chemical action.

Optical observation of a liposomal suspension which has released 75 percent of entrapped glucose during a freeze thaw cycle reveals no gross disruption, liposomal structure re-maining upon thawing (Plate 1). Four mechanisms by which glucose leakage may occur with the liposomes appearing morphologically intact upon thawing are:-

1) An alteration in the permeability coefficient to glucose of the bilayers, resulting in an increased rate of solute release compared with the controls.
2) A non-selective loss of solute with water during freeze-induced dehydration.
3) The formation of a 'lesion' in some bilayers, which remains open after thawing.
4) A transient lesion which forms during freezing and thawing but reseals immediately or soon after thawing.

Kinetics of Glucose Release upon Thawing

Following freezing to and thawing from -10°C the release of
glucose is complete within 3 min and there was no significant
difference between liposomes incubated at 4 or 25°C (Morris
and McGrath, 1981a). Under these conditions the long-term
permeability of the lipid bilayer is therefore not modified
by freezing and thawing. Leakage appears to be an all or
nothing event, involving either the total release of marker
from some of the liposomes with others remaining intact or
damage to the peripheral lamellae of all liposomes. Following
multiple freezing and thawing a similar increment of glucose
release occurred following each cycle (Table 1). This indicates that damage is a statistical event, a given proportion
releasing glucose at each freeze-thaw cycle, and not that
there exist sub-populations which differ in their sensitivity
to freezing injury.

In cells damaged by freezing there is a loss of membrane
selective permeability which may (Morris *et al*., this volume)
or may not (Palta *et al*., 1977) increase with subsequent
incubation. This suggests that the primary damage following
freezing and thawing to both cells and liposomes is of a
similar nature. In some cell-types this then triggers a
complex series of events which results in death.

Hypertonic Simulation of Freezing Injury

The possibility that glucose was released along with water
during freeze-induced dehydration was investigated by exposure
of liposomes to hypertonic solutions of NaCl at 20°C. There
was little loss of glucose into sodium chloride solutions of
less than 1.5 M, but at higher concentrations the loss of
glucose into the hypertonic solution increased. Following
resuspension into 0.1 M NaCl there was a further large loss
of glucose (Fig. 6). Similar results have been reported by
Siminovitch and Chapman (1974). Within the range 0.1 to
1.5 M NaCl the combined stresses associated with shrinkage
and rehydration are therefore necessary for the loss of membrane integrity.

Resealing upon Thawing

Following freezing and thawing liposomes are osmotically
sensitive, and their shrinkage in response to hypertonic
sodium chloride is identical to that of unfrozen liposomes
(Fig. 7). Thus the lesion which allows a leakage of glucose
during freezing and thawing reseals upon thawing. This can

Table 1

The loss of intraliposomal glucose following multiple cycles of freezing to and thawing from $-10°C$.

Cycle	Loss of glucose (%)	Increment (%)
1	37.2	37.2
2	55.2	28.6
3	72.7	39.1
4	82.6	36.2

be confirmed by a reverse loading process. The amount of glucose loaded into liposomes during freezing and thawing was proportional to the final temperature attained in the range -5 to $-15°C$, with no further increase occurring at lower temperatures (Fig. 8). There was little glucose loading into liposomes following incubation at $0°C$, but if glucose was added to a liposome suspension immediately upon thawing from $-25°C$ uptake did occur, approximately 20 percent of that

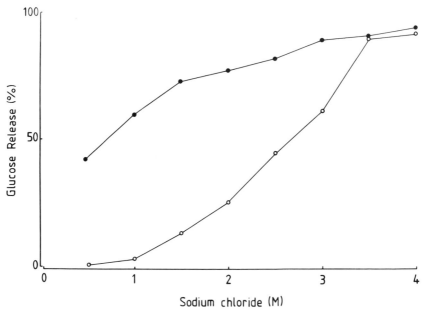

Figure 6. *Glucose release (%) from liposomes following exposure to hypertonic solutions of NaCl (○) and subsequent resuspension into isotonic NaCl (●).*

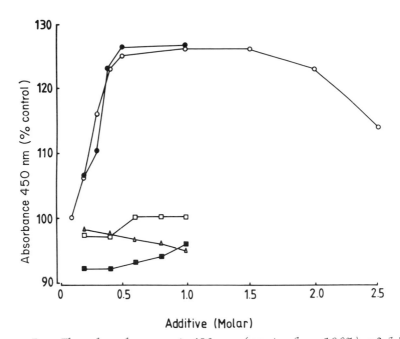

Figure 7. *The absorbance at 450 nm (control = 100%) of liposomes in hypertonic solutions of sodium chloride (O), dimethylsulphoxide (■), glycerol (□) or methanol (△). Liposomes
frozen to and thawed from −25°C and then exposed to hypertonic
sodium chloride (●).*

following freezing and thawing in the presence of glucose.

This phenomenon of reverse loading is important in the
critical evaluation of liposomal damage. The lesion in the
bilayer is not analogous to a one-way valve, for re-entrapment
of marker may occur during freezing and in the immediate post-
thaw incubation. The observed release of marker following
freezing and thawing must therefore always be an underestimate of total release.

. In summary, the damage to the liposomal bilayer appears to
be a discrete lesion which forms during freezing and thawing,
through which intraliposomal markers are lost by diffusion.
This lesion also allows an exchange of material with the
extraliposomal compartment before it reseals immediately or
soon after thawing.

AN HYPOTHESIS OF INJURY AT LOW RATES OF COOLING

An hypothesis of membrane damage at low rates of cooling is

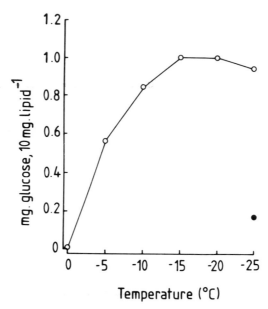

Figure 8. *Glucose entrapment during freezing to and thawing from different temperatures (O). Liposomes were prepared in 0.1 M NaCl and suspended in 0.1 glucose prior to freezing; upon thawing non-liposomal glucose was removed by dialysis. In one experiment following freezing to -25°C in 0.1 M NaCl, glucose was added immediately upon thawing (●).*

presented schematically in Figure 9. This is derived in part from Steponkus and Wiest (1979); Williams and Hope (1981); McGrath, Steponkus *et al.* (this volume). Exposure to hypertonic solutions induces osmotic shrinkage and if the cell remains spherical a decrease in the interfacial membrane tension occurs. At a maximum packing density within the membrane any further increase in applied pressure can be

Figure 9. (opposite) *Schematic representation of cellular damage at low rates of cooling. The following assumptions are made; a spherical cell of diameter 20 μm with one limiting membrane which is composed entirely of phospholipid. During shrinkage and expansion the cell remains spherical and there is no physical limitation to shrinkage or any alteration in the intracellular solute concentration. The number of molecules per unit area of membrane is arbitrarily taken to be 100 in the isotonic condition. The maximum packing density is 110 molecules per unit area.*

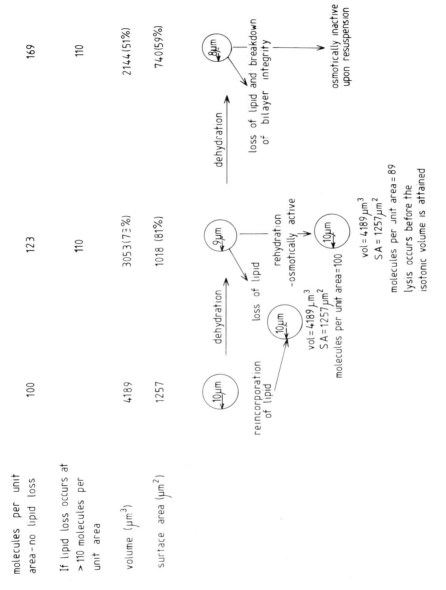

molecules per unit area – no lipid loss   100   123   169

If lipid loss occurs at >110 molecules per unit area   110   110

volume (μm^3)   4189   3053 (73%)   2144 (51%)

surface area (μm^2)   1257   1018 (81%)   740 (59%)

10 μm

reincorporation of lipid

dehydration

loss of lipid

10 μm
vol = 4189 μm^3
S A = 1257 μm^2
molecules per unit area = 100

9 μm

rehydration – osmotically active

10 μm
vol = 4189 μm^3
S A = 1257 μm^2
molecules per unit area = 100

dehydration

loss of lipid and breakdown of bilayer integrity

8 μm

osmotically inactive upon resuspension

vol = 4189 μm^3
S A = 1257 μm^2
molecules per unit area = 89

molecules per unit area = 89 lysis occurs before the isotonic volume is attained

accommodated either by an alteration in the bilayer structure
or the loss of membrane components. Release of membrane
material may occur into the intracellular compartment (Steponkus *et al.*, this volume) or into the surrounding aqueous
solution (Araki, 1979; Fujikawa, this volume). If the loss
of membrane material does not exceed a critical value the
cell is osmotically active during thawing. However, because
of the loss of membrane material it cannot return to its
isotonic volume which is determined by the intracellular
solute concentration and lysis may occur. Of obvious relevance is the potential to reincorporate material released from
the membrane during shrinkage (Williams and Hope, 1981;
Steponkus *et al.*, this volume). At lower temperatures a loss
of membrane selective permeability may occur as the result of
either the release of an excess amount of membrane or a second
independent mechanism of injury such as membrane fusion or a
mesomorphic phase change. Such cells will not be osmotically
responsive during thawing.

Evidence for this hypothesis and its relationship to other
proposed mechanisms of injury are summarised elsewhere
(McGrath, Steponkus *et al.*, this volume). In the next section
data from liposomes will be examined with respect to this
concept. Because the sequence of events outlined in Figure
9 will be complicated by interactions with the dynamic processes of freezing and thawing, this hypothesis is examined
initially in terms of shrinkage and rehydration at a constant
temperature (20°C) then the additional stresses associated
with freezing and thawing are discussed.

Intramembrane Pressure

Interfacial membrane tension has two components, a surface
tension and a surface pressure (McGrath, this volume). The
first term is the tension within the membrane caused by the
hydrophobic effect and is relatively insensitive to membrane
surface area changes. Simplistically, it is a measure of the
tendency for water to repel the hydrocarbon portion of the
phospholipid from the aqueous phase independent of whether
the molecules are loosely or tightly packed once they are in
the membrane. Surface pressure is associated with the lateral
momentum within the membrane and at a constant temperature is
a function of the area per molecule.

In liposomes containing the probe 1,6-diphenyl-1,3,5-
hexatriene (DPH) a direct relationship exists between the
fluorescence anisotropy and surface pressure of the bilayer
(Fulford and Peel, 1980). Osmotic shrinkage of liposomes
containing DPH induces a decrease in fluorescence depolariz-

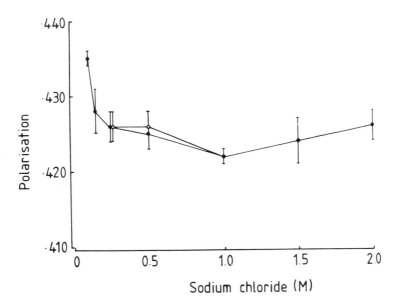

Figure 10. *Steady-state fluorescence depolarization of the probe 1,6-diphenyl-1,3,5-hexatriene in liposomes of dipalmitoyl phosphatidylcholine:cholesterol:dicetyl phosphate at a mole ratio of 7:2:1 following exposure to hypertonic NaCl at 20°C. The liposomes were prepared in 0.1 M NaCl. A range of liposome concentrations was examined at each sodium chloride level and the depolarization at zero absorbance was calculated using a maximum likelihood programme.*

ation (Fig. 10). This is consistent with either an isothermal decrease in fluidity or an increase in lateral pressure. Currently, these results can only be expressed in terms of steady-state fluorescence depolarization, since the DPH fluorescence half-life may also be affected by lateral pressure. However, this demonstrates qualitatively an increase in lateral pressure which attains a maximum following shrinkage in 1.0 M NaCl. This change occurs within 15 seconds of exposure and remains constant for 15 min. The phenomenon is reversible, as liposomes shrunken in 1.0 M NaCl and then diluted exhibit a decrease in lateral pressure. Similar results for hypotonic swelling of liposomes have been reported (Borochov and Borochov, 1979). At higher levels of NaCl where direct hypertonic damage occurs there is a relaxation in lateral pressure.

Response of the Membrane during Shrinkage

There are several potential reactions of the membrane to an
increase in lateral pressure (Morris, in press b). A homeostatic release of lipid may occur when a maximum value is
attained. This does not invoke a novel or pathological
mechanism, a dynamic equilibrium exists between a lipid bilayer and its environment. Liposomes prepared with radio-
active labelled phospholipid and cholesterol exhibit an
exchange of label into isotonic medium (Clarke and Morris,
unpublished results). The stresses associated with shrinkage may alter the position of the equilibrium (McGrath, this
volume). However, many technical problems exist in measuring
the loss of lipid in response to shrinkage, as any material
released would reseal to form small liposomes or micelles
which cannot be separated from pre-existing liposomes.

Shrinkage and Rehydration

Following exposure of liposomes to hypertonic solutions of
NaCl there is a correlation between the release of intraliposomal glucose (Fig. 6), optical absorbance (Fig. 7)
alterations in morphology and increase in surface pressure
(Fig. 10). The steady state absorbance of a liposome suspension increases (the liposomes shrink) following exposure
to a hypertonic solution of a non-penetrating compound. A
maximum in absorbance occurs in 0.5 M NaCl, increase to
1.5 M had no further effect. Within this range the bilayer
lateral pressure attained a maximum value. In the shrunken
state the liposomal bilayers are metastable. Prolonged
exposure to the hypertonic solution did not result in any
swelling and significant loss of glucose occurs only upon
resuspension into 0.1 M NaCl.

At higher concentrations, 1.5 to 2.0 M, loss of bilayer
selective permeability develops in the shrunken state.
Following an initial shrinkage a reswelling of the liposomes
occurs, this is accompanied by a relaxation in the surface
pressure and the release of glucose. Finally at 3.0 M NaCl
there is a disruption of large liposomes within 15 seconds
of exposure and a total release of intraliposomal glucose.

These results are compatible with the model presented in
Figure 9. However, exposure to and dilution from 1.0 M NaCl
at 20°C results in the release of 60 percent of intraliposomal
glucose; although this stress is equivalent osmotically to
that occuring during freezing to and thawing from -4°C, the
later treatment releases only 20 percent of the entrapped
glucose. A similar disparity exists between the response to

osmotic stress and freezing in plant protoplasts (Steponkus
et al., this volume). During freezing and thawing the imposed stress is not simply osmotic, other variables including
membrane phase separations, pH, gas and solute solubility and
viscosity are all changing simultaneously (Franks, this
volume). In addition, the rate of any chemical process will
decrease at lower temperatures.

Freezing and Thawing

Freeze-induced dehydration results in an ordering of the
phospholipid polar regions of liposomes; this was not a temperature effect as there was no equivalent change in undercooled samples (Singh and Miller, 1980). A similar decrease
in rotational freedom near to the hydrophilic regions of
phospholipids has been reported during the drying of liposomes
(Griffith *et al.*, 1974; Jost *et al.*, 1971). The decrease in
fluorescence anisotropy observed during shrinkage (Fig. 10)
may reflect this molecular ordering. Alternatively, at low
water contents (< 20 percent), a change in the conformation of
phospholipids from the lamellar to hexagonal phase may occur
(Simon, 1978) and there is evidence for such a lipid phase
change in model systems (Chapman, 1975) and during the drying
of sarcoplasmic reticulum vesicles (Crowe *et al.*, in press).
However, in biological systems which remain viable at low
water content the results are contradictory. In seed membranes
only lamellar bilayers could be detected by X-ray diffraction
(McKersie and Stinson, 1980), whilst freeze-fracture studies
reveal hexagonal phases (Toivo-Kinnucan and Stushnoff, 1981).
    Another effect of freeze-induced dehydration is to induce
membrane fusion. Repeated freezing of small unilamellar
vesicles induces an aggregation into large unilamellar structures (Gerritsen *et al.*, 1978; Hui *et al.*, 1981; Yu and Branton,
1976). In multilamellar vesicles the concentric bilayer
structure is disrupted by the fusion between bilayers within
the liposome and a parallel decrease in average lamellar
spacing indicating dehydration is observed. In addition a
significant proportion of the bilayer structure was disrupted
(Hui *et al.*, 1981).
    During slow cooling crinkling of some liposomes surfaces
can be seen (Plate 1). This may be equivalent to the collapse
observed in monolayers when a critical surface pressure has
been exceeded (Reis, 1979; Williams and Harris, 1977). During
thawing dense bodies can be seen to be associated with the
surfaces which previously exhibited crinkling. It is assumed
that membrane lesions are formed during this process of lipid
extrusion which then allows the loss of intraliposomal markers

to the external medium. These lesions reseal during warming
so that upon thawing the liposomes are osmotically active.
The lipid released from the multilamellar liposomes form small
vesicles and micelles upon warming and extraliposomal solutes
may be entrapped during this process. The above sequence
explains both the nature of the freeze-induced lesion and the
entrapment of extraliposomal material during thawing.

However, the molecular nature of the lesion, whether a
bilayer hexagonal phase change, membrane fusion or an extrusion of lipid and its relationship to membrane phase separation
is not yet apparent. In addition the specific cryoprection
afforded by Me_2SO but not other "cryoprotective additives"
and the sensitising effect of cholesterol are unresolved.
Further studies on liposomes will be expected to provide
insights into these and other problems of freezing injury.

FREEZING DAMAGE TO THYLAKOID MEMBRANES IN VITRO AND IN VIVO

U. Heber and J.M. Schmitt

*Institute of Botany and Pharmaceutical Biology,
University of Würzburg, West Germany.*

G.H. Krause, R.J. Klosson and K.A. Santarius

Institute of Botany, University of Düsseldorf, West Germany.

INTRODUCTION

Freezing of metabolically active and fully hydrated cells, tissues or organisms is often lethal. This chapter deals with an important cause of freezing damage and outlines some strategies of resistant cells.

Although lowering the temperature is a simple physical process, it produces several effects that are potentially damaging (Franks, this volume). Individual metabolic reactions are slowed down at different rates and this will change the overall pattern of metabolism. In biomembranes, phase changes of lipids or lipid-protein complexes may alter membrane activity and function (Pringle and Chapman, this volume). While such events may lead to damage in chilling-sensitive plants at temperatures above freezing (Levitt, 1980; Lyons, 1973; Lyons *et al.*, 1979; Wolfe, 1978; Wilson and McMurdo, this volume), undercooling is not detrimental to the viability of many cell-types if ice formation is avoided (Burke *et al.*, 1976; Levitt, 1980). Rather, reduced temperatures are known to stabilize biological systems and are often used for preservation.

At slow rates of cooling extracellular ice crystals form and upon further lowering of the temperature these crystals grow at the expense of intracellular water. This shrinks the cell osmotically and in addition, mechanical stress may be exerted by the accumulating ice. However, when damage occurs under these conditions, it is rarely attributed to

mechanical causes. Only when ice forms intracellularly at
rapid rates of cooling has mechanical membrane rupture been
postulated as a cause of cell death (Levitt, 1980;
Mazur, 1969, 1970, 1977; Meryman, 1966; Fujikawa, this volume).
Extracellular ice formation, on the other hand, may or may
not kill a cell depending on its physiological state. Plant
cells which are genetically incapable of hardening cannot
withstand the dehydration that accompanies ice formation,
unless they are supplied with compounds that can act as
cryoprotectants. Whether or not potentially hardy cells
survive freezing depends on the extent of dehydration and
on the degree of hardiness. Freezing resistance of plants
is usually acquired before the onset of winter and is lost
in the spring.

In terms of water potential, dehydration during freezing
is solely a function of temperature and is independent of the
solute. Since a sensitive cell can be protected against
freezing damage by an additive, dehydration in itself (that
is a reduction in water potential) cannot be the cause of
damage. In this chapter, we will present evidence from
studies on thylakoid membranes, which show that the extent
of cell damage during freezing depends on the composition of
the complex cellular solutions in contact with cellular
membranes, and which undergo concentration during freezing.

CELLULAR SITES OF FREEZING INJURY

In attempts to define which cellular constituents are sensitive to freezing or to those effects such as solute accumulation brought about by freezing, it becomes immediately
apparent that the activity of many enzymes is decreased
during dehydration or exposure to increased solute concentrations. Photosynthesis of isolated spinach chloroplasts,
which is optimal at water potentials close to -10 bar (-1MPa)
is severely curtailed at water potentials between -14 and
-40 bar (Kaiser *et al.*, 1981a,b). Also depending on the
solute environment, soluble chloroplast enzymes are significantly inhibited at water potentials between -20 and -70 bar
(Kaiser and Heber, 1981). For comparison, a solution in
equilibrium with ice at $-15^{\circ}C$ has a water potential close to
-180 bar. However the same enzymes which are inhibited by
a decrease in water potential at room temperature, are
preserved for long times when stored at freezing temperatures.
The inhibitory effects of low water potentials on enzyme
activity are usually reversible, and low temperatures, even
though they are accompanied by low water potentials, protect
unstable systems such as enzymes by decreasing reaction rates.

Transfer of enzymes to room temperature and physiological water potentials restores activity (Santarius, 1969). The available evidence thus does not support the view that the resistance of cells to freezing is limited by the sensitivity of soluble enzymes.

Biomembranes represent a higher organizational level than soluble proteins and their response to low temperatures and the low water potentials accompanying freezing is complex. In some cases, temperature-dependent phase transitions are known to alter membrane activity and function (Pringle and Chapman, this volume) but no direct evidence is available that reversible changes in the state of membrane lipids are significant factors in producing freezing injury (Levitt, 1980). Chilling injury of sensitive plants requires metabolic activity and often a considerable time span to become effective. In contrast, at freezing temperatures metabolic rates are very slow, and freezing damage develops rapidly.

At room temperature, electron transport of thylakoid membranes in intact chloroplasts is less sensitive to decreased water potentials than the activity of soluble chloroplast enzymes (Kaiser et al., 1981b). Under severe water stress, the composition of the suspending medium drastically influenced the membrane response. In the presence of high concentrations of polyols such as sucrose and glucose, photophosphorylation was inhibited but electron flow was maintained (Santarius and Ernst, 1967). After rehydration of the membranes, ATP synthesis was restored. However, when a salt such as NaCl was a major solute at low water potentials, irreversible membrane alterations were observed.

Similarly, when thylakoids were frozen to $-20^{\circ}C$ with glucose, sucrose or raffinose as major solutes, the membranes remained intact and after thawing photosynthetic electron transport and phosphorylation were unimpaired (Lineberger and Steponkus, 1980a; Santarius, 1973b; Steponkus et al., 1977). However, when NaCl or a similar salt was a major solute, freezing led to membrane precipitation, phosphorylation was lost and, depending on conditions, electron transport was stimulated or inhibited (Heber and Santarius, 1964; Santarius, 1969; Santarius and Heber, 1970; Heber et al., 1979). Severely damaged membranes had lost their osmotic properties and were unable to swell in hypotonic solutions and to shrink in hypertonic media (Garber and Steponkus, 1976; Heber, 1967, 1968; Uribe and Jagendorf, 1968; Williams and Meryman, 1970).

Freezing injury to biomembranes appears to be a general phenomenon (many authors, this volume). Besides chloroplast membranes, mitochondrial membranes are frost-sensitive (Araki, 1977; Heber, 1968; Heber and Santarius, 1964; Porter *et al.*, 1953; Thebud and Santarius, 1981). Injury observed in frozen-thawed spinach protoplasts has been correlated with alterations in plasma membranes (Steponkus and Wiest, 1978, 1979; Wiest and Steponkus, 1978) and tonoplast instability appears to be an important factor in freezing damage to plant cells (Morris and Clarke, 1978; Zeigler and Kandler, 1980).

THYLAKOID MEMBRANE DAMAGE

Biochemical Activity

During freezing of biological membranes lesions may be produced at various sites. This will be illustrated with thylakoids, which are closed membrane vesicles capable of oxidizing water in the light and transferring electrons *via* an electron transport chain to suitable acceptors. Electron transport is accompanied by phosphorylation of ADP and ATP. The permeability of the membranes to small neutral solutes such as glycerol is high, but permeability to larger molecules such as sucrose or raffinose and to many electrolytes is very low. The permeability coefficients of a related membrane system, the chloroplast envelope, have been determined (Gimmler *et al.*, 1981). In consequence, the vesicles will shrink when the water potential is lowered by adding a non-permeating solute, and expand on dilution (Gross and Packer, 1965).

The capacity of thylakoids for photophosphorylation is a function of the composition of the suspending medium and sub-zero temperature (Fig. 1). Photophosphorylation is dependent on electron transport, the formation of ion gradients across the membrane, a light-induced membrane potential and a functional coupling enzyme. At 0°C, the membranes tolerate considerable concentrations of a mildly chaotropic salt such as NaCl. Concentrations of NaCl up to 200 mM did not affect membrane activity within 3 hours. Freezing to -6°C, and the parallel increase in the solute concentration of the residual unfrozen solution did not inactivate membranes at low ratios of NaCl to sucrose. However, when this ratio was increased, photophosphorylation decreased and was finally abolished at an initial NaCl to sucrose ratio of unity. Lowering the temperature to -12°C increased injury and caused loss of photophosphorylation at a decreased ratio of NaCl to sucrose, because further concentration of NaCl occurs at the lower temperature. If NaCl activities in the unfrozen solutions are calculated, it becomes apparent that the NaCl concentration producing 50 percent membrane inactivation increases as the temperature is decreased (Table 1). This is explained by

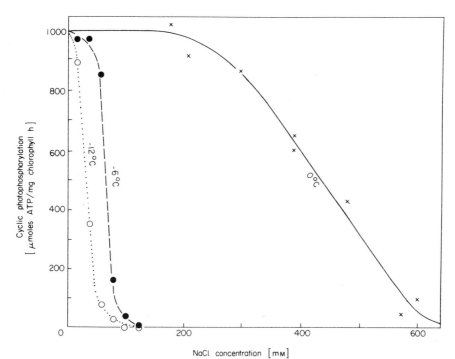

Figure 1. *Effect of temperature on inactivation of thylakoids suspended in solutions containing 100 mM sucrose and various levels of NaCl and maintained for 3 hours at 0°C, -6°C and -12°C. From Heber et al., (1979).*

two opposing effects of temperature. Lowering the temperature increases ice formation and consequently solute accumulation in the unfrozen part of the system, which tends to increase membrane damage. On the other hand, lower temperatures decrease the rate of any inactivation processes (Santarius and Heber, 1970). Sucrose can be replaced as a cryoprotective agent in the thylakoid system by other compounds such as raffinose, glucose, glycerol, dimethylsulfoxide, γ-aminobutyric acid, proline or threonine. NaCl can be substituted by other potentially membrane-toxic solutes that cause membrane damage when concentrated during freezing (Heber and Santarius, 1973).

The membrane inactivation shown in Figure 1 was not caused by mechanical effects exerted by ice crystals. This can be shown by varying the proportion of unfrozen solution to ice, which is possible by varying initial solute concentrations without changing the ratio of salt to cryoprotectant (Heber and Santarius, 1964). Still, mechanical effects such as in eutectic crystallization in membrane suspensions are possible,

U. HEBER *et al.*

Table 1

*Concentration of NaCl producing 50 percent loss of photophos-
phorylation capacity in thylakoids at 3 different temperatures.*

Temperature ($^{\circ}$C)	NaCl concentration (Molar)
0	0.44
-6	0.98
-12	1.4

*Values were calculated from data in Figure 1 and tabulated
values in Weast (1974).*

and when they occur, they are destructive (Santarius, 1973a;
Santarius and Heber, 1970).

The various biochemical activities of thylakoids respond
differently to freezing and the increasing ratio of salt to
cryoprotectant (Fig. 2). When this ratio is increased, first
photophosphorylation of the thylakoids decreases and simultaneously, electron transport to ferricyanide is stimulated.
This is a direct consequence of uncoupling of phosphorylation
from electron transport. After this initial increase, ferricyanide reduction declines together with the rise in variable
fluorescence indicating progressive loss of photosystem II
activity as the ratio of NaCl to sucrose is further increased.
While photosystem II activity decreases, photosystem I activity
of the membranes (which is expressed by electron flow from
ascorbate/dichlorophenolindophenol to methyl-viologen) is
stimulated, indicating a pathological alteration of membrane
function. This is probably caused by increased accessibility
of photosystem I to the electron donor/acceptor system. Final
loss of photosystem I activity is evidence of extensive
membrane destruction.

Membrane Permeability

While photophosphorylation begins to decline, membrane permeability to large neutral solutes such as sucrose is not
immediately altered (Jensen *et al.*, 1981). The membranes still
respond osmotically to changes in water potential. However,
the photoinduced proton gradient necessary to support photophosphorylation can no longer be fully maintained, because of
an increased leakiness to protons. As the ratio of salt to

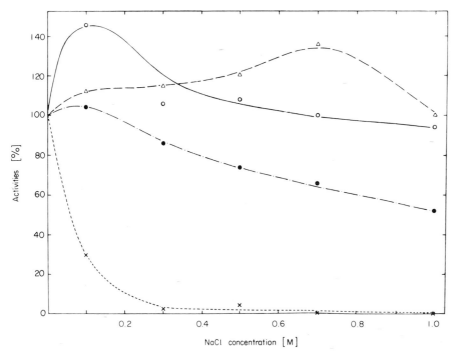

Figure 2. *Inactivation of biochemical activities of washed thylakoids from spinach leaves during freezing for 3 hours to about -20°C as a function of the ratio between salt and sucrose in the suspending medium. Sucrose concentration was 75 mM prior to freezing. To allow direct comparison, the activities of frozen samples that were fully protected by sucrose in the absence of added NaCl served as a reference (100%). Ferricyanide reduction (O); methylviologen reduction (Δ); cyclic photophosphorylation (×); rate of rise of variable fluorescence (—●—). From Heber et al., (1973).*

sucrose is increased, freezing-induced leakiness becomes more obvious and the membranes finally lose osmotic responses even to solute molecules as large as sucrose (Heber, 1967).

Loss of Proteins

The ratio of lipid to protein in the thylakoids is lower than in many other biomembranes (Douce and Joyard, 1979). Loss of membrane function during freezing is usually accompanied by membrane aggregation which indicates a decrease in the charge density of the membrane surface (Morris *et al.*, this volume). During inactivation, many different polypeptides dissociate from the membrane (Garber and Steponkus, 1976; Heber *et al.*, 1979; Lineberger and Steponkus, 1980b; Santarius *et al.*, 1979; Volger *et al.*, 1978), including components of the enzyme complexes responsible for electron transport and ATP synthesis. However, while it is possible to restore ATP synthesis by adding frozen and thawed, isolated CF_1 to unfrozen membranes from which CF_1 had been removed (Heber, 1967), attempts have failed to reconstitute severely frost-damaged membranes (Steponkus *et al.*, 1977). It appears that in such membranes, leakiness which results in loss of ATP synthesis, is not only caused by pores left in the membrane after the coupling factor has been lost.

It is therefore valid to ask whether solubilization of membrane proteins is the cause or a consequence of membrane damage. It is presently thought that dissociation follows primary events which weaken intramembrane interactions and finally result in loss of membrane function (McGrath; Steponkus, this volume).

The amount of protein detached from the membranes during freezing depends on the solute environment. Loss of protein is insignificant and membrane function is preserved when the ratio of salt to cryoprotectant is low but both loss of protein and membrane damage increase as the ratio of salt to cryoprotectant is increased. Up to 15% of the total membrane protein has been observed to be lost from the membranes during freezing (Mollenhauer, unpublished results). The ATPase activity of coupling factor released from thylakoids during freezing is lower than expected from the amount of CF_1 released (Lineberger and Steponkus, 1980b) and this suggests that some denaturation has occurred.

Stresses other than freezing will induce the release of protein from thylakoids. Dissociation from the membrane also occurs on exposure to high concentrations of salts at $0°C$ (Kamienietzky and Nelson, 1975; Volger *et al.*, 1978). Ice formation thus serves only to concentrate solutes in the vicinity of the membranes, and dissociation actually increases with temperature. In the presence of 0.75 M NaCl, dissociation was negligible at $0°C$; it was considerable at $15°C$ and even more pronounced at $30°C$ (Fig. 3). At supraoptimal temperatures, protein release almost paralleled the inactivation of the photosynthetic reactions of the thylakoid membranes (Volger and Santarius, 1981).

Only water soluble extrinsic (peripheral) membrane proteins can be expected to leave the membrane phase. The electrophoretic polypeptide pattern of intact thylakoids is, except for band intensities, very similar to that of thylakoids damaged by freezing. This shows that many proteins are

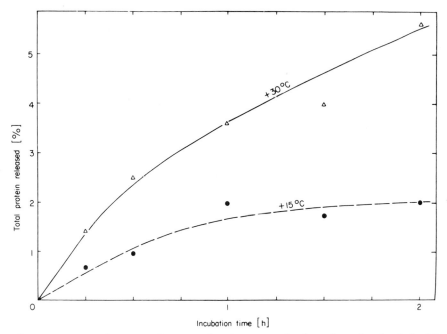

Figure 3. *Release of protein from washed spinach thylakoids suspended in 0.75 M NaCl at +15° and +30°C as a function of incubation time.*

retained in damaged thylakoids. Sulphydryl groups are thought to play an important role in frost injury (Levitt, 1980) but titration provided no evidence for significant SH-group oxidation during freeze-induced membrane inactivation (Heber and Santarius, 1964; Schmitt, unpublished results).

Chlorophyll is an intrinsic fluorescence probe of the thylakoid membrane. Upon excitation with blue light (400 nm) at $77°K$, red light is emitted from photosystem II and far-red light from photosystem I. Changes in the ratio of band peaks indicate changes in excitation energy distribution between the photosystems. Such changes would be expected if freezing induces alterations at the sites of the chlorophyll protein complexes situated in the membrane. No changes were observed upon freeze inactivation of thylakoid membranes *in situ* (Klosson and Krause, 1981a). Thus it appears that the interaction of major intrinsic protein components remains unaltered during membrane inactivation.

The Lipid Phase

Lipids rather than proteins are commonly believed to be
responsible for the barrier properties of biomembranes.
Increased membrane leakiness after freezing therefore suggests
that loss of protein from the membranes during freezing damage
is a corollary, not a cause of injury. Electron spin resonance (ESR) studies failed to reveal irreversible changes in
the lipid phase of freeze-damaged thylakoids (Jensen *et al.*,
1981). The temperature profile of lipid mobility obtained
from ESR spectra of spin-labelled fatty acids after incorporation into spinach thylakoids suggested a phase transition
of membrane lipids at $+15^\circ$C and another one at -10°C, but
these transitions were inconclusive. A small increase in
lipid mobility parallel to the increase in membrane damage
was observed during freezing. However, room temperature ESR
spectra from freeze-damaged thylakoids were undistinguishable
from spectra of intact membranes, when the membranes were
suspended in a hypotonic medium. Differences were observed
only under hypertonic conditions, when only the intact membranes were able to shrink osmotically.

MECHANISM OF MEMBRANE DAMAGE

The extent of membrane damage during freezing is influenced
not only by the ratio of salts to cryoprotectants but also by
the nature of the salts (Fig. 4). Similar results were
obtained at 2°C with isolated thylakoid membranes exposed to
high concentrations of sodium salts (Santarius, 1969) and
with spinach protoplasts subjected either to osmotic volume
changes or to a freeze-thaw cycle (Steponkus and Wiest, 1978).
     The electropherogram of polypeptides released from thyl-
akoid membranes during freezing in the presence of sorbitol
and of different sodium salts is presented in Plate 1. In
a control maintained at 0°C, only minor amounts of protein
were found in the supernatant after centrifugation. Freezing
resulted in the appearance of different polypeptide bands.
Some of them have been identified (Garber and Steponkus, 1976;
Volger *et al.*, 1978). The two most intense bands show the
α and β subunits of the coupling factor CF_1. Except for the
nitrate experiment, in which more protein was released from
the membranes than would be predicted from the data in Figure
3, the intensity of the bands increased in the order of increasing anion toxicity. This order is identical with a
Hofmeister lyotropic power series (Abernethy, 1967a,b; Heber
et al., 1979; Hofmeister, 1888). Toxicity increases as the
Stoke's law hydrated radius of the anions decreases and field

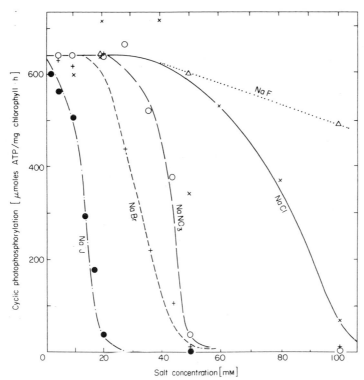

Figure 4. *Cyclic photophosphorylation of thylakoids following 'slow' cooling to -20°C. The thylakoids were suspended in 100 mM sucrose and various sodium salts. From Heber et al., (1979).*

strength exerted by the anions increases.

The order of membrane toxicities of cations during freezing was also similar to the order of cation effects in a Hofmeister power series (Heber *et al.*, 1979). However, in this case, strongly hydrated cations such as K^+ were more effective in inactivating thylakoids than weakly hydrated cations such as Cs^+, and divalent cations were more effective than monovalent cations. Similar observations have been made in freeze-thaw experiments with isolated spinach protoplasts (Steponkus and Weist, 1978).

There are two main hypotheses to explain freezing damage caused by solutes. The first one proposes a direct interaction between membranes and potentially toxic solutes as the cause of membrane damage. The second suggests that a decrease in water structure brought about by the accumulation of chaotropic solutes is responsible for freezing injury. This

decrease is thought to facilitate increased interaction between the membrane and water and therefore to cause membrane
destabilization (Hatefi and Hanstein, 1974).

Hydrophobic interactions and Coulomb effects make major
contributions to the stability of biomembranes (McGrath, this
volume). The accumulation of solute ions during freezing
will, at sufficiently high concentrations, suppress intramembrane ionic interactions and lead to protein dissociation.
Coulomb interaction between solute and membrane will depend
on field strength and should be stronger for anions with a
small hydrated radius than for larger ones (Larsen and Magid,
1974); this would explain the varying toxicity of different
anions. Direct interaction can also explain the toxicity of
anions possessing apolar side chains, such as phenylpyruvate
or decenylsuccinate, and of neutral solutes such as phenol
(Heber *et al.*, 1971, 1973, 1979; Santarius *et al.*, 1979).
When concentrated during freezing, these solutes will contribute to destabilization of the membrane by intramembrane
hydrophobic interactions.

Different cation effects on the membranes can in part be
explained by charge neutralization (Hauser and Dawson 1967;
Träuble and Eibl, 1974). Divalent cations are more effective
in neutralizing negative surface charges on the membrane than
the less toxic monovalent cations. Charge neutralization
facilitates increased anion interaction with the membrane
during freezing.

The monovalent cations may also act in a different manner
since by reducing the surface potential they may exert a
stabilizing effect. Possibly, a lowering of the surface
potential leads to increased ionization of acidic surface
groups (Träuble and Eibl, 1974), which will decrease interaction with toxic anions. Less freezing damage would result
in the presence of weakly hydrated monovalent cations such
as Cs^+, because these are more effective in reducing the

Plate 1 (opposite). *Effect of anions on protein release from
thylakoid membranes during freezing. Spinach thylakoids were
suspended in a medium containing 50 mM. sorbitol, 5 mM $MgCl_2$,
5 mM NaCl and 2.5 mM KH_2PO_4, pH 7.8. The sodium salts of the
anions indicated on the figure were added to give a final
concentration of 45 mM. The samples were frozen for 3 hours
at $-20°C$. Upon thawing, the thylakoids were centrifuged and
the supernatants were applied to a 10-18% polyacrylamide gel
and separated in the presence of sodium dodecyl sulphate.
Migration was from top to bottom. Co is a control containing
180 mM sorbitol for protection (Mollenhauer, unpublished
results).*

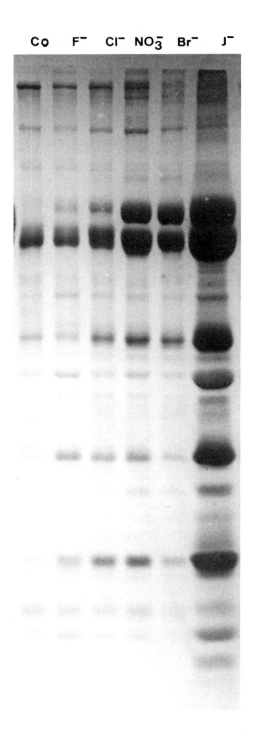

surface potential than strongly hydrated species such as Li^+.
In addition, the anion activity is lower in the presence of
weakly hydrated cations.

Hydrophobic bonds between membrane components form as a
consequence of the repulsion of apolar groups from the
aqueous phase rather than as a result of mutual attraction
(Tanford, 1980; McGrath, Pringle and Chapman, this volume).
Repulsion depends on water structure and since chaotropic
reagents decrease water structure, they increase interaction
of partially polar membrane components with the aqueous phase
and eventually facilitate dissociation reactions.

These mechanisms need not be considered as alternatives.
They may both contribute to membrane destabilization during
freezing.

Thylakoid Inactivation in vivo

Freezing experiments with isolated biomembranes *in vitro*
permit factor analysis, but raise the question whether results are related to freezing damage occurring to cellular
systems *in vivo*. Comparison of the biochemical activities
and physical properties of thylakoid membranes frozen *in
vitro* isolated rapidly from frost killed leaves revealed
differences in the inactivation pattern (Heber *et al.*, 1973).

In spinach leaves which were slowly cooled, maintained
for several hours at a sub-zero temperature and then thawed
slowly, injury occurred at a threshold temperature (Klosson
and Krause, 1981a). Photosynthetic activities of thylakoids
isolated from frost-treated leaves were decreased in parallel
to the inactivation of photosynthesis in the leaves (Fig. 5).
In contrast to freezing damage to thylakoids *in vitro*,
photophosphorylation and electron transport were affected by
freezing to a similar extent, and evidence for uncoupling of
phosphorylation from electron transport was observed only
when damage was extensive. In thylakoids isolated from
partially damaged leaves, membrane permeability to protons
was not increased compared to unfrozen controls (Fig. 6).
Partial reactions involving either photosystem I or II were
decreased in injured leaves. However, similar to the effect
of freezing on thylakoids *in vitro*, the water splitting
mechanism was more inhibited in thylakoids from severely
frost-damaged leaves than partial reactions of the two
photosystems.

Cold acclimation of the plants lowered the threshold temperature of freezing damage by about 6^0C, but did not alter
the pattern of injury (Figs. 5,6).

That discrepancies between freezing injury observed *in vivo*

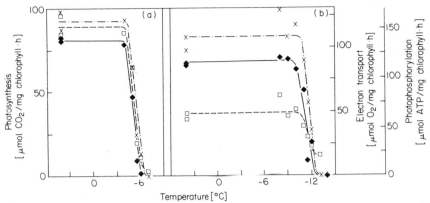

Figure 5. *Effect of freezing of spinach leaves on non-cyclic electron transport and photophosphorylation of chloroplast membranes. Thylakoids were isolated from (a) sensitive and (b) frost-hardened leaves after maintaining the leaves at +5°C or subjecting them to a freeze-thaw cycle; the rates of cooling and warming were 3°C h⁻¹. The leaves were exposed to the minimum temperatures for about 2 hours. Electron transport mediated by methylviologen, (O₂ uptake) (□); photophosphorylation (x); rates of photosynthesis (◆) of the whole leaves measured at 20°C. From Klosson and Krause (1981a).*

and *in vitro* were not caused by the procedure of thylakoid isolation from injured leaves, was shown by a study of physical phenomena in frost-damaged leaves, such as chlorophyll fluorescence emission and light scattering reactions (Klosson and Krause, 1981b). By these criteria, injury of the thylakoid system can be detected, although indirectly, *in situ*. For instance, the photoinduced change in 535 nm light scattering by leaves which has been shown to be an indicator of thylakoid energization *in vivo* (Heber, 1969; Krause, 1973,1974) is inhibited in parallel to photophosphorylation of the isolated thylakoid membranes.

There are several possibilities to explain the different pattern of thylakoid membrane inactivation *in vivo* and *in vitro*. Amphiphilic solutes such as phenylalanine, phenylpyruvate or caprylate exhibit a much higher membrane toxicity during freezing of thylakoids *in vitro* than salts such as NaCl. The profile of damage caused by this class of compounds is different from that caused by inorganic salts (Heber *et al.*, 1979; Volger *et al.*, 1978; Heber *et al.*, 1971, 1973). Damage to thylakoid membranes isolated rapidly from frost-killed spinach leaves is similar to that produced by amphiphilic

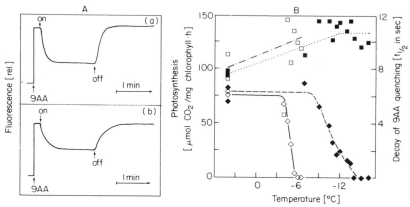

Figure 6. *Effect of freezing of spinach leaves on the permeability of thylakoids to protons. Thylakoids were isolated after cold treatment of sensitive or frost-hardened leaves and illuminated in the presence of 5 μM 9-aminoacridine. (A) Time course of fluorescence in a thylakoid suspension from hardened leaves kept at $+5°C$ (a) and from leaves previously frozen to $-13°C$ (b). 9-aminoacridine is thought to be trapped in the intrathylakoid space, when this becomes acidic during illumination. Trapped amine is non-fluorescent. Proton efflux after darkening results in 9-aminoacridine release and fluorescence increase. (B) Half time of the fluorescence increase after darkening which is a measure of the permeability of the thylakoids to protons (■, □) and rates of photosynthesis of the leaves after cold treatment (♦, ◇). Open symbols: frost-sensitive leaves; closed symbols: hardened leaves. Experimental conditions as in Figure 5. From Klosson and Krause (1981a).*

solutes during freezing *in vitro* (Heber *et al.*, 1973). This suggests that in addition to inorganic salts other potentially toxic solutes play a role in membrane inactivation during freezing *in vivo*.

This raises the question whether *in vivo* damage to thylakoids is primarily an effect of freezing. In order to minimize effects of slow thawing, chloroplasts were isolated as rapidly as possible from frozen leaves (Klosson, unpublished). When the leaves had been kept just below the threshold temperature of injury, the rapidly isolated thylakoids showed no indications of damage, whereas strong injury occured *in situ* when leaves were thawed slowly. Inactivation of rapidly isolated thylakoids was apparent only 4 to 5°C below the threshold temperature of leaf survival. The pattern of inactivation was similar to that observed in thylakoids isolated after slow thawing of leaves which had been frozen just beyond the limit of survival. These results show that membrane injury occurs not only during freezing but also during thawing as has been found with isolated spinach protoplasts (Wiest and Steponkus, 1978). Solute stress may be considerable during thawing when the concentration of solutes is still high and increased temperatures favor inactivating reactions. Release of protein from thylakoids frozen *in situ* is very similar to the release observed during freezing *in vitro* and indicates that solute effects are indeed involved in membrane inactivation *in vivo* (Mollenhauer, unpublished results).

Primary Freezing Injury in Leaf Cells

Solute injury to thylakoids during freezing of leaf cells, while certainly sufficient to cause cell death, does not necessarily imply that cell death is actually caused by damage to thylakoids. If another biomembrane is more sensitive to the solute stress produced during freezing, it would be damaged first and this might lead to a variety of secondary damaging effects. Such a view is supported by the complexity of damage observed *in vivo* (Klosson and Krause, 1981a,b). In spinach leaves, photosynthesis, dark respiration and the capability of the cells to reabsorb water lost during freezing were decreased to a similar extent. Chlorophyll fluorescence and light scattering measurements here indicated that in partially damaged leaves photosynthetic gas exchange was slightly more inhibited than thylakoid activity. The percentage of chloroplasts retaining functional envelopes was smaller when they were isolated from these leaves than from unfrozen controls (Fig. 7). This suggests that the chloroplast envelope is also damaged during freezing. This membrane system plays an important role in the control of reactions of the Calvin cycle (Heber and Heldt, 1981). Increased leakiness of the envelope would interfere with pH regulation of photosynthesis causing deactivation of Calvin cycle enzymes, although such enzymes are not affected directly by freezing. However, damage to the chloroplast envelope also does not necessarily indicate primary injury. Indirect evidence is available of freezing damage to other cell membranes such as plasmalemma and tonoplast (Senser and Beck, 1977; Steponkus and Wiest, 1978, 1979; Wiest and Steponkus, 1978; Yoshida, 1978; Ziegler and Kandler, 1980). In view of their different functions it is unlikely that the various cellular membrane systems possess comparable sensitivity to

U. HEBER *et al.*

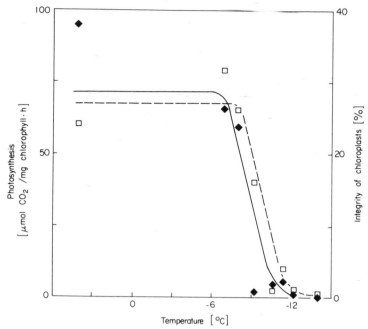

Figure 7. *Effects of freezing on the integrity of the chloroplast envelope. Percentage of intact chloroplasts present in the leaf homogenate (□); rates of photosynthesis at 20°C after cold treatment of leaves (◆). From Klosson and Krause (1981a).*

solute effects. Indeed, under comparable isolation and freezing conditions, spinach leaf mitochondria were less sensitive to freezing *in vitro* than thylakoids (Thebud and Santarius, unpublished results; Morris *et al.*, this volume).

Solutes are unevenly distributed in the cell. Assymetry of membrane structure and compartmentation of solutes must lead to different solute effects at the individual sides of a membrane. Thus, depending on membrane properties and on the composition of solutes in contact with the membrane surface, one or the other of the membranes will first be damaged during freezing producing secondary effects that limit survival of the cell.

MEMBRANE PROTECTION

Membrane protection has been discussed previously (Heber *et al.*, 1979; Heber and Santarius, 1973, 1976; Santarius, 1978; Santarius and Heber, 1972) and will be considered here only in broad outline.

Colligative Protection

It has been mentioned above that the ratio of potentially membrane-toxic solute to cryoprotectant dictates whether or not membrane injury occurs during freezing. In the case of thylakoids, a wide range of compounds show protective properties; including polyols which are generally considered to be non-permeating such as glucose, sucrose and raffinose (Heber and Santarius, 1964; Santarius, 1973a), permeating alcohols such as methanol, glycerol or ethylene glycol (Santarius, unpublished results), amino acids such as proline, threonine or γ-amino butyric acid (Heber *et al.*, 1971; Tyankova, 1972) and even weakly chaotropic salts such as sodium citrate and sodium acetate (Santarius, 1971).

All that is necessary for a compound to protect a membrane against solute injury during freezing is that it does not damage the membrane and does not crystallize when concentrated during freezing. The mechanism of protection by these solutes is simple (Lovelock 1953). During freezing, the temperature determines the total solute concentration in the residual unfrozen part of the system. If at the beginning of freezing only one solute is present and this is capable of damaging the membrane at high concentrations, freezing will produce such concentrations and thereby damage the membrane. On the other hand, if non-toxic solutes are present in addition to the potentially toxic one, freezing to the same temperature will produce the same decrease in water potential, but now the concentrated solution existing in equilibrium with ice will contain several solutes, and the concentration of the potentially cryotoxic solute will be only a fraction of that reached before. Protection will result, if that decreased level of the solute is not yet damaging (Heber and Santarius, 1973, 1976).

Specific Protection

Colligative protection predicts that different cryoprotectants should, on an osmolar basis, afford comparable protection to the membranes. However, deviations from this expectation are often observed and these can be interpreted in several ways. Various cryoprotectants differ in osmotic coefficient and when this is taken into account, the effect of many cryoprotectants can be explained exclusively by colligative action (Lineberger and Steponkus, 1980a; Meryman *et al.*, 1977). However, in some cases specific interactions between cryoprotectants and membrane may add to colligative protection enhancing or decreasing membrane stability. It is also

necessary to consider the assymetry of membranes and the
permeability properties of cryoprotectants and toxic solutes.
Clearly, protection cannot be expected if a non-permeating
cryotoxic solute is localized on one side of a membrane and
a non-permeating cryoprotectant is added to the other side.

Little information is available on specific protection of
biomembranes. When it was noticed that thylakoids isolated
from frost-hardened spinach were more resistant to freezing
than thylakoids from non-hardy leaves, it was initially
thought that hardening changed membrane structure. However,
washing abolished the differences in membrane resistance, and
a search for the factor responsible for membrane protection
yielded a protein fraction which, when added to thylakoids,
increased their resistance to freezing (Heber, 1968, 1970;
Heber and Ernst, 1967; Heber and Kempfle, 1970). This active
fraction was heterogeneous, containing polypeptides of molecular weight between 10,000 and 20,000 Daltons. An amino
acid analysis of two of the polypeptides revealed high per-
centages of polar amino acids (Volger and Heber, 1975). The
proteins partially protected thylakoids against freezing
damage at a concentration lower by a factor of almost 1000
than that of an equally effective sucrose solution. This
makes protection on a colligative basis unlikely. Full
protection of thylakoids by the proteins was not observed,
possibly because the inside of the thylakoid vesicles was
not accessible to the proteins (Santarius *et al.*, 1979).

Alteration of Membrane Structure

There is considerable literature (Levitt, 1980; Santarius,
1978) on changes in membrane composition during frost hardening, but little evidence is available as to whether these
changes actually result in increased frost tolerance of the
membranes. Senser and Beck (1977) reported for spruce
chloroplasts large seasonal changes in thylakoid properties,
but their data do not show that thylakoids from hardy winter
needles were more resistant to freezing than thylakoids from
summer needles. With spinach chloroplast membranes, it was
observed that in the presence of very low (millimolar) concentrations of amphiphilic anions such as decenylsuccinate
or phenylpyruvate resistance of thylakoids to freezing was
increased (Heber *et al.*, 1979; Santarius *et al.*, 1979).
Possibly, the reported increase in the cellular contents of
phospholipids which are also amphiphilic serves to increase
freezing tolerance of biomembranes *in vivo* (Siminovitch *et
al.*, 1975; Yoshida, 1976).

SUMMARY

Cells contain and require for their physiological function a
large variety of electrolytes. These ions interact with
charged and polar groups at the surfaces of biomembranes; at
temperatures above freezing such interaction is necessary for
maintenance of membrane function. Accumulation of electrolytes during freezing, however, leads to membrane destabilization and irreversible damage. Strategies of living organisms
to adapt to freezing conditions involve the accumulation of
both colligatively and specifically acting cryoprotectants.
Isolated thylakoids have served as a model system to elucidate
the basic mechanism of freezing injury by solute stress and
preservation by cryoprotectants.
    Since green leaves are complex systems, it is hardly
surprising that many harmful effects of freezing and thawing
are observed *in vivo*. Thylakoid damage has been found to
increase during thawing of leaves and the pattern of thylakoid
damage *in situ* is different from that caused by inorganic
salts *in vitro*. Injury *in vivo* may possibly be caused by
endogenous toxic solutes acting on the chloroplasts in a similar manner as observed with amphiphilic substances *in vitro*.
Alternatively, injury *in vivo* may be the consequence of
freezing damage to more sensitive membrane systems elsewhere
in the cell. However, there is good reason to assume that
damage to and protection of these membranes during freezing
are governed by the same principles that have been outlined
for thylakoids.

ACKNOWLEDGEMENTS

We are grateful to Dr. U. Schreiber, M. Jensen and A. Mollenhauer for helpful discussions. Work from our laboratories
which is discussed in this review was supported by the Deutsche
Forschungsgemeinschaft.

FREEZING INJURY IN CHLAMYDOMONAS:
A SYNOPTIC APPROACH

G.J. Morris, G.E. Coulson, K.J. Clarke

Institute of Terrestrial Ecology, Culture Centre of Algae and Protozoa, Cambridge, England.

B.W.W. Grout

Department of Biology, North East London Polytechnic, London, England.

A. Clarke

British Antarctic Survey, Cambridge, England.

INTRODUCTION

Vegetative cells of the unicellular green alga *Chlamydomonas reinhardii* are extremely susceptible to freeze-thaw injury and are a useful system for investigating the biochemistry of freezing damage (Grout *et al.*, 1980; Morris *et al.*, 1979). In this article the response of the mutant CW15+ (Hymans and Davies, 1972) to the stresses of freezing and thawing will be examined in detail. CW15+ is a cell wall-less mutant, analogous to a free-living protoplast, and hence has many advantages for investigating freezing injury (Steponkus *et al.*, this volume). An additional feature is that in *C. reinhardii* CW15+, the cell wall components are synthesised but not polymerized, in contrast to the protoplasts of higher plant cells which are prepared enzymatically and are metastable, continually resynthesising and depositing cell wall material.

THE ECOLOGY OF CHLAMYDOMONAS

The ecology of an organism should first by considered for an understanding of its response to freezing and thawing.

However, with the exception of discussions of cold-hardening
in higher plants, such an ecological approach has been
ignored.

Whilst it is generally assumed that damage to the cellular
membranes occurs at an early stage of both freeze-thaw and
osmotic injury, it is not known to what extent the bulk chemical composition of the membrane determines the response to
such stresses. *C. reinhardii* is a mesophilic, freshwater
species sensitive to both freeze-thaw and shrinkage-expansion
injury. Some related species however, have evolved resistance
to either diurnal freezing (*C. nivalis*) or saturated salt
solutions (*Dunaliella salina*) and this enables interspecific
comparisons of membrane composition and stress resistance to
be made.

Temperature

C. reinhardii has an optimum growth temperature of *ca* $20^{\circ}C$;
At both higher and lower temperatures the rate of cell division
decreases, and at $10^{\circ}C$ it is inhibited completely. However,
cultures transferred to $4^{\circ}C$ show no reduction in viability
and are motile for at least 21 days, indicating that there is
no metabolic imbalance at this temperature. There is no
response analogous to chilling injury of higher plants (Wilson
and McMurdo, this volume). At $4^{\circ}C$ there is a decrease in
both the average number of double bonds per phospholipid
fatty acid molecule and also in the ratio of unsaturated to
saturated fatty acids (Table 1). This response is in the
opposite direction to that predicted by the concept of strict
homeoviscous adaptation of phospholipid fatty acids (Sinensky,
1974). But as the lipids are extracted from viable, metabolically active cells, the membranes of such cells must be
functional, suggesting that factors other than phospholipid
fatty acid composition are important in determining membrane
fluidity and function at low temperatures (Clarke, this
volume).

The rate limiting step for cell division in *C. reinhardii*
at low temperatures is not apparent. Lipid phase separations
in membranes would provide such a mechanism (Pringle and
Chapman, this volume). However, all the temperature-sensitive
mutants of *C. reinhardii* so far analysed are limited at the
restrictive temperature by steps in protein synthesis (Cross
and McMahon, 1976; Grossman and Togasaki, 1979; Hansen and
Bogorod, 1978; McMahon, 1971). The same pattern occurs in
temperature sensitive mutants of *Escherichia coli* (Simon,
this volume).

Like many other freshwater organisms, *C. reinhardii* responds

Table 1

Fatty acid composition of phospholipids from three species of unicellular alga

Fatty acid	Chlamydomonas reinhardii CW15+ 7d 25°C	21d 4°C	Dunaliella salina 7d 25°C	Chlamydomonas nivalis 7d 25°C
14:0	0.22	–	0.11	0.17
15:0	0.76	3.47	5.28	0.11
16:0	15.46	19.03	12.42	13.39
16:1	1.34	–	1.79	1.37
16:2	1.31	1.88	2.83	1.61
16:3	2.52	2.49	2.87	10.23
16:4	19.27	21.50	12.60	15.73
17:0	1.31	0.35	2.28	0.35
18:0	1.40	1.11	0.51	0.12
18:1	5.75	4.00	7.01	7.68
18:2	3.87	3.26	21.12	1.67
18:3	35.11	27.92	17.29	37.51
18:4	3.70	5.68	5.44	8.14
minor components [a]	0.08	1.66	2.54	–
unknowns	7.90	7.65	5.91	1.92
db mol^{-1} [b]	2.41	2.32	2.01	2.59
unsat/sat. [c]	3.80	2.85	3.56	5.93

(a) *Mainly branched chain and minor C_{14} and C_{15} unsaturated acids.*
(b) *Average number of double bonds per fatty acid molecule.*
(c) *Ratio of unsaturated to saturated fatty acids.*

to certain adverse environmental conditions by initiation of sexual reproduction (Sager and Granick, 1975). Fusion of the gametes with those released from the opposite mating type results in the formation of zygotes. These structures are the most resistant stage of the life cycle; vegetative cells

are very sensitive to freezing injury whilst the recovery of
zygotes from -196°C exceeds 50 percent (Bennoun, 1972). In
the environment there is no selection pressure to evolve
freezing resistance in the vegetative stage as the species
can overwinter in the form of zygotes. No process similar
to that of cold-hardening of higher plants can be induced in
vegetative cells of *C. reinhardii* (Morris *et al.*, 1979).
This is in contrast to species of algae which do not form
zygotes, in which the vegetative cells are either intrinsically
resistant to freezing injury or mechanisms for increasing
freezing tolerance exist (Morris, 1980).

Chlamydomonas nivalis is one of the organisms responsible
for the 'red' and 'green' snows which occur in some environments where snow persists through the summer (Kol, 1968).
It has been suggested that the red colour is due to the
intracellular accumulation of carotenoids synthesised under
conditions of nitrate limitation (Czygan, 1970). A high
water potential is necessary for algal growth (Fogg, 1967),
and vegetative cells do not appear until liquid meltwater
has been in the snow for several days; complete melting,
however, is inhibitory. Isolates of *C. nivalis* exhibit a
high photosynthetic activity at 0°C, although the temperature
optima for various strains vary within the range 0 to 20°C.
In samples isolated from the Beartooth Mountains, U.S., there
was no correlation between the optimal temperature for
photosynthesis and the elevation of the sampling site. Natural populations are thus composed of a mixture of psychrophilic and psychrotrophic strains (Mosser *et al.*, 1977).

C. nivalis overwinters in the form of resistant zygotes.
Once growth has been initiated in spring the vegetative cells
are often exposed to a diurnal freeze-thaw cycle. In some
strains photosynthesis occurs in the frozen state at -3°C
and following freezing to -8°C photosynthetic activity returned to control values immediately upon thawing (Mosser
et al., 1977). Resistance to freezing injury is therefore
essential for the survival of vegetative cells of *C. nivalis*
and some isolates survive upon thawing from -196°C (Morris
et al., 1979). When compared with *C. reinhardii* there was a
small increase in the average number of double bonds per
phospholipid fatty acid molecule, due largely to a higher
content of 16:3 and 18:4 fatty acids. There are, however,
major differences in the ratio of unsaturated to saturated
fatty acids, *C. nivalis* having a much higher proportion of
fatty acids with at least one double bond (Table 1). The
physical properties of phospholipid bilayers are not a linear
function of the average degree of unsaturation of the fatty
acids, the most dramatic change occuring with the insertion

of one olefinic bond into a saturated fatty acid. Addition
of subsequent double bonds has progressively smaller effects
(Demel *et al.*, 1972; Ghosh *et al.*, 1971).

It is tempting to speculate that the freezing tolerance of
C. nivalis is directly related to the large proportion of
unsaturated fatty acids in the membrane. However, a major
difficulty with this hypothesis is that the molecular composition of membranes is not homogeneous and differences in
composition exist between the various membrane systems in a
cell and also within individual membranes (Clarke; Pringle
and Chapman, this volume). The cellular response to freezing
may be determined by a critical microenvironment within a
membrane and bulk lipid extraction and analysis would not be
expected to provide any insight into the local composition
of these regions. There is no correlation between the cellular response to freezing and either the average degree of
phospholipid unsaturation or the ratio of unsaturated to
saturated fatty acids for a variety of protists cultured
under different conditions (Spearman rank correlation, all
$P > 0.05$: Table 2). Similarly, there is no simple relationship between phospholipid fatty acid composition and stress
tolerance for a number of wheat cultivars (de la Roche *et al.*,
1975).

Osmotic Response

C. reinhardii is cultured in a medium of 45 mOsm; the cell
wall-less mutant CW15+ tolerates this low osmolality medium
due to the presence of a contractile vacuole, which regulates
cell volume by active pumping. Any treatment which inactivates
the contractile vacuole but still allows osmotic uptake of
water by the cell will cause lysis. It is of interest that
several other organisms which possess contractile vacuoles,
Tetrahymena, Amoeba and *Paramecium*, are also very sensitive
to freeze-thaw damage. However, it cannot be concluded that
the contractile vacuole is the primary target of such injury;
indeed some cell-types (*e.g. C. nivalis*) have contractile
vacuoles and are also resistant to freezing injury. It could
equally be argued that the possession of a contractile vacuole
is a necessary adaptation of some cell-types to freshwater
environments and that freshwater organisms are susceptible
to freezing injury for reasons independent of the possession
of this organelle. The role of the contractile vacuole in
chilling and freezing injury is thus unclear and requires
further investigation.

In media supplemented with different compounds cell growth
is dependent both on the additive and the mode of nutrition

Table 2

A comparison of the cellular response to either hypertonic or freezing stress with the average number of double bonds per fatty acid molecule (Db mol^{-1}) and the ratio of unsaturated to saturated fatty acids (unsat:sat) in selected unicellular protists.

Organism	Method of culture	Db mol^{-1}	unsat:sat	Freeze/thaw (%) (a)	Hypertonic NaCl (M) (b)
Chlamydomonas reinhardii	7d 20°C	2.41	3.80	< 1	0.24
"	21d 4°C	2.32	2.85	< 1	-
Dunaliella salina	7d 20°C	2.01	3.56	40	saturated
Chlamydomonas nivalis	7d 20°C	2.59	5.93	60	1.05
Chlorella emersonii	14d 20°C	2.22	4.09	< 1	1.25
"	35d 20°C	2.42	4.90	50	-
"	14d 20°C nitrate limited }	2.29	3.80	30	1.60
"	14d 4°C	2.44	4.29	25	-
Chlorella protothecoides	7d 20°C	2.07	4.32	98	2.0
"	7d 20°C light heterotrophic }	1.51	5.50	95	-
"	7d 20°C dark heterotrophic }	1.25	4.43	60	2.0
Tetrahymena pyriformis	3d 20°C	1.17	4.46	< 1	0.51
"	2d 15°C	1.44	4.83	< 1	-

(a) Recovery of cells from −196°C. Due to the interactions between rate of cooling and cell size the maximal survival obtained is given rather than the recovery following any single rate of cooling.
(b) The molar concentration of NaCl which reduces cell viability by 50% following a 5 min exposure at 20°C and dilution back into normal growth media.

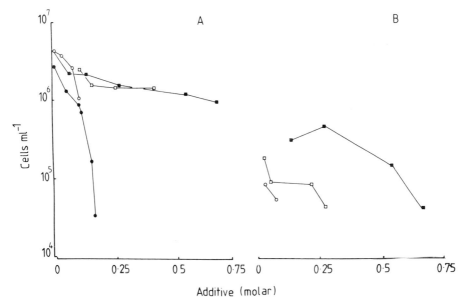

Figure 1. *Cell yield of Chlamydomonas reinhardii CW15+ foll-*
owing incubation at 20°C in growth medium supplemented with
either glycerol (■), glucose (□), sodium acetate (○) or
sodium chloride (●).   Cultures were maintained either in
constant light (A) or dark (B).

(Fig. 1). In the light, sodium chloride and sodium acetate
inhibit growth at concentrations exceeding 0.1 M, whereas
with glycerol and glucose, higher concentrations are tolerated.
Under dark heterotrophic conditions the cell yield is lower,
significant growth occurring only at low concentrations
of glycerol (0.25 M or less).

Not all *Chlamydomonas* species are sensitive to solutions
of high osmolality, for many marine species are known and a
closely related genus (*Dunaliella*) contains representatives
which are extreme halophiles. The main taxonomic feature
which distinguishes *Dunaliella* from *Chlamydomonas* is the lack
of a cell wall (Butcher, 1959) and *Dunaliella* may thus be
regarded as analogous to *C. reinhardii* CW15+. *D. salina* has
been isolated from the Dead Sea and the Great Salt Lake, Utah
and will grow in media containing from 0.3M to fully saturated
NaCl, the optimum being 2.0M. It is the most halotolerant
eukaryotic microorganism known (Brock, 1975). The tolerance
to hypertonic medium is not specific to NaCl for growth also
occurs in high concentrations of sucrose.

D. salina is not in ionic equilibrium with its environment
as it excludes NaCl and will shrink and swell in response to

changes in external NaCl concentration. The cells can tolerate extensive swelling or shrinkage and osmotic lysis is
rarely observed (Brown and Borowitzka, 1979). As discussed
elsewhere (McGrath; Steponkus *et al.*, this volume), these
observations suggest that an effective mechanism exists for
the release and reincorporation of cell membrane components
to accommodate changes in the effective membrane surface
area. Unlike halotolerant bacteria (Hunter *et al.*, 1981;
Kates, 1978; Langworthy, 1977) *Dunaliella* phospholipids
contain no unusual fatty acid although in comparison with
C. reinhardii and *C. nivalis* the average number of double
bonds per fatty acid molecule is low, due primarily to the
reduced content of 16:4 and 18:3 acids. The ratio of unsaturated to saturated acids is similar to that of *C. reinhardii*
(Table 1). There is thus no correlation between the cellular
phospholipid fatty acid composition and resistance to sodium
chloride solutions (Table 2: $P > 0.05$, Spearman r).

Long-term adaptation of *Dunaliella* to hypertonic conditions
involves the synthesis of glycerol as a compatible solute
(Brown and Borowitzka, 1979). Glycerol is a well known cry-
oprotective additive, and it is thus not suprising that
D. salina is resistant to freezing to and thawing from $-196^{\circ}C$
(Morris, unpublished results).

RESPONSE OF CHLAMYDOMONAS REINHARDII TO FREEZING AND THAWING

Cell viability following freezing and thawing is determined
by complex interactions between factors such as growth
conditions prior to freezing, the rates of cooling and warming
and the final temperature attained. To allow a detailed
analysis of freezing injury the response of cells from the
early stationary phase of culture (7 days at $20^{\circ}C$) to a
single freeze-thaw protocol was examined (Grout *et al.*, 1980).

Physics of Freezing

C. reinhardii CW15+ is cultured in a medium of low osmolality
(45 mOsm) and this has important consequences for the properties of solutions during freezing, properties which are not
observed in media of higher osmolality. The physical stresses
to which the cells are exposed during freezing are thus of a
different type to those experienced by cells suspended in
media of higher osmolality.

At any sub-zero temperature above the eutectic point the
osmolality of an ideal, dilute aqueous solution in equilibrium
with ice is

$$\frac{\Delta FPt \ ^{\circ}C}{1.86} \quad (\text{Osmoles Litre}^{-1})$$

The portion of unfrozen solution is therefore directly related to the molar concentration before freezing. Assuming ideal solution behaviour for the *C. reinhardii* growth medium (equivalent to approximately 0.025 M NaCl) the fraction of unfrozen medium following the formation of ice will be < 1 percent at $13^{\circ}C$ (Morris and McGrath, 1981b). The cells are thus restricted to a very small aqueous compartment, since the majority of the system exists as ice.

In addition, during cooling the solubility of dissolved gases increases and will double between 0 and $-20^{\circ}C$ (Hobbs, 1974). However, when a portion of the liquid water is removed as ice, the gases are then concentrated in the residual solution. The maximum solubility of air in ice is at least one thousand times less than the solubility of air in water and inclusion of gas in the ice crystal lattice may therefore be neglected (Scholander *et al.*, 1953). The dissolved gases will be concentrated by a factor of about 100 in the residual aqueous solution at $-3^{\circ}C$ and under these conditions it is probable that the gases will come out of solution in the form of bubbles. The cells may thus be exposed to a three phase system of ice, gas bubbles and an aqueous solution saturated with gases and containing concentrated solutes.

Cell Viability

C. reinhardii CW15+ can be undercooled to at least $-7.5^{\circ}C$ without loss of viability. In the presence of ice however, the cells are damaged at high sub-zero temperatures and at a rate of cooling of $0.25^{\circ}C \ min^{-1}$ the median lethal temperature is $-5.3^{\circ}C$ (Fig. 2). The response of the cell wall-less mutant is not significantly different from that of the wild type (Morris *et al.*, 1979), which indicates that in this organism damage to the plasmalemma:cell wall interface is not a primary cause of freezing injury. Similar conclusions have been made with protoplasts from higher plant cells (Steponkus *et al.*, this volume). With all other mutants (70) and strains (5) of *C. reinhardii* examined a similar sensitivity to freezing injury was observed. In addition, there is no selection of freeze-resistant mutants following freezing of cultures pre-treated with mutagens, for any clones derived from cells surviving freezing are still freeze-sensitive.

Cryomicroscopy

At a rate of cooling of $0.25^{\circ}C \ min^{-1}$ large extracellular ice crystals form and the cells are restricted to narrow aqueous channels in which they undergo osmotic dehydration (Plate 1).

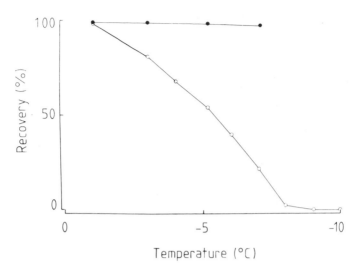

Figure 2.   *Recovery (%) of Chlamydomonas reinhardii CW15+ following cooling at 0.25°C min⁻¹to different final temperatures.   Cell suspensions were either undercooled (O) or frozen (●).*

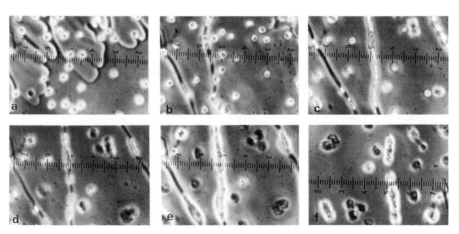

Plate 1.   *The response of Chlamydomonas reinhardii CW15+ during freezing and thawing (magnification x630).   (a-d) During cooling at 0.25°C min⁻¹ (a) -1°C, (b) -3°C, (c) -4°C, (d) -5°C.   (e-f) During warming at 10°C min⁻¹ (e) -2.5°C, (f) +5°C.*

At faster rates of cooling there is less time for dehydration
to occur and the probability of intracellular ice nucleation,
as determined by the phenomenon of 'black flashing', increases
(Fig. 3). When intracellular ice nucleation occurred it did
not become apparent until $-8.1 \pm 2.7^{\circ}C$ (McGrath and Morris,
unpublished observations). It can therefore be assumed that
under the conditions used in this study the cellular injury
observed is the result of the stresses associated with shrinkage and rehydration. The rate of cooling is too slow and the
temperature at which injury occurs is too high for primary
damage to be caused by intracellular ice. It is however
possible that ice is propagated into the cells following damage to the plasmalemma and that organelles may then be in
direct contact with ice.

During freeze-induced dehydration changes in the colour
of the cells were frequently observed, the sequence being,
dark green - bright green - orange - red; upon warming and
subsequent rehydration the reverse pattern occurred (McGrath
and Morris, unpublished observations). These changes may
indicate an alteration in the relative position of the photosynthetic pigments within the thylakoids, such as a vertical
movement of carotenoids within the plane of the membrane, and
are a consequence of freeze-induced dehydration; they are
not observed either during undercooling or at rapid rates of
cooling.

Upon thawing cells which have been damaged by slow cooling
swell beyond their original volume and this is accompanied
by a vesiculation of the cytoplasm and a loss of cellular
definition (Plate 1). It is thus apparent that the plasmalemma is damaged at an early stage of freezing injury.

BIOCHEMISTRY OF CELLS DAMAGED BY FREEZING

Plasmalemma Selective Permeability

As with many other cell-types the reduction in cell viability
is associated with an alteration in plasmalemma function, as
demonstrated here by the release of intracellular enzymes.
Both the total amount of enzyme released and the kinetics of
this release, however, differ from that observed with the
wild-type (Fig. 4). More enzyme is released from the cell
wall-less mutant, this loss occuring within 5 min of thawing
with no evidence of the two-phase kinetics observed in the
wild-type. These differences probably do not reflect two
mechanisms of injury but merely the presence of a cell wall
in the wild-type which will retain the cellular constituents
even when the plasmalemma is damaged. Also the rates of

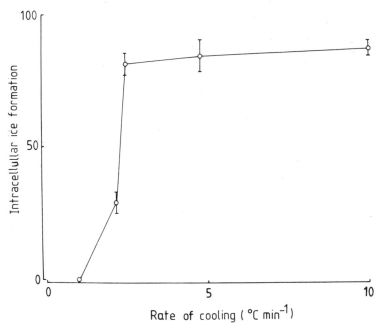

Figure 3. *The probability of intracellular ice formation in Chlamydomonas reinhardii CW15+ at different rates of cooling.*

diffusion will be altered as the cell wall matrix will behave as a molecular sieve (Thomas and Syrett, 1976). Cell lysis is observed during the thawing of *C. reinhardii* CW15+, whereas in the wild-type although morphological alterations (*e.g.* vesiculation of the cytoplasm) are apparent, the cells remain intact.

Metabolic Activity

Changes in the mitochondrial and photosynthetic activity of *C. reinhardii* (after rewarming to $20^{\circ}C$) are observed following both undercooling and freezing (Table 3).

Both respiration and the Hill reaction are sensitive to low temperatures *per se*. In mitochondria this could be due to either a direct effect on the enzymes of the respiratory chain or the effects of low temperature lesions in the cell that have an indirect effect on respiration. The results from the modified Hill reaction must relate to an altered function of the thylakoid membranes rather than to lesions within the cell, for activity is assayed in a cell-free system. In contrast cellular ^{14}C-fixation in the light was only

Table 3.

Metabolic activities (percentage control values) of C. rein-
hardii CW15+ measured at 20°C following either chilling or
freezing and thawing (Grout et al., 1981).

Temperature (°C)	Respiration	Hill reaction	C^{14}-fixation light	C^{14}-fixation dark
0	96	93	99	151
-2.5 undercooled	86	82	95	174
-5.0 undercooled	48	69	90	108
-2.5 frozen/thawed	41	52	0.8	15
-5.0 frozen/thawed	28	28	1.7	16

slightly reduced, whilst the dark reaction was stimulated.
At low temperatures both chloroplast and mitochondrial membranes may undergo phase separations (Lyons, 1973), although
lipid phase changes are completely reversible and thus cannot
account directly for the reduction in metabolic activity upon
warming to +20°C. If, however, the phase separations induced
effects such as segregation of membrane proteins or a redistribution of ions they could account for the observed reduction in metabolic activity. Whatever the mechanism of inhibition following exposure to low temperatures it must be either
readily repairable or due to inhibition which is reversible,

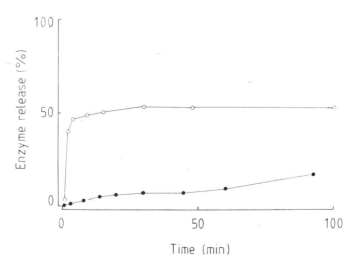

Figure 4. *Release (%) of Glutamic-oxaloacetic transaminase*
from wild-type Chlamydomonas reinhardii (●) and C. reinhardii
CW15+ (O) at different times following freezing to -5°C.

for such short-term undercooling does not reduce cell viability (Fig. 2).

    Following freezing and thawing there is a further reduction
in both respiration and Hill activity. Photosynthesis, which
is insensitive to temperature reduction *per se* shows an
extreme sensitivity to the stresses associated with freezing
and thawing. This is inconsistent with the observed cell
viability since on thawing from $-2.5^{\circ}C$ some 85 percent of
the cell population has the potential to divide and form
colonies in agar, whereas upon thawing C^{14}-fixation is < 1
percent of the control value. This indicates that non-lethal
metabolic injury occurs at low temperatures in cells, but
that this is repairable upon incubation.

Surface Charge

Whole cell electrophoresis is a technique which provides
direct information on the effective surface charge at the
plasmalemma (Jones, 1975). The electrophoretic mobility of
cells undercooled to $-5^{\circ}C$ and rewarmed was not significantly
different from that of control cells. Following freezing
and thawing a marked alteration in electrophoretic mobility,
measured at $20^{\circ}C$, was observed in cells cooled below $-2.5^{\circ}C$
(Table 4) and there was no further effect with time after
thawing. In addition, the reduction in electrophoretic mobility, recorded following cooling to $-196^{\circ}C$ at various rates
from 0.3 to 800 $C^{\circ}min^{-1}$, was not significantly different from
that upon thawing from $-2.5^{\circ}C$ (Grout *et al.*, 1980). In cells
thawed from $-1^{\circ}C$, which have a high recovery (Fig. 2), there
was little effect on electrophoretic mobility. The presence
of extracellular ice *per se* is therefore not directly responsible for the change in surface charge. The changes
observed represent a fundamental alteration in the plasmalemma
which seems to be directly related to freeze-thaw damage. A
similar decrease in surface change density accompanies freezing
injury in thylakoids (Heber *et al.*, this volume).

    The changes in electrophoretic mobility represent alter-
ations to the density of charged groups at the membrane surface or to the neutralisation of charged species by intracellular compounds leaking as the result of cellular injury.
The carrier of negative charges at the membrane surfaces is
predominantly protein and following freezing and thawing of
thylakoids there is a release of membrane proteins (Heber
et al., this volume). Electrostatic repulsion is important
in preventing membrane fusion or aggregation and the loss of
charged groups may then result in membrane fusion. Indeed
such membrane aggregation has been proposed as a cause of

Table 4

Electrophoretic mobilities of Chlamydomonas reinhardii CW15+ 5 min after thawing from a range of subzero temperatures.

Temperature ($^{\circ}$C)	Mobility (μ sec^{-1} V cm^{-1})
20 (control)	-4.14 ± 0.92
0	-3.66 ± 0.99
-1	-3.16 ± 0.58
-2.5	-0.94 ± 0.43
-5	-1.02 ± 0.08
-10	-0.78 ± 0.01
-15	-0.83 ± 0.05
-20	-0.79 ± 0.08
-5 (undercooled)	-3.96 ± 0.82

freezing injury of thylakoids (Jensen *et al.*, 1981). Similar freeze-induced membrane fusion occurs in liposomes both in the presence (Gerritsen *et al.*, 1978; Yu and Branton, 1976) and absence (Hui *et al.*, 1981; Strauss and Ingenito, 1980) of integral proteins.

Fatty Acid Composition

As well as changes in the membrane function, alterations to the composition of the membrane lipids occur following freezing and thawing. With cells frozen to -5°C and then maintained for between 5 and 100 min before lipid extraction, there is a progressive change in the phospholipid fatty acid composition (Table 5). The overall trend is for a relative loss of polyunsaturated acids, notably 16:4 and 18:3ω3, with a corresponding increase in the more saturated fatty acids. Consequently the mean number of double bonds per fatty acid molecule was reduced. This effect on fatty acid composition became more apparent with increasing time at 25°C. In parallel with these alterations in fatty acid composition there was a release of free fatty acids (Morris *et al.*, 1979).

These changes in fatty acid composition are due to either the activation of intracellular phospholipases or a peroxidation process. Whilst the specific mechanism is not yet apparent, the relationship between the observed changes in phospholipid composition and reduction in cellular viability is of critical importance for an understanding of freezing injury. Yoshida (1976, 1978, 1979) has proposed that the activation of phospholipase D is the primary mechanism of

Table 5.

Phospholipid fatty acid composition of Chlamydomonas reinhardii CW15+ following freezing to $-5°C$; upon thawing cells were incubated for different times at $25°C$.

Fatty acid	Unfrozen control	5 min	15 min	30 min	100 min
14:0	0.33	0.42	0.40	0.49	0.47
15:0	0.07	0.07	0.18	0.07	0.23
16:0	14.54	19.17	18.07	23.11	23.03
16:1ω9	0.40	0.50	0.45	0.55	0.55
16:1ω7	4.75	6.07	5.71	7.20	7.72
16:2ω6	2.56	2.96	2.94	3.24	3.27
16:3	6.18	5.68	5.65	4.30	3.94
16:4	13.07	10.35	10.67	7.26	6.39
18:0	2.06	1.55	1.80	2.02	1.81
18:1ω9	4.96	6.23	6.08	7.94	8.80
18:1ω7	3.83	4.79	4.69	5.61	6.02
18:2ω6	10.48	10.72	10.46	10.51	10.13
18:3ω6	8.60	6.89	6.93	6.33	6.43
18:3ω3	21.33	18.27	17.89	14.36	12.84
18:4ω3	2.54	2.06	1.89	1.68	1.66
18:5	1.61	1.76	3.40	2.71	3.64
minor components	0.15	0.12	0.16	0.11	0.16
unknowns	2.54	2.31	2.63	2.51	2.91
db mol^{-1}	2.22	1.98	2.05	1.75	1.72

freeze-thaw damage in woody twigs. However, in *C. reinhardii*
CW15+ it is suggested that the alteration in fatty acid
composition is a secondary event and not the cause of freezing
injury.

Several independent determinants of cell function, including
the loss of mobility, release of intracellular enzymes (Fig.
4), alterations in cellular ultrastructure (see below) and
reduction in cell surface change, occur immediately upon
thawing from -5°C. However, the changes in lipid composition
become apparent only with increasing time at 20°C (Table 5).
In addition, when lipids are extracted immediately from cells
frozen to different final temperatures, with a range of
survival from 100 to 0 percent, there was no difference in
fatty acid composition between any of the frozen and thawed
cell suspensions (Clarke *et al.*, in press). Following cooling
to the lower temperatures cell disruption is observed during
thawing by cryomicroscopy (Plate 1) and thin section electron
microscopy (see below). These major differences in cell
viability and morphology are not however reflected in changes
in fatty acid composition.

Whilst the alterations in phospholipid fatty acid composition are not the primary cause of freezing injury the release
of lipid breakdown products from damaged cells may affect
potentially viable cells. Free fatty acids, which are released
following freezing and thawing (Morris *et al.*, 1979) are
potent inhibitors of photosynthesis (Krogman and Jagendorf,
1959) and this may account for some of the differences observed between metabolic activity and cellular viability.

ULTRASTRUCTURE OF CELLS DAMAGED BY FREEZING

Following freezing to and thawing from -5°C there are numerous
changes in the cellular ultrastructure visible by thin-section
electron microscopy (Plate 2). These pathological alterations
become more extreme the longer the cells are maintained at
-5°C. Vesicles appear in the cytoplasm and increase progressively in size. Mitochondria can be detected after short
times at -5°C, but following longer periods of incubation
they apparently degenerate and are not recognisable in thin
section. The thylakoids which are uniformly stacked in control become 'baggy' upon thawing and the cytoplasmic ground
matter becomes steadily less granular and contains densely
staining bodies which are possibly lipid. After extended
periods at -5°C evagination of the plasmalemma occurs, there
is a loss of cell definition and eventually only membrane
vesicles remain. Upon thawing from -10°C a similar, but
more rapid, sequence of events occurs.

Great care must however be taken in the interpretation of

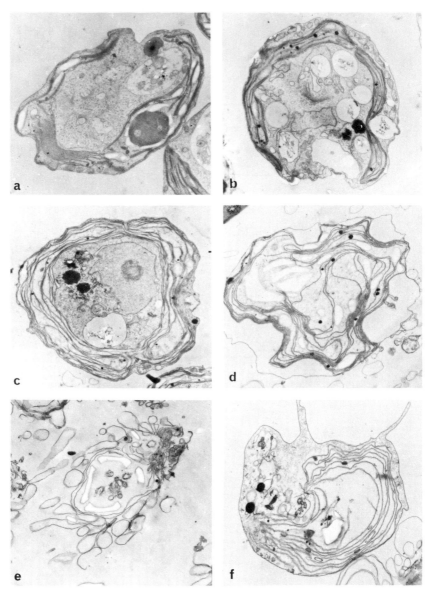

Plate 2. *Effect of freezing and thawing on the ultrastructure of Chlamydomonas reinhardii CW15+. (a) Control x4,500, (b-e) following freezing to and thawing from $-5^{\circ}C$, cells were maintained at $-5^{\circ}C$ for different times before thawing, (b) 5 min x4,500, (c) 15 min x4,500, (d) 30 min x5,940, (e) 1,000 min x6,750, (f) cells frozen to and maintained at $-10^{\circ}C$ for 5 min x5,625.*

this and other studies of freezing injury by thin section
electron microscopy, when a fixation and embedding protocol
satisfactory for unfrozen cells is applied, without modification, to thawed material. It is tacitly assumed that the
fixation characteristics of cells damaged by freezing are
the same of those of unfrozen cells. However, freezing modifies both the membrane lipid (Table 5) and protein (Heber
et al., this volume) components and causes release of intracellular proteins and lipid degradation products (Morris *et
al.*, 1979). Any or all of these factors may modify the
chemical fixation processes. Thus it is not apparent whether
the observed alterations in morphology are due to freezing
and thawing *per se* or to the indirect effects of freezing
damage upon the fixation characteristics of cells.

MECHANISM OF FREEZING INJURY

At the low rate of cooling examined in this study (0.25°C
min^{-1}) cell damage is associated with the stresses of freeze-
induced dehydration and resuspension upon subsequent thawing.
It is often assumed that the response of cells to freezing
and thawing may be simulated under isothermal conditions by
hypertonic dehydration followed by resuspension in isotonic
medium. However, during freezing the median lethal temperature for *C. reinhardii* CW15+ is -5.3°C (Fig. 2) at which
temperature the cells would be exposed to a solution equivalent to 1.42 M NaCl. However, at 20°C a 50 percent reduction
in viability occurs following exposure to and resuspension
from 0.24 M NaCl, equivalent to a solution in equilibrium
with ice at 0.9°C. In other systems, isothermal shrinkage:
rehydration studies also overestimate the damage which occurs
during freezing.

   C. reinhardii is extremely sensitive to hypertonic media
and the recovery upon resuspension is dependent on the nature
of the hypertonic medium (Fig. 5). Compounds that are non-
penetrating (NaCl, glucose) or slowly permeating (glycerol)
are damaging, whilst rapidly permeating compounds (dimethyl
sulphoxide, methanol) are relatively non-toxic since they
induce only a transient volume change. The toxicity observed
at high concentrations of penetrating additives may be the
result of the direct effects of high intracellular levels of
these additives. The toxic effects of non-penetrating compounds however, cannot be ascribed solely to dehydration, as
a significant difference is observed between ionic and non-
ionic additives at the same molarity; other factors must
therefore contribute to cellular injury.

   During osmotic shrinkage of plant protoplasts there is a

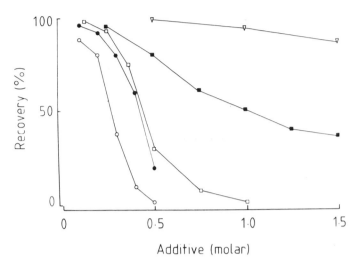

Figure 5. *Recovery (%) of Chlamydomonas reinhardii CW15+*
following exposure to methanol (▽), dimethylsulphoxide (■),
glycerol (□), glucose (●) or sodium chloride (○) for 5 min
at 20°C and dilution into isotonic medium.

reduction in the effective surface area and thus upon resuspension lysis occurs before the isotonic volume is attained.
Evidence for this has come largely from the observation that
following hypertonic exposure cell survival is modified by
the osmolality of the resuspension medium (Steponkus *et al.*,
this volume). *C. reinhardii* CW15+, however, does not behave
in an analagous manner in that 50 percent cell damage occurs
following exposure to 0.25 M NaCl at an average cell volume
of 56 percent of control value and viability is not affected
by dilution into either hypertonic medium or distilled water
(Morris and Grout, unpublished data). Direct examination of
cells exposed to and then fixed in 0.25 M NaCl reveals many
pathological features (Plate 3), including vesiculation of
the cytoplasm and numerous vesicles associated with thylakoids
in many cells. There was no direct evidence of extrusion of
membrane from the plasmalemma.

In *C. reinhardii* CW15+ it thus appears that cellular injury
occurs directly in the shrunken state. The mechanism of this
injury, and its relation to freeze-thaw damage require further
investigation.

CONCLUSIONS

It is obvious that in *C. reinhardii* CW15+, as with many other

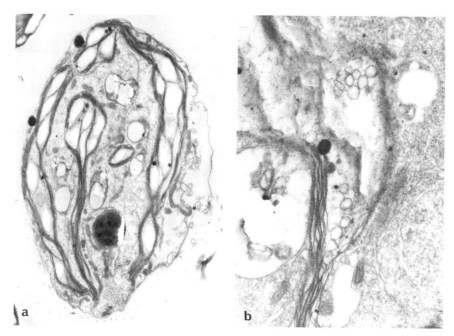

Plate 3. *Ultrastructure of Chlamydomonas reinhardii CW15+ following exposure to 0.25 M NaCl for 5 min at 20°C, for control see plate 2(a). (a) x7,650, (b) x24,750.*

cell-types, early events of freeze-thaw injury include alterations to the structure and function of the cellular membranes. However, despite extensive research (Many authors, this volume), the biochemistry of this injury is not understood. The data presented in this article highlight many of the problems associated with this type of research. All the properties of the membranes that were examined in this study were modified in populations which were damaged by freezing; it is thus impossible to distinguish the primary site of membrane injury (if one exists), from secondary, pathological events.

It is apparent that in *C. reinhardii* CW15+ non-lethal or 'metabolic' injury may occur, *e.g.* a > 95 percent reduction in ^{14}C fixation is observed immediately upon thawing in a population with a 50 percent survival. Currently only averaged properties of the population can be measured and compared with cell survival which is obtained some considerable time after thawing. A critical appraisal of freezing injury will be obtained only when it is possible to separate lethally from non-lethally injured cells and study the nature of injury and repair in the two populations separately.

There is no simple relationship between the bulk phospholipid fatty acid composition of a cell and the response to either hypertonic or freeze-thaw stress. If the membrane is the primary site of injury factors other than the fatty acid composition must be important, such as the direct effects of temperature on proteins (Franks, this volume) and the destabilization of the interactive forces between the constituent phospholipids, sterols and proteins that ultimately determine the mechanical properties of membranes.

STRESSES INDUCED BY CONTRACTION AND EXPANSION DURING A FREEZE-THAW CYCLE: A MEMBRANE PERSPECTIVE

P.L. Steponkus, J. Wolfe and M.F. Dowgert

Department of Agronomy, Cornell University, USA.

INTRODUCTION

During a freeze-thaw cycle, biological cells are subjected
to numerous stresses. Proposed mechanisms of injury frequently derive from either theoretical or experimental attempts
to resolve the freeze-thaw stress complex into the component
thermal, mechanical, and chemical stresses. As osmotic stress
has been long invoked in freezing injury (Scarth *et al.*, 1940),
resolution of the chemical versus mechanical effects, *i.e.*
solute concentration versus volumetric contraction, has been
a major concern. For example, Lovelock (1953a,b, 1954b)
suggested that injury is a consequence of the concentration
of electrolytes. Whereas, Meryman (1967, 1968, 1974, Meryman
et al., 1977) proposed that injury is a direct result of
volumetric contraction *per se*. Although Rall *et al.* (1978)
concur with Lovelock (1953a,b) that injury is associated with
the concentration of salts, and not the result of volumetric
reductions, they nevertheless differ with Lovelock by considering injury as a consequence of extracellular rather than
intracellular salt concentration. Despite the claims that
"nearly perfect covariance of the two phenomena (chemical *v.*
mechanical) has been dissected" and that injury cannot be
"ascribed to injurious levels of electrolyte" (Williams and
Shaw, 1980), the interpretations remain equivocal.

This dilemma is, in part, due to the concept of injury
being the result of a single critical stress, *e.g.* either
volumetric contraction or solute concentration. The demise
of most hypotheses based on a single stress occurs when a
given hypothesis does not apply under all conceivable conditions. This is not to say, however, that the basic tenets

of any of the hypotheses arc necessarily erroneous. Such a
conclusion is only implied in the concept that a single stress
is universally critical. In contrast, Mazur (1977) and
Steponkus (1978, 1980) envisage a sequence of potentially
lethal stresses, although they differ as to order (Steponkus
and Wiest, 1979). However, even when experimental results
are embodied in such a sequential framework (Mazur, 1977), no
precise explanation of injury at the molecular level has emerged.

Also contributing to the dilemma is that few hypotheses
have evolved from a consideration of membrane lesions, with
most being based solely on measurements of gross survival.
Although freezing injury is commonly inferred to be the
consequence of membrane damage, there have been few attempts
to characterize specific membrane lesions, especially to the
plasma membrane. It is difficult to envisage how a mechanism
of injury can be proposed without insight into the nature of
injury. As the freeze-thaw stress complex is composed of
several individual stresses, freeze-thaw injury may result
from any one of several potentially lethal lesions. Lovelock
(1953a,b, 1954b) attempted to provide insight into specific
lesions resulting from chemical stresses, but recent studies
have not been as complete, and multiple forms of injury are
only inferred. For instance, Rall *et al.* (1978) infer that
in addition to salt damage, survival is diminished by an increased susceptibility to osmotic shock caused by glycerol.
In other cases, multiple forms of injury are apparent in the
data but are not discussed. For example, in the report of
Williams and Shaw (1980), two forms of injury diminish survival;
that which occurs upon exposure to hypertonic conditions
(approx. 30%) and that which occurs upon subsequent hypotonic
exposure (approx. 40%).

A greater understanding of freezing injury would develop
if both the freeze-thaw stress complex and the freeze-thaw
injury complex were resolved into specific stresses and
lesions and viewed as a sequence of events. If the tolerance
of the plasma membrane could be quantified then genotypic
differences in cold hardiness, the effects of cold acclimation,
or the efficacy of various cryoprotectants could be ascribed
to changes which either mitigate or preclude the various
cellular stresses or increase membrane tolerance to these
stresses. It is even possible that an enduring hypothesis
of injury might emerge. Therefore, we have attempted to
identify specific lesions in the plasma membrane and to
examine their incidence as a function of the various stresses
arising during a freeze-thaw cycle. This report discusses
the influence of stresses associated with contraction and
expansion of the plasma membrane of protoplasts isolated from
higher plants.

The Isolated Protoplast

A major impediment in the characterization of plasma membrane lesions is the inability to isolate sufficiently pure fractions from higher plants in a state which can be shown to be identical to their state *in vivo*. Such limitations led to the termination of our early efforts using isolated plasma membrane enriched fractions (Steponkus, 1973; Steponkus and Wiest, 1973; Wiest and Steponkus, 1975, 1977). Alternatively, we elected to study the effects of a freeze-thaw cycle on the plasma membrane *in situ* using isolated protoplasts. The first reports of such an approach were reported simultaneously by two independent groups (Siminovitch *et al.*, 1976; Wiest and Steponkus, 1976: Also, see Singh, 1979a,b; Siminovitch, 1979).

   Since damage to the plasma membrane is of prime concern, the isolated protoplast is a most appropriate arena in which to study freezing injury in higher plants. It is unfettered by a cell wall and unencumbered by neighbouring cells. A protoplast stands alone, naked in its environment, with strengths and weaknesses in full view. Isolated protoplasts behave as ideal osmometers over a wide range of osmolality, 300-3,000 mOsm (Steponkus and Wiest, 1978, 1979; Wiest and Steponkus, 1978a). Over this range, the protoplasts remain spherical; this facilitates quantitative microscopic studies of cell dehydration. Calculations of both volume and surface area are thus possible. Finally, isolated protoplasts provide an arena in which the effects of a freeze-thaw cycle on higher plant cells can be compared directly with those of mammalian cell types exposed to similar environments.

EXPANSION-INDUCED LYSIS DURING A FREEZE-THAW CYCLE

Scarth *et al.* (1940) were among the earliest to note that injury to the plasma membrane of plant cells was related to plasmolysis and deplasmolysis which occurred during freezing and thawing and that lysis occurred during deplasmolysis. This phenomenon has been reported or inferred to occur in mammalian cell types, especially erythrocytes (Lovelock, 1953a, b, 1954b; Ponder, 1955; Zade-Oppen, 1968). It is largely from studies of erythrocytes that proposals for such behaviour have developed. Lovelock (1953a,b) proposed that red blood cells become permeable to sodium ions under hypertonic conditions, and the uptake of solute causes the cells to lyse when the cells were subsequently returned to isotonic conditions. Similarly, Zade-Oppen (1968) and Meryman (1971) concluded that the influx of solutes under hypertonic conditions resulted in the cell reaching a haemolytic

volume at a higher osmolality upon return to isotonic
conditions. While agreeing in the outcome of such an osmotic
excursion, the proposals differ in the manner by which the
additional solute uptake is effected. Lovelock (1953a, b)
proposed that the altered electrolyte permeability is
attributable to increased electrolyte concentration; Meryman
(1967, 1968) attributed it to the physical stress associated
with the attainment of a minimum critical volume; Mazur
(1977) invoked a transient hydrostatic tension producing a
driving force for a net influx of solutes.

     None of these proposals, based on studies of erythrocytes,
are however, applicable to isolated plant protoplasts. All
of the proposals suggest that solute influx occurs during the
hypertonic exposure and lysis occurs during subsequent hypotonic exposure. Implicit is the concept of a fixed maximum
critical volume at which lysis occurs and that this is merely
achieved at higher osmolalities. Such an event would lead
to a hysteresis in the osmotic behaviour upon exposure to
hypotonic conditions following a hypertonic exposure. No such
hysteresis is observed in the osmotic behaviour of isolated
protoplasts exposed to hypertonic conditions and subsequently
diluted (Wiest and Steponkus, 1978a). Instead, protoplast
lysis occurs at a lower volume and surface area. Although
osmotic behaviour is best described as a function of volume,
expansion-induced lysis is a function of the area over which
the plasma membrane must be distended. This observation
shifts attention from factors which alter the solute content
of the cell to factors which directly affect the plasma membrane.

THE TOLERABLE SURFACE AREA INCREMENT (TSAI)

Isolated protoplasts behave as ideal osmometers within the
range of 300 to 3,000 mOsm (Steponkus and Wiest, 1978, 1979;
Wiest and Steponkus, 1978a) and remain spherical to the
resolution of light microscopy. These volumetric changes
necessitate substantial changes in the surface area of the
plasma membrane. The upper limits of expansion are not fixed,
for expansion-induced lysis does not occur at an absolute
critical surface area. In protoplasts isolated from non-
acclimated tissues, surface area contractions are incompletely
reversible, and the expansion potential is limited by an
absolute tolerable surface area increment. We refer to
this constant as the TSAI value and use $TSAI_{50}$, a population
parameter, to denote the surface area increment at which 50
percent of the cells will lyse.

     The basic tenets of the TSAI concept are that surface
area contractions are incompletely reversible and that the

TSAI value is independent of the extent of contraction. For example, for protoplasts isolated from leaves of 'Puma' rye (*Secale cereale*) the $TSAI_{50}$ value is 1000 μm^2. In 530 mOsm solutions, the average surface area is 3750 μm^2, and 50 percent of the population will lyse if the suspending solution is diluted to 360 mOsm when a surface area of 4750 μm^2 is achieved. If, however, the protoplasts are first contracted in a hypertonic medium (910 mOsm) so that the average surface area is 2750 μm^2, 50 percent of the population will lyse when returned to the isotonic medium (530 mOsm) - when a surface area of 3750 μm^2 is achieved.

We have therefore proposed that during a freeze-thaw cycle, injury to protoplasts isolated from non-acclimated tissues is caused by an alteration in plasma membrane resilience (Steponkus and Wiest, 1979; Wiest and Steponkus, 1978) rather than an alteration in the internal solute concentration as proposed for red blood cells (Lovelock, 1953a; Mazur, 1977; Meryman, 1971; Zade-Oppen, 1968). Specifically, an alteration in the membrane as a result of contraction decreases the maximum critical surface area, and an expansion-induced dissolution of the plasma membrane occurs when the maximum critical surface area is exceeded.

A contraction-induced alteration in protoplast resilience has been observed previously. As early as 1939, Tornava observed that protoplasts could double their initial surface area before lysis occurred. It was added, however, that if they were first contracted, the original surface area could not be regained. Scarth *et al.*, (1940) also observed that the extent of plasmolysis influenced the point of lysis of intact cells, the point of lysis of strongly plasmolyzed cells being lower than that of cells which were weakly plasmolyzed. The significance of these observations has been overlooked in most contemporary explanations of freeze-thaw injury.

Given that surface area contractions of the plasma membrane are potentially lethal, characterization of both the $TSAI_{50}$ and the osmotic stresses arising during a freeze-thaw cycle increase the ability to establish cause and effect. This approach enables the influence of factors which alter the sensitivity of the plasma membrane to contraction and expansion to be examined separately from those factors which alter the extent of osmotically-induced contraction and expansion. These factors may be a result of genotypic differences, the addition of various cryoprotectants, or cold acclimation.

Genotypic Differences

The $TSAI_{50}$ value differs with species, being 900 μm^2 in

spinach protoplasts, *cv.* Winter Bloomsdale (Wiest and
Steponkus, 1978a); 400 μm^2 for wheat, *cv.* Yorkstar (Wiest
and Steponkus, 1978b); 600 μm^2 in 'New York Common' rye and
1000 μm^2 in 'Puma' rye. One cannot, however, infer a ranking
of hardiness based solely on the $TSAI_{50}$. Because TSAI value
is an absolute value, it must be considered in relation to
the osmotic behaviour during a freeze-thaw cycle which will
be influenced by the internal solute concentration and the
initial cell size.

Effect of Cryoprotectants

Studies to directly determine the effect of specific cryoprotectants on the $TSAI_{50}$ value have not yet been conducted.
It is to be expected, however, that cryoprotectants will
influence the plasma membrane since the chemical nature of
the suspending medium influences the $TSAI_{50}$ value (Steponkus
and Wiest, 1978). For instance, the ionic species influence
the sensitivity of protoplasts to a freeze-thaw cycle and
follows the series $Li^+ = Na^+ < K^+ = Rb^+ = Cs^+$ and $Cl^- < Br^-$
$< I^-$. Thus, in wheat protoplasts the $TSAI_{50}$ is 600 μm^2 in
$NaCl + CaCl_2$ but only 400 μm^2 in $KCl + CaCl_2$ (Wiest and
Steponkus, 1978a). In addition, there is an influence of pH
on the expansion potential (Steponkus and Wiest, 1978).
Hence, although the mechanical stresses of contraction and
expansion are the primary causes of expansion-induced lysis,
chemical influences cannot be eliminated, as has been suggested by Meryman *et al.* (1977) and coworkers (Williams and
Shaw, 1980).

Effect of Cold Acclimation

In 1941, Siminovitch and Levitt reported that the plasma
membrane of hardy cells is less easily ruptured during
deplasmolysis than that of unhardy cells - an observation
which has also been overlooked in contemporary hypotheses of
injury. The TSAI concept confirms this observation and allows
for its quantification. The $TSAI_{50}$ for protoplasts isolated
from non-acclimated leaves of Puma rye is 1000 μm^2, the
$TSAI_{50}$ for protoplasts isolated from acclimated leaves is
3000 μm^2 (Steponkus *et al.*, 1979). Thus, one component of
the cold acclimation process is to increase the tolerance
of the plasma membrane to contractile stresses.

ELABORATION OF THE TSAI CONCEPT BY CRYOMICROSCOPY

The TSAI concept was originally developed from studies of
populations of protoplasts contracted and expanded by either

direct osmotic manipulation or by exposure to a freeze-thaw
cycle (Wiest and Steponkus, 1978a). In such studies, cell
volumes and corresponding surface areas were determined
before and after the manipulation. With recent advances in
cryomicroscopy (Diller and Cravalho, 1970; McGrath et al.,
1975), our attention has been directed to individual protoplasts during a freeze-thaw cycle. The specifications of
our particular cryomicroscope have been previously described
(Levin et al., 1978; Steponkus et al., 1981). Elaboration
of the TSAI concept by cryomicroscopy has considered a
number of aspects. These include the incidence of expansion-
induced lysis relative to other freeze-thaw induced lesions,
determination of the TSAI value of individual cells exposed
to a freeze-thaw cycle, and quantitative accountability for
freeze-thaw injury.

Incidence of Expansion-Induced Lysis Relative to other Lesions

Visual observation of non-acclimated protoplasts during
a freeze-thaw cycle has revealed several different lesions
as manifested in cellular volumetric responses. At fast
rates of cooling to low temperatures ($>5 min^{-1}$ to $-20^{\circ}C$)
immediate physical disruption of the plasma membrane is
associated with the formation of intracellular ice (Dowgert
and Steponkus, 1979, 1981). When cooled at slow rates to
relatively warm subzero temperatures ($< 3.1 min^{-1}$ to $-5^{\circ}C$)
protoplasts exhibit characteristic osmometric behaviour
during cooling and warming. During warming, however, lysis
occurs at a surface area which is always less than the initial surface area. When protoplasts are cooled to temperatures
lower than $-5^{\circ}C$ (-10 to -20°) they exhibit characteristic
osmometric behaviour during cooling but are osmotically
inactive during warming and remained contracted. The incidence of these two forms of injury at slow cooling rates is a
function of the minimum temperature imposed. At temperatures
of $-3^{\circ}C$ and $-5^{\circ}C$, expansion-induced lysis accounts for all of
the injury; at $-15^{\circ}C$ and $-20^{\circ}C$, all of the injured cells are
osmotically inactive; at $-10^{\circ}C$, the incidence of both types
of injury is approximately equal.

Thus, injury of protoplasts may be due to any one of
several lesions, depending on the freeze-thaw protocol. At
relatively warm subzero temperatures, injury to non-acclimated
protoplasts is the result of expansion-induced lysis. At
lower temperatures, however, injury is manifested as a loss
of osmotic responsiveness. On this basis, it is argued that
the mechanical stresses of contraction and expansion are the
first potentially lethal stress encountered by non-acclimated

protoplasts during a freeze-thaw cycle. Similar studies with
protoplasts isolated from acclimated tissue indicate that
the incidence of expansion-induced lysis is reduced considerably. In acclimated protoplasts, the $TSAI_{50}$ value of
3000 μm^2 is not exceeded even at temperatures below $-26^{\circ}C$,
the LT_{50} for the population. Thus, it is predicted that
injury to acclimated tissue is not the result of contraction-
induced alterations in the plasma membrane.

*Direct Measurement of the TSAI Value During a Freeze-Thaw
Cycle*

The $TSAI_{50}$ value is a population parameter calculated from
the average surface area of the population at any given
osmolality. Since there is a range of protoplast sizes
(radius = 10 to 20 μm at 530 mOsm), absolute changes in volume
and surface area during osmotic manipulation differ for the
various cell sizes within the population. Although the $TSAI_{50}$
is an absolute constant for a population of protoplasts and is
independent of the extent of contraction, it is assumed that
the TSAI value for any particular cell is proportional to
its initial size (see Steponkus and Wiest, 1979). Cryo-
microscopic studies confirm this assumption, that is,
larger cells have a larger TSAI value. Measurements of
individual protoplasts were made before cooling, in the
contracted state, and before lysis during warming. TSAI
(lytic surface area - contracted surface area) was related
to initial surface area, A, (in μm^2) according to the
equation: TSAI = 45 + 0.26A. A cell with an initial
surface area of 1500 μm^2 thus had a TSAI value of 435 μm^2
whereas a cell with an initial surface area of 6000 μm^2 had
a TSAI value of 1605 μm^2. While these values were as predicted from osmotic manipulation experiments, variability
was rather large because mechanical deformation of the
protoplasts by surrounding ice resulted in a poor estimate
of the contracted surface area in the frozen state.
Deformation was minimal at warmer subzero temperatures, but
increased at lower temperatures. Since a sphere has the
minimum surface area possible for a given volume any
deviation from sphericity due to mechanical deformation
would decrease the extent of the surface area contraction
for a given reduction in volume.

*Expansion-Induced Lysis: Quantitative Accountability for
Freeze-Thaw Injury.*

It was reported previously that the amount of injury
incurred by non-acclimated spinach protoplasts during a

freeze-thaw cycle could be accounted for quantitatively by
lysis that occured when protoplasts were induced to contract
and expand osmotically at room temperature (Wiest and
Steponkus, 1978a). In subsequent studies with wheat protoplasts the quantitative agreement was less exact (Wiest, 1979).
The amount of injury following a freeze-thaw cycle was 30
percent less than predicted on the basis of the TSAI value
determined at room temperature. This discrepancy was ascribed
to a temperature dependency of the expansion process, as less
injury occurred if the protoplasts were expanded at 0°C
instead of 25°C. Such a temperature dependency was opposite
of what one might have expected intuitively, $i.e.$ more injury
if expansion occurred at low temperatures. In the studies
with spinach and wheat, the protoplasts were suspended in
varied concentrations of osmoticum and frozen to -3.9°C.
Hence, the extent of contraction was similar in all of the
treatments and only the extent of expansion was varied. The
range of survival was between 10 and 90 percent.

    Subsequently, studies with rye protoplasts have also
indicated that the extent of injury incurred during a freeze-
thaw cycle is less than that predicted from the TSAI value.
The LT_{50} for a population of rye protoplasts frozen at
$2.8 hr^{-1}$ was -6°C. On the basis of TSAI value, the LT_{50} would
have been predicted to be only about -2°C. In the case of
rye protoplasts, the $TSAI_{50}$ values were determined at 0°C,
and a thermal dependency could not be invoked to explain the
difference. Again, however, we ascribe this discrepancy
to the fact that during freezing protoplasts do not necessarily remain as perfect spheres due to mechanical deformation
by the surrounding ice. Survival from a freeze-thaw cycle
might be greater than predicted if it is assumed that the
protoplasts remain spherical during freezing.

UNCOUPLING VOLUMETRIC FROM SURFACE AREA CONTRACTION

Meryman (1967, 1968, 1971) proposed the "minimum critical
volume" hypothesis to account for freeze-thaw induced lysis
in a wide range of biological organisms. However, we do not
consider that it is applicable to the expansion-induced lysis
of isolated protoplasts. Specifically, the hypothesis invokes a mechanical resistance to volumetric reduction which
is manifested as anomalous osmotic behaviour at high osmolalities. No such behaviour is observed in isolated protoplasts, and that inferred for red blood cells has been
questioned (Wiest and Steponkus, 1979). Also the hypothesis
invokes an alteration in permeability so that solute influx
occurs and is manifested as anomalous behaviour upon exposure

to hypotonic conditions. Again, no such behaviour is observed in isolated protoplasts (Steponkus and Wiest, 1978, 1979; Wiest and Steponkus, 1978a).

The unequivocal determination of whether injury is the result of a reduction in cell water volume or a reduction in the surface area of the plasma membrane requires the uncoupling of volumetric from surface area contraction. Upon first consideration, such a task would seem formidable; in fact, it is not. With a micropipette connected to a manometer, individual protoplasts can be "towed" along an osmotic gradient in order to effect a volume change. By applying a greater negative pressure to the pipette and allowing a portion of the protoplast to intrude into the pipette, the shape can be controlled and in this way surface area contractions can be uncoupled from volumetric contractions. This approach has provided the most direct and unequivocal demonstration of the TSAI concept (Wolfe and Steponkus, 1981b). Protoplasts can be "towed" along a continuous concentration gradient and allowed to contract in both volume and surface area. Such protoplasts cannot survive a sufficiently slow and large increase in surface area upon reexpansion and lyse when "towed" back to a region of low osmolality. In contrast, protoplasts allowed to contract volumetrically while an appropriate pressure is applied, change their shape and intrude into the pipette while maintaining their surface area constant. When such cells are returned to isotonic conditions, they do not lyse. On this basis, we conclude that the minimum critical volume hypothesis is not applicable to isolated protoplasts.

STRESS-STRAIN RELATIONSHIP (SSR) OF THE PLASMA MEMBRANE

Experimental and theoretical studies of the mechanical properties of biological membranes have been limited almost exclusively to sea-urchin eggs (Cole, 1932; Mitchison and Swann, 1954) and erythrocytes (Norris, 1939; Rand and Burton, 1964; Waugh and Evans, 1979). These membranes are atypically rigid (Gomperts, 1977) and such studies have only considered small changes in area. The mechanical properties of erythrocytes have been considered in several sophisticated analyses (Chien *et al.*, 1978; Deuling and Helfrich, 1976) and the classical treatment of visco-elasticity has been extended to describe the stress-strain relationship (Evans and Skalak, 1979).

Several of the observations which led to the formulation of the TSAI concept suggest that the SSR of the plasma membrane of isolated plant protoplasts is distinctly different from the previously characterized systems.

Specifically, over a large range of osmolality, isolated
protoplasts behave as ideal osmometers and cell volume changes
uniformly with the protoplast remaining spherical within the
limits of resolution of the light microscope. The contraction
may be as large as threefold and very large contractions are
irreversible (inelastic). The maintenance of sphericity over
a large range of volumes suggests that the membrane is, at
all volumes, tending to contract in surface area. As such,
it is under tension at all values of area and the equilibrium
state has a resting tension (γ_r), which is not determined by
the extent of contraction or expansion.

Measurement of the Resting Tension

Using the 'elastimeter' of Mitchison and Swann (1954), *i.e.*
micropipette aspiration of the plasma membrane, the resting
tension (γ_r) of the plasma membrane of isolated protoplasts
has been determined (Wolfe and Steponkus, 1981a,b). The
analysis is based on the fact that a spherical membrane with
a radius of R, withstanding a hydrostatic pressure difference
of P_i must have a tension of

$$\gamma = P_i R/2$$

For protoplasts isolated from non-acclimated 'Puma' rye
leaves in 530 mOsm sorbitol, γ_r is typically 100 $\mu N.m^{-1}$,
although a considerable range of values has been measured (0
to 1000 $\mu N.m^{-1}$). If the protoplasts are subsequently
equilibrated in either 410 or 700 mOsm sorbitol solutions,
the protoplasts remain spherical and the same range of γ_r is
observed. The spherical shape and the ability to sustain a
tension suggest that the contraction or expansion is associated with an effective area contraction or expansion of the
plasma membrane. Simple homogeneous contraction or expansion
alone cannot explain the approximate constancy of γ_r (Wolfe
and Steponkus, 1981a). Instead, we infer the existence of
a reservoir into which membrane material is deposited during
contraction and from which material is retrieved during
expansion.

Time Dependence of the Stress-Strain Relation

The SSR of isolated rye protoplasts depends strongly on the
time scale of deformation. For small deformations over
short periods of time (seconds), the amount of material in
the plane of the membrane is conserved and the deformations
follow a simple elastic relation:

$$\gamma = k_A (A-A_o)/A$$

where γ is the membrane tension, A is the area of the membrane, A_o is the area which would be occupied by the material currently in the membrane if it were not subject to a tension, and k_A is the modulus of elasticity. Over longer periods of time (minutes), a large imposed tension produces a plastic change in the area, *i.e.* a change in A_o. Since the unstretched area per molecule is unchanged by tension, this change must be due to exchange of membrane material between the plane of the membrane and the inferred reservoir. If the tension is rapidly relaxed to γ_r, this change is apparent as a change in membrane area after the excursion. Over long time scales (minutes), however, the tension reverts to its resting tension (γ_r), independent of the induced change in area. Thus, the deformation on this time scale follows a surface energy law, where $\gamma = \gamma_r$. Since the rate at which A_o changes depends on the tension, the change in area (ΔA) may be written as:

$$\Delta A = \delta(A-A_o) + \delta A_o(\gamma,t)$$

$$= \frac{A}{k_A} (\gamma-\gamma_r) + \delta A_o(\gamma,t)$$

Experimental determinations (Wolfe and Steponkus, 1981b) gave the value $k_A = 230$ mN.m^{-1} and $\gamma_r = 100$ μN.m^{-1}.

From analysis of the SSR of isolated protoplasts, it is apparent that contraction and expansion of the plasma membrane involves both elastic and plastic responses (Wolfe and Steponkus, 1981a,b). Both an elastic law and a surface energy law must therefore be invoked in describing this behaviour. A small change in γ_r following large surface area changes would require a very low value of k_A, on the order of 100 μN.m^{-1} as opposed to the measured value of 230 mN.m^{-1}. Such a value if unrealistic for biological membranes (Wolfe and Steponkus, 1981a). Thus, only a small change, on the order of 2 percent, in membrane area may be achieved by intrinsic contraction or expansion. In contrast, the SSR analysis suggests that large surface area changes in the plasma membrane involve subduction of membrane material into a reservoir during contraction.

Resolution of the TSAI Value.

Application of the SSR data to the TSAI concept requires the introduction of a parameter γ_c, the 'critical tension' for lysis, which we loosely define as the tension necessary to

greatly increase the probability of lysis. The plasma membrane of isolated protoplasts usually ruptures at tensions
greater than 4 mN m^{-1}. Using γ_c, TSAI may be defined as the
amount by which the area of the plasma membrane may be increased (under defined conditions) before $\gamma = \gamma_c$. TSAI$_{50}$ is
a population parameter, $i.e.$ the value of the average area
increase in a population of cells of various sizes where 50
percent lyse. Note that TSAI$_{50}$, the median of TSAI, is not
necessarily equal to the mean TSAI. The value of A-A$_o$ at
which $\gamma = \gamma_c$, $i.e.$ the amount by which the membrane is
stretched before lysis, may be defined as TSAI^o. In principle, it may be measured as the limit of TSAI for short
periods. Thus, there are some relations which are true by
definition, viz:

$$A_r - A_o = \frac{\gamma_r A_r}{k_A}$$

$$TSAI^o = \frac{(\gamma_c - \gamma_r)A}{k_A}$$

$$and\ TSAI = \frac{(\gamma_c - \gamma_r)A}{k_A} + \delta A_o(\gamma, t)$$

Hence, a change in the TSAI value may be due to a change in
the elastic modulus k_A; the critical lysing tension, γ_c; the
resting tension, γ_r; or the function $\delta A_o(\gamma, t)$ which describes
the transfer of material between the membrane and the reserv-
oir.

     A molecular scenario for the apparent irreversible volum-
etric responses which are quantified by the TSAI concept may
therefore be proposed in terms of the SSR. Upon exposure to
hypertonic conditions, the volume of the cell contracts and
the tension in the membrane is quickly relaxed to zero. A
very small volume change is sufficient and this occurs within
milliseconds. Since at zero tension the membrane can contract
without doing work, the reservoir-membrane equilibrium is
altered slightly (only approximately 0.01 times the thermal
energy per molecule, but for an aggregate of 1000 molecules
this is considerable) and so material is subduced into the
reservoir on a time scale of several minutes until the cell
regains sphericity, with a small resting tension (γ_r). When
the cell is returned to isotonic conditions, water begins to
enter rapidly. This stretches the membrane elastically and
creates a large ($< mN.m^{-1}$) tension, inducing reincorporation
of reservoir material. The excursion will not be totally
reversible, however, if the reincorporation does not proceed
as rapidly as water influx, if all of the subduced membrane
material is not readily accessible, or if the elastic

expansion exceeds γ_c. Initially, the reaction proceeds faster
on expansion because $\gamma > \gamma_r$, and appears to be an exponential
function of time. Thus, within the first seconds, the excursion is partially reversible, but over the range of minutes
(1 to 60), the excursion appears irreversible. The initial
amount retrieved is manifested as the TSAI value and represents the elastic component plus the rapid phase of the plastic
component. The cell lyses, however, because of the limited
retrieval of the entire amount deleted. The above explanation
may include homogeneous (intrinsic) contraction and closer
molecular packing, microscopic buckling of the membrane,
vesiculation, and subduction of the membrane into the associated reservoir.

MICROSCOPIC OBSERVATIONS OF SUBDUCTION OF THE PLASMA MEMBRANE

Subduction of plasma membrane components during contraction
is predicted by the TSAI concept (Steponkus and Wiest, 1979)
and elaborated by the SSR analysis (Wolfe and Steponkus,
1981a). It was thus of obvious interest to determine whether
such a membrane alteration could be seen by light microscopy.
Initial experiments used non-acclimated protoplasts labelled
with Concanavalin-A conjugated with fluorescein. When
labelled in 530 mOsm sorbitol, a uniform fluorescence of the
plasma membrane was observed. Following contraction in 1000
mOsm sorbitol, regions of the plasma membrane appear to
protrude into the protoplast interior, although the outer
perimeter appears smooth. Upon greater contraction in 3000
mOsm sorbitol, fluorescent regions appear within the protoplast. Following expansion upon subsequent dilution, fluorescent material remains observable within the protoplast.
These observations suggest that contraction of non-acclimated
protoplasts results in endocytotic vesiculation. Similar
studies with acclimated protoplasts have been precluded
because of difficulties in acclimating etiolated leaves which
are necessary when using Concanavalin-A fluorescein to minimize autofluorescence from chlorophyll.
    Comparisons of the behaviour of the plasma membrane of
acclimated with non-acclimated protoplasts during a freeze-
thaw cycle have been attempted without the use of a specific
plasma membrane marker. During freezing of non-acclimated
protoplasts, vesiculation of the plasma membrane cannot be
discerned with light microscopy. In some instances,
mechanical constriction of a portion of the plasma membrane
by the ice mass has resulted in "pinching off" of the cir-
cumscribed region. During freezing of acclimated protoplasts,
extensive exocytotic vesiculation of the plasma membrane is

clearly visible. This phenomenon has also been observed in acclimated protoplasts exposed to extremely high osmolalities (3000 mOsm). Most often the extruded material appears as either tethered spheres or long filamentous strands. Although a noticeable proportion of these exclusions become detached from the membrane during thawing, the majority remain attached and are gradually reincorporated into the plane of the membrane. While a systematic study has not yet been completed, exocytotic vesiculation of the plasma membrane appears to occur quite frequently in acclimated protoplasts in contrast to non-acclimated protoplasts. In some instances, however, vesiculation is not observed in some of the acclimated protoplasts, and invariably such protoplasts lyse upon subsequent expansion. These observation suggest that those protoplasts which undergo extensive exocytotic vesiculation during contraction are most likely to regain their initial surface area during expansion; whereas those which do not exhibit such vesiculation lyse during expansion.

The ability of the acclimated protoplasts to survive subsequent expansion, in spite of the extensive vesiculation, suggests that this deleted membrane material is reincorporated into the membrane. There is, however, a proportion which is detached and not reincorporated into the membrane suggesting that additional material is brought into the membrane during expansion. Before freezing, acclimated protoplasts are characterized by an extensive amount of membraneous material within the cell, usually just beneath the plasma membrane and quite often surrounding the vacuole. This network of membraneous material is not observed in non-acclimated protoplasts. It is quite tempting to suggest that such material is readily available for incorporation into the plasma membrane during expansion. Such a possibility, however, awaits experimental documentation.

Previously, Siminovitch *et al.* (1967a,b) have reported that substantial membrane augmentation occurs during cold acclimation. Cytological studies have revealed a substantial increase in cellular membranes and chemical studies have documented a large increase in the phospholipid content. Subsequent electron microscopic studies (Pomeroy and Siminovitch, 1971) demonstrated alterations in the cellular membranes, especially "membrane-bound vesicles derived from invaginations and folding of the plasmalemma" and "a vesicular form of endoplasmic reticulum". The authors suggested that "the folded condition of the membrane would....alleviate the stresses of contraction and expansion". Later, Singh *et al.* (1975) considered membrane augmentation as one of the major mechanisms of resistance to freezing, but the mechanism by

which augmentation enabled the cell to better accomodate the
stresses was not elaborated. While the impact of membrane
augmentation on the ability of a cell to endure the contraction and expansion stresses during a freeze-thaw cycle remains
to be demonstrated directly, it is clearly consistent with
the TSAI concept of freezing injury. More importantly, the
TSAI concept now allows for the significance of augmentation
to be assessed, so that it need not remain merely as one of
the many correlative observations in the study of cold
acclimation.

SUMMARY

Freezing injury may result from any one of several stresses
arising during a freeze-thaw cycle. The impact of the
mechanical stresses of contraction and expansion incurred by
isolated protoplasts during a freeze-thaw cycle is obvious
when considered from a membrane perspective. The plasma
membrane does not respond in a simple elastic manner to the
large surface area deformations resulting from osmotically
induced volumetric changes of the protoplast. Instead,
membrane material is deleted from or incorporated into the
membrane during contraction and expansion. The extent of
this exchange and the resultant tolerance of the plasma
membrane to surface area deformations is characterized by the
TSAI value. This parameter, which is readily quantified, is
influenced by the chemical environment, altered by cold
acclimation, and varies among genotypes. Analysis of the
global stress-strain relationship of the plasma membrane
allows for resolution of the TSAI value into its individual
components, γ_r, γ_c, k_A and $\delta A(\gamma, t)$, and allows for a molecular
description of the behaviour of the plasma membrane during
surface area deformations. With such a detailed understanding,
the significance of compositional changes in the plasma membrane and other cellular components resulting from cold acclimation may be determined directly. Similarly, the influence
of cryoprotectants may be determined directly. Clearly, an
analysis of both freeze-thaw induced stresses and the resultant lesions is essential for a detailed understanding of
freezing injury and the process of cold acclimation.

THE EFFECT OF DIFFERENT COOLING RATES ON THE MEMBRANE OF FROZEN HUMAN ERYTHROCYTES

S. Fujikawa

*Institute of Low Temperature Science, Hokkaido University,
Japan.*

INTRODUCTION

The survival of cells following freezing and thawing is
dependent both upon cell-type and the method of freezing and
thawing. The rates of cooling and warming, the type and
concentration of cryoprotective additive are all factors
which influence cell survival (Mazur 1970; Rapatz *et al.*,
1968; Franks, this volume). The loss of cellular viability
upon thawing is related to structural changes within the
membrane during freezing and thawing (Levitt, 1966; Lovelock, 1953; Mazur *et al.*, 1970; Meryman, 1968; McGrath;
Morris; Steponkus *et al.*, this volume). Haemolysis, upon
rapid warming, of unprotected human erythrocytes is dependent
upon the rate of cooling (Fig. 1). In the present study the
technique of freeze-fracture electron microscopy is used to
examine the structure of membranes in erythrocytes cooled
at equivalent rates. The ultrastructural alterations
observed in the membranes are compared with the haemolysis
which occurs upon thawing and possible mechanisms of injury
are discussed.

SLOW RATES OF COOLING

At rates of cooling slower than $1,800\,^{\circ}C\ min^{-1}$ intramembraneous particle-free (IMP-free) patches formed on the fracture
faces (Plate 1). The size and frequency of these patches
were dependent upon the rate of cooling. As the rate
increased from 3 to $1,800\,^{\circ}C\ min^{-1}$ the number of these patches
increased whilst their size decreased. At rates of cooling
in excess of $3,500\,^{\circ}C\ min^{-1}$ these IMP-free patches were not

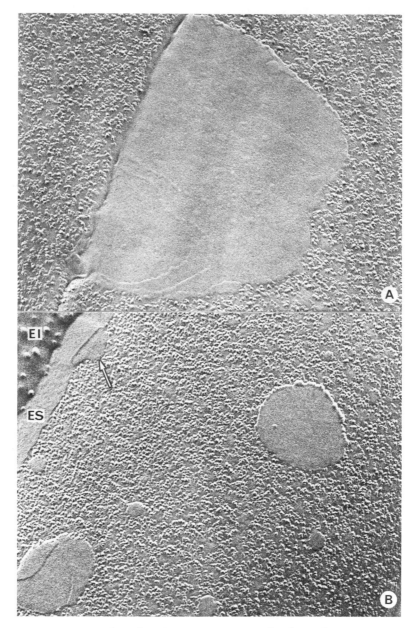

Plate 1. *Protoplasmic fracture faces (PF) of human erythrocytes cooled at rates of $3^\circ C$ min^{-1} (A) and $140^\circ C$ min^{-1} (B) to $-196^\circ C$. An arrow in B shows an intramembraneous particle-free (IMP-free) patch. X 54,000*

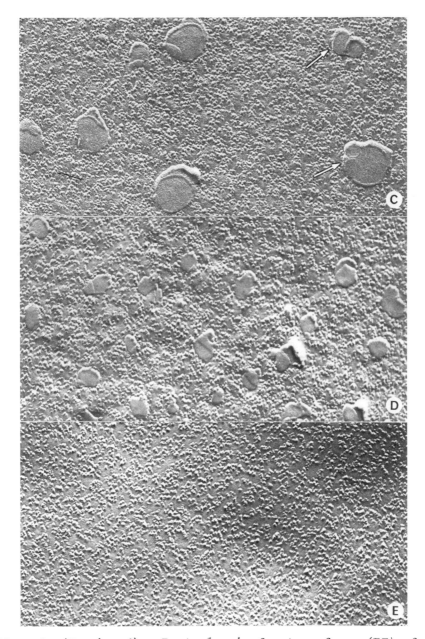

Plate 1. (Continued). *Protoplasmic fracture faces (PF) of human erythrocytes cooled at rates of 700°C min⁻¹ (C) and 1,800°C min⁻¹ (D) to -196°C. Arrows in C show fused patches, EI extracellular ice.   X   54,000*

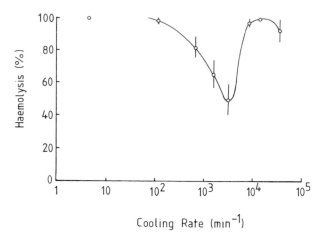

Figure 1. *Haemolysis (%) of unprotected human erythrocytes cooled at different rates to −196°C, all treatments were rewarmed by immersion of the specimens into isotonic saline at 20°C (a rate of warming of approx. 10,000°C min⁻¹). The different rates of cooling were obtained as described elsewhere (Fujikawa, 1981). The vertical lines show the range of variation.*

observed. The haemolysis upon thawing was at a minimum value of 50% at a cooling rate of 3,500°C min⁻¹.

The IMP-free patches which formed at low rates of cooling were elevated on the protoplasmic fracture face (PF) and depressed on the exoplasmic fracture face (EF). Their distribution and size were identical on both fracture faces and replica pairs demonstrated that they were of complimentary structure (Plate 2). It may be significant that when abnormal lipoprotein vesicles were fused with erythrocyte membranes a similar freeze-fracture morphology was observed (Verkleij *et al.*, 1976). Membrane associated vesicles were also observed by conventional thin-section electron microscopy (Plate 3). Their frequency and size were similar to the patches revealed by freeze-fracture. Therefore the IMP-free patches observed in fracture faces of erythrocytes at low rates of cooling are believed to be membrane derived vesicles. Formation of these vesicles was highly dependent upon the rate of cooling and was not affected by extensive washing of the erythrocytes before cooling. It is assumed that they result from structural changes which occur within the erythrocyte membrane at low rates of cooling (Pringle and Chapman, this volume).

The IMP-free vesicles which were derived from within the membrane at low rates of cooling were observed to pinch off

from the membrane during warming (Plate 4).   At room
temperature no such structures were observed within frozen
and thawed erythrocyte membranes.

Following slow rates of cooling, cholesterol-enriched,
protein-free vesicles have been isolated and characterised

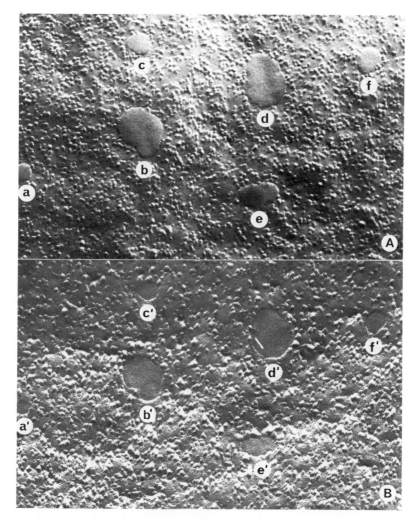

Plate 2.   *A double-replica pair of freeze-fractured*
membranes of human erythrocytes cooled at 700°C min⁻¹
-196°C.   Intramembrane particle-free (IMP-free) elevated
patches on PF (A) and IMP-free depressed patches of EF (B)
form a complementary structure.   A-F on PF are complimentary
with A'-F' on EF respectively.   X 73,000

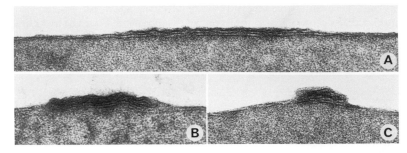

Plate 3. *Ultrathin sections of human erythrocytes cooled at (A) 3°, (B) 140°, and (C) 700°C min⁻¹ to -196°C. These specimens were then freeze-dried at room temperature and fixed with osmium tetroxide vapour. X 90,000*

(Araki, 1979). Similar vesicles are also released from erythrocytes following exposure to hypertonic solutions of sodium chloride at sub-zero temperatures (Araki, 1979). Erythrocytes of different cholesterol to phospholipid ratio were frozen at a low rate of cooling and the areas of IMP-free patches were determined (Table 1). The percentage area of the membrane exhibiting IMP-free patches increased with increasing cholesterol to phospholipid ratio. This suggests that the cholesterol-enriched, protein-free vesicles

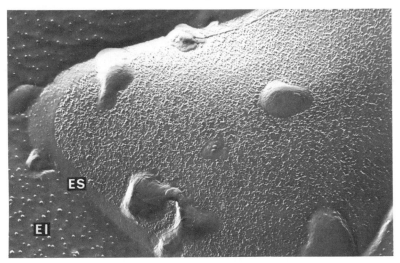

Plate 4. *The membrane structure of human erythrocytes cooled at 140°C min⁻¹ to -196°C, rewarmed to -10°C, and rapidly re-cooled to -196°C to fix the membrane structure at -10°C. X 27,000*

TABLE 1

The relationship between cholesterol content and the formation of intramembraneous particle-free (IMP-free) vesicles in slowly cooled erythrocytes*

Cholesterol:phospholipid ratio	Total areas of measured fracture faces (μ^2)	Total areas of IMP-free vesicles in measured fracture faces (μ^2)	Area of IMP-free vesicles/measured fracture faces
0.768	87.321	6.523	0.075
0.528	110.309	4.325	0.039
0.363	97.528	3.184	0.033

* The erythrocytes were prepared by the methods of Grunze and Deuticke (1974) and Deuticke and Ruska (1976). Erythrocytes were cooled at a rate of $700^{\circ}C$ min^{-1} to $-196^{\circ}C$ and areas of IMP-free patches in about 100 μ^2 fracture faces were measured using an image analyser.

released upon thawing (Araki, 1979) originage from the IMP-
free patches visualised by freeze-fracture electron microscopy.

   At low rates of cooling it is assumed that cellular
injury is a consequence of prolonged exposure to hypertonic
solutions (Mazur, 1966; Franks, this volume). Although the
mechanism of membrane vesicle formation is unknown (McGrath;
Morris; Steponkus *et al*., this volume), the exposure of
erythrocytes to hypertonic solutions during slow cooling
causes the segregation of cholesterol from intramembraneous
particles and the resultant formation of cholesterol-
enriched, protein free vesicles. Formation of these vesicles
could be avoided by cooling the erythrocytes at rates faster
than $3,500^{\circ}C$ min^{-1}; this would reduce the time of exposure
to hypertonic solutions. The gradual increase in the
percentage of unhaemolysed cells as the rate of cooling
increased to $3,500^{\circ}C$ min^{-1} (Fig. 1) was correlated with the
decrease in the formation of IMP-free vesicles.

RAPID RATES OF COOLING

At rates of cooling faster than $8,000^{\circ}C$ min^{-1} an abrupt
increase in lysis was observed (Fig. 1). This phenomenon
has been related to the nucleation of intracellular ice
crystals. In most cell-types examined the abrupt decrease
in viability at some intermediate rate of cooling
corresponds with the onset of intracellular freezing (Bank,
1973, 1974; Bank and Mazur, 1973; Mazur *et al*., 1970; Nei,
1976). In erythrocytes cooled at 8,000 and $11,500^{\circ}C$ min^{-1}
intracellular ice crystals ranging from 100 nm to 2 μm in
diameter were formed. Following these rates of cooling the
fracture plane through the interior of the membrane was
found to be interrupted at several regions, associated with
etchable ice crystals (Plate 5a). After etching such ice
crystals were removed from the fracture faces and altered
membrane regions resembling "worm eaten spots" were
observed on PF and exoplasmic surfaces (Plate 5b). The
presence of these worm eaten spots was restricted to regions
in which intracellular ice crystals were in direct contact
with the cell membrane (Plate 5c). It is probable that at
the regions in direct contact with intracellular ice crystals,
the fracture plane was deviated from the interior of the
membranes. In these areas the fracture plane passed extra-
cellularly. The removal, by etching, of ice crystals from
such regions revealed worm eaten spots on the membrane. The
fracture seen in Plate 5a was a typical feature of cells
containing intracellular ice; such deviations in the plane

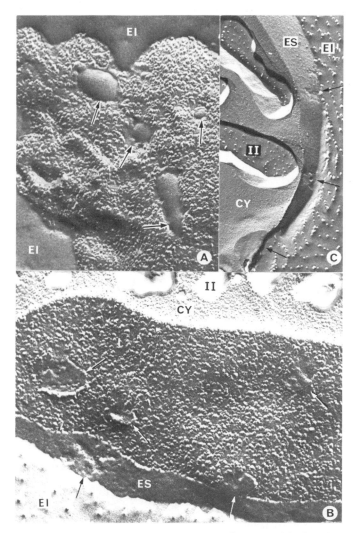

Plate 5. *The membrane appearance of intracellularly frozen human erythrocytes cooled at 8,000°C min⁻¹ to -196°C. (A) freeze-fractured PF before etching was interrupted at several etchable ice crystals (arrows). X 68,000. (B) after etching, the ice crystals were removed and revealed the structure resembling worm-eaten spots (arrows) on PF and ES X 68,000. (C) these worm-eaten spots (arrows) on the membrane corresponded with the regions in direct contact with intracellular ice crystals. X 28,000. II: intracellular ice crystals. CY: cytoplasmic region.*

of fracture were not observed at lower rates of cooling.

In the technique of freeze-fracture the plane of fracture passes through the hydrophobic interior of cellular membranes (Branton, 1966, 1969; Deamer and Branton, 1967), this no longer occurs when the membrane organization is destroyed (de Smet et al., 1978; James and Branton, 1971; Nermut and Ward, 1974). This alteration in the fracture plane from within the interior to the exterior of the membrane may thus reflect an alteration in the organization of the lipid bilayer. It is suggested that this loss of membrane structure results from being physically squeezed between growing intra- and extracellular ice crystals (Fig. 2). Sometimes not only worm eaten spots but also membrane holes were seen on the etched PF face, resulting from a deviation of the fracture plane into the intracellular compartment (Fujikawa, 1980). Thus the deviation of the fracture plane to either the extra-cellular or intracellular compartment depends upon the surface irregularity of the altered membrane.

Following spray freezing of erythrocytes the haemolysis upon thawing was again observed to decrease (Fig. 1). At ultrarapid rates of cooling intracellular water is converted into very small ice crystals or amorphous ice and the recovery of some cell-types increases (Bachmann and Schmitt, 1971; Moor, 1965). Erythrocytes have a relatively high water content and at the rate of cooling obtained by spray freezing intracellular ice crystals ranging from 50 to 100 nm were observed in many cells. A small percentage of red cells contained ice crystals of less than 50 nm in diameter. In cells containing crystals in the range 50 to 100 nm membrane fracture faces were rarely formed. Ice crystals of this size have numerous direct contacts with the membrane and in

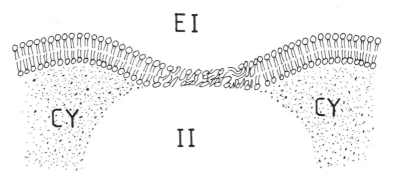

Figure 2. *Hypothetical scheme indicating the alteration of molecular organization of a lipid bilayer by the formation of intracellular ice crystals.*

consequence the fracture plane does not pass through the interior of the membrane. In erythrocytes containing ice crystals less than 50 nm in diameter membrane fracture faces were observed and the membranes were not morphologically altered (Plate 6).

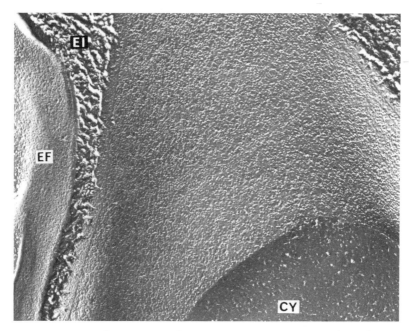

Plate 6. *PF of human erythrocytes spray frozen by the method of Nakanishi et al. (1979). X 27,000*

In summary, a correlation between the membrane ultra-structure in the frozen cell and haemolysis upon thawing has been demonstrated. The decrease in lysis observed as the rate of cooling increased from 3 to $3,500^{\circ}C$ min^{-1} corres-ponded with a reduction in the percentage of cells which formed IMP-free vesicles. Recovery was maximal at a rate of cooling of $3,500^{\circ}C$ min^{-1}. At rates of cooling faster than $8,000^{\circ}C$ min^{-1} the formation of intracellular ice crystals resulted in a decrease in cell recovery and damage to the cellular membranes induced by the formation of intracellular ice was observed. Following ultrarapid cooling the intra-cellular ice crystals formed were less than 50 nm in diameter and did not damage the membrane. Lysis decreased in parallel with the percentage of cells with normal membrane morphology. Thus the retention of normal membrane morphology during freezing is a prerequisite for cell recovery and the form-

ation of IMP-free vesicles and worm eaten spots seems to be
directly correlated with the occurence of lysis upon thawing
following slow and rapid freezing respectively.

ACKNOWLEDGMENTS

This paper was presented during the author's temporary stay
in the Department of Biochemistry and Chemistry, Royal Free
Hospital School of Medicine, University of London, London,
WC1N 1BP. The author is indebted to Professor J.A. Lucy
for the opportunity to present this study and to Dr Q.F.
Ahkong for critical reading of the manuscript.

THERMODYNAMIC MODELLING OF MEMBRANE DAMAGE

J.J. McGrath

Bioengineering Transport Processes Laboratory, Mechanical Engineering Department, Michigan State University, U.S.A.

INTRODUCTION*

This paper describes the development of a thermodynamic model to account for the haemolysis of human erythrocytes following freezing at sub-optimal rates of cooling. Several major difficulties were associated with this modelling process, namely:-

1) There are currently no satisfactory explanations of the mechanism of haemolysis following freezing and thawing.
2) Freezing involves a reduction in temperature and a con-comitant increase in solute concentration. Exposure to hypertonic conditions generally results in cell damage, whilst a reduction in temperature will diminish the rate of this destructive process. Thus during freezing two competing tendencies, and possibly others, will determine the fate of the erythrocyte. It was not obvious how to account for both factors during freezing when they occur simultaneously.
3) In some cases the damage resulting from hypertonic expos-ure is not apparent until the cells are returned to iso-tonic conditions.
4) Temperature reduction *per se* can haemolyse erythrocytes.
5) The model must be kinetic and not static since the recovery of erythrocytes is dependent on the rates of both cooling and warming.

The specific model presented in this chapter is less

---

* *Symbols used in this chapter are listed at the end of the text.*

important at its present stage of development than the method
used to derive it. It is not to be expected that this model
is accurate in every detail, for too many assumptions used in
its development remain uncertain at this time. However, the
model is expected to be a useful tool with a potential for
computer simulation of freezing injury. Two of its most
important features are that it is a non-equilibrium model
capable of yielding rate information, and that classical
thermodynamic principles can be applied to specify the expected inter-relationships between the thermal, chemical, mechanical and electrical properties of the system.

The paper begins with a review which places this model in
the context of previous thermodynamic modelling of freezing
injury. Assuming that the membrane is the primary site of
freezing damage and knowing that in erythrocytes damage can
be induced by a variety of means the biophysics of membranes
is discussed to characterise their thermal, chemical, mechanical and electrical properties.

Thermodynamic terminology and techniques are then applied
to the membrane as the system of interest. Finally to
demonstrate the potential and the limitations of the present
model a specific example is defined including all thermodynamic and kinetic details. The results of computer simulations
of the kinetics of hypertonic haemolysis are presented to
demonstrate the predictive potential of this method.

A REVIEW OF RELEVANT CRYOBIOLOGY

Over the years successful cryopreservation protocols have
been defined for many cell-types. Unfortunately these
successes currently cannot be extended to more complex biological systems. This is largely because of the lack of an
understanding of the mechanism of freezing injury.

Although many causes of freezing injury have been suggested,
the 'two-factor' hypothesis (Mazur, 1970) can be considered
the classic theoretical basis for interpreting experimental
data in the field of cryobiology. This hypothesis proposes
that rapid cooling produces cellular injury by the formation
of intracellular ice, whereas damage to cells at slow rates
of cooling is by so-called 'solute effects'.

It has been suggested that the nucleation of intracellular
ice produces holes in erythrocyte membranes (Nei, 1976).
Recently, a more subtle suggestion has been proposed namely
that the formation of intracellular ice in direct contact
with the plasma membrane causes a molecular disorganisation
of the normal bilayer structure. Haemolysis then occurs at
these altered membrane sites during thawing (Fujikawa, 1980,
this volume).

With the formation of extracellular ice the residual solution becomes increasingly concentrated as the temperature is
reduced (Franks, this volume). Water is lost osmotically
from cells and the intracellular solute concentration thus
increases. Slow rates of cooling therefore expose cells to
concentrated solutions for extended periods and destructive
events are thereby given a substantial period in which to
occur. To minimise exposure times to concentrated solutes
faster rates of cooling are required. Unfortunately, at these
faster rates of cooling an increased intracellular undercooling is to be expected (Mazur, 1963). This undercooling
enhances the probability of intracellular ice nucleation and
diminishes the likelihood of cell recovery. The outcome of
having these two major factors determining cell recovery is
that an optimal cooling rate is often observed (Mazur *et al.*,
1970).

It is important to emphasise that dynamic events will play
a large, and possibly dominant, role in the outcome of a
typical freeze-thaw protocol, both at sub- and supra-optimal
cooling rates. Thus, whilst 'equilibrium' experiments are
important in defining more of the detailed mechanisms involved
in freezing injury, 'non-equilibrium' experimental and analytical research must also be pursued. A complete picture of
freezing injury must account for the sensitivity of cell
recovery to both cooling and warming rate.

An analytical model which attempted a quantitative description of the kinetics of water transport during cellular
freezing has been published (Mazur, 1963). This non-
equilibrium model could predict cell water loss during cooling
as well as the degree of undercooling of intracellular water,
cellular shrinkage and the occurence of intracellular ice.
The importance of surface area to volume ratios and the water
permeability were clearly specified, and the lack of vital
data such as the temperature dependence of water permeability
was clearly illuminated.

Other modelling efforts have built upon this initial
analysis. Some are refinements of the original model while
others represent complementary aspects of freezing damage.
Thus a non-equilibrium model has been developed which considered such effects as extracellular supercooling, solute
concentration effect on water permeability and warming response (Silvares *et al.*, 1974). This model was linked to
nucleation theory in order to predict heterogeneous nucleation
temperatures (Toscano *et al.*, 1975). Subsequently thermodynamic models for membrane water permeability (Levin *et al.*,
1976a), intracellular solution behaviour (Levin *et al.*, 1976b)
and concentration polarization effects (Levin *et al.*, 1977)

have been published. Finally solute influx has been incorporated in order to explain anomolous volume changes during
warming (Knox *et al.*, 1980).

These thermodynamic models are important for a number of
reasons. Firstly, the classic two-factor hypothesis of
freezing injury and the examples of optimal cooling and warming rates make it clear that dynamic events determine cellular recovery and destruction. Dynamic modelling is therefore necessary and these models are dynamic models. Secondly,
the process of modelling requires a rigorous delineation of
the 'building blocks' of the model, an indication of whether
these building blocks are facts or assumptions, and a clear
description of how they are assembled. It is thus a relatively
easy task to re-examine any model, and to test its sensitivity
to alterations in the assumptions and their interactions.
Finally, these non-equilibrium thermodynamic models offer
the possibility of quantifying many of the changes which occur
during freezing and thawing.

Although dynamic modelling is a powerful tool, none of the
existing models, with the exception of that of Knox *et al.*
(1980), consider what happens to the membrane during freezing
and thawing. Rather these models focus on what occurs within
a cell, or in the extracellular volume. Hence the major use
of these models has been aimed at an understanding of the
water transport process at sub-zero temperatures.

There is much evidence to suggest that dramatic alterations
of the membrane occur during freezing and thawing (many
authors, this volume). There is thus a need to account for
dynamic alterations occuring within the membrane during
freezing and thawing. It is known that cell damage is related
to a number of factors including hypertonic exposure (Lovelock,
1953a), thermal shock (Lovelock, 1954b) and post-hypertonic
dilution damage (Zade-Oppen, 1968), but it is not apparent
how much of the total freeze-thaw damage is due to each
category (Farrant and Morris, 1973).

The rationale of the model presented in this chapter is as
follows:- In many cases of freezing injury the membrane
appears to be the site of damage. Methods of modelling
freezing injury on a quantitative, dynamic basis had been
applied successfully in cryobiology, but not all of the three
major modes of damage discussed above can be modelled on the
first attempt. Hypertonic exposure is one of the first
stresses a cell encounters during freezing, and so a non-
equilibrium thermodynamic model is developed for the membrane
during hypertonic stress, in the hope that a basis for understanding aspects of 'solute effect' damage can be established.
Further improvements on this initial effort may prove to be

useful in understanding problems in other areas such as
thermal shock, dilution shock or possibly even intracellular
ice formation at supraoptimal cooling rates.

MEMBRANE MODELS

General Considerations

The modern view of biological membranes centres on the fluid
mosaic model proposed by Singer and Nicolson (1972). A
major contribution of this model was the postulate that the
membrane is a dynamic structure with lateral diffusion of
constituent molecules. The biological membrane is seen as a
bilayer matrix of lipid with peripheral and integral proteins
embedded in the membrane. This view of the membrane allows
for asymmetry of membrane components so that carbohydrates
may extend from the external surfaces of cells for cell-cell
recognition and immune response, *etc*. This model also maintains that component flip-flop or tumbling does not occur at
significant rates.
     Lateral diffusion of lipid soluble spin-labels and membrane
proteins indicate that the apparent membrane viscosity is of
the order of 1.0 Poise, ranging from 0.1 Poise to 10 Poise
depending upon the measuring technique and the biological
system (Poo and Cone, 1974). For comparison the viscosity
of water at 0°C is approximately 0.01 Poise. Although special
mechanisms may exist for flip-flop of membrane molecules from
one side of the bilayer to the other these events are normally
taken to be infrequent ($\sim 10^{4}$-10^{5} sec) and controlled in order
that the normal membrane asymmetry necessary for proper
functioning is not compromised (Kornberg and McConnell, 1971).
     Although for the most part, components are preferentially
sequestered in the membrane it is important to recognize that
the membrane is in dynamic equilibrium with its environment.
Thus, there is a partitioning of lipid and other components
between the membrane and its surroundings. Changes in the
environment would be expected to produce changes in the
membrane and vice versa. Since the hydrophobic components
will largely be in the membrane, the extramembranous component or changes in it are often ignored.
     Following an administration of heavy water a rise in
deuteruim content of cholesterol in human erythrocytes and
plasma occurs (London and Schwarz, 1953). *In vitro* synthesis
of cholesterol by erythrocytes is insignificant and therefore
membrane cholesterol must be in a state of dynamic equilibrium
with plasma cholesterol.
     Crowley *et al*. (1965) cite the evidence that erythrocyte

cholesterol and certain phospholipids exchange with corresponding plasma lipids. The lipid differences between foetal
and adult erythrocytes are parallel to those differences seen
in the lipid of foetal and adult plasma. In addition, the
idea of exchange between membrane and plasma is also apparent
in studies of blood disease in which cell lipids may be
normal until placed in contact with abnormal lipid distrib-
utions in the plasma of a diseased person. Normally phosphatidylcholine and sphingomyelin are exchangeable but
phosphatidylethanolamine and phosphatidylserine are not (Zwaal
et al., 1973).

The thermodynamic basis for the stability and properties
of the membrane have been reviewed (Tanford, 1978, 1980). This
is not meant to produce dramatic advances or to provide
answers to all questions but to provide a framework for
organizing existing information and defining future problems.
After biosynthesis the organization of living matter is
dictated by thermodynamic considerations: in all cases molecules in the system seek a configuration which represents
the state of lowest chemical potential. In particular, the
stability of biological membranes results from the "hydrophobic effect". Amphiphilic molecules form membrane systems
not because of exceptionally strong attractive forces between
amphiphilic molecules in general, but because water has such
a strong cohesive self-attraction that it repulses the hydrocarbon chains of the amphiphilic molecules and, in effect,
squeezes them out. Thus bilayers, micelles and other such
structures formed are generally quite fluid due to this lack
of strong cohesive forces between the membrane components.
Although in some cases stronger attractive forces do exist
and create cohesive structures within biological membranes.

If the membrane is to be considered in equilibrium with
its environment then the chemical potential (μ) of any
(uncharged) species, j, must be equal in the membrane (m)
and environment (e):

$$\mu_j^m = \mu_j^e \tag{1}$$

where each chemical potential is defined in terms of a
reference state (normally taken as unit concentration) and a
term related to the activity (or mole fraction) of species j:

$$\mu_j = \mu_j^o + RT \ln a_j \tag{2}$$

where μ_j^o is a function of temperature and pressure only
(Tanford 1978, 1980).

Since the activity and mole fraction can be considered equal for ideal, dilute solutions, one can write $a_j = X_j$ and:

$$\mu_j = \mu_j^o + RT \ln X_j \qquad (3)$$

Hence,

$$\mu_j^{o,m} + RT \ln X_j^m = \mu_j^{o,e} + RT \ln X_j^e \qquad (4)$$

A reorganization of this equation reveals how cholesterol, phospholipid, or any other membrane constituent may be partitioned between the membrane and its environment in the equilibrium state:

$$\left(\frac{X_j^e}{X_j^m}\right) = \exp\left[\frac{(\mu_j^{o,m} - \mu_j^{o,e})}{RT}\right] = -\exp\left[-\frac{\Delta\mu_j^o}{RT}\right] \qquad (5)$$

Since there is much more amphiphilic material in the membrane than in the environment, $X_j^m \gg X_j^e$, and it is expected that $\mu_j^{o,e} > \mu_j^{o,m}$. This is consistent with the assumption made above that molecules in the system will seek a state of lowest chemical potential. For linear alkyl chains the term $\Delta\mu_j^o = \mu_j^{o,e} - \mu_j^{o,m}$ can be estimated from the total number of CH_2 groups in the membrane molecule since the free energy required to exchange a CH_2 group between a hydrocarbon phase and water is approximately 850 calories per mole of CH_2 group (Tanford, 1978, 1980). Thus a hydrocarbon "membrane" molecule with 20 CH_2 groups would be expected to partition such that:

$$\frac{X_j^e}{X_j^m} = 5 \times 10^{-13} \qquad (6)$$

Thus almost all of the hydrocarbon molecules will be in the membrane, but a finite amount would be in the environment.

The state in which external membrane components exist is a separate question. If the concentration of these components is small, then they will exist in solution as monomers. At concentrations greater than the critical micelle concentration they may exist as micelles or aggregates. Estimates of the critical micelle concentration of membrane phospholipids are

very low: $10^{-10}M$ (Jones, 1975). An account of these factors
and other interfacial phenomena is given by Davies and
Rideal (1963). Israelachvili *et al.* (1980) have published
an extensive review of the physical principles of membrane
organisation which addresses the thermodynamic aspects of
self-assembly of membrane components.

Recently the traditional fluid mosaic bilayer model has
been challenged. A "metamorphic mosaic" model of biological
membranes has been proposed as a more satisfactory explanation
of such phenomena as membrane component flip-flop and cell
membrane fusion. Using new NMR techniques phase transitions
from bilayer to hexagonal structures have been detected with
an intermediate structure thought to be an inverted micelle
(Cullis and de Kruijff, 1979). This inverted micelle structure is postulated for erythrocyte membrane "blebbing".
Such a model could be relevant as a mechanism for the formation of the microvesicles observed during freezing and
thawing (Araki, 1979; Fujikawa, this volume).

Chemical Properties of Membranes

The chemical properties of the membrane system must be
considered in any thermodynamic analysis because part of the
total system energy will be in chemical form. The model
developed in this chapter arose primarily from studies on the
human erythrocyte. The major constituents of this are shown
in Table 1. The membrane contains one major neutral lipid
and four major phospholipids, 35-40 types of protein and a
number of glycoproteins (Wintrobe, 1976). It remains to be
seen how important, thermodynamically speaking, the trace
components will be or how to best synthesize a mixture model
of the membrane.

Although the model presented here has not been developed
to the point of accounting for membrane asymmetry, this
feature of erythrocyte membranes (and other membranes) should
be appreciated. For example, phosphatidylcholine and
sphingomyelin appear to reside on the outside surface of the
bilayer while phosphatidylethanolamine and phosphatidylserine
appear to reside on the cytoplasmic surface of the membrane
(Finean *et al.*, 1978; Pringle and Chapman, this volume). It
is not clear at this time whether there is a preferential
location of cholesterol.

The effects of the major membrane components on membrane
bilayer stability are presented in Table 2. Israelachvili
et al. (1980) argue that membrane aggregation and asymmetry
must be explained in part by the geometric packing constraints
of the molecules involved.

Table 1

Major constituents of the human erythrocyte

Approximate Composition	(% dry weight)
Protein	40–50
Lipid	35–45
Carbohydrate	7–15

Lipid Composition	μmole per 10^{10} cells	mg per 10^{10} cells
Neutral		
Cholesterol	3.2	1.3 (1.1–1.4)
Free Fatty Acids	< 0.1	
Phospholipids		
Phosphatidylcholine	1.2	1.0
Phosphatidylethanolamine	1.1	0.9
Sphingomyelin	1.0	0.8
Phosphatidylserine	0.6	0.4
Lysphosphatidylcholine	0.04	
Others	0.07	
Glycolipids	0.2	0.1
Total lipid (Range in parentheses)		4.5 (3.9–5.2)

From Wintrobe, (1976).

TABLE 2

Characteristics of Major Membrane Lipid Components

Phosphatidylcholine:	membrane bilayer stabiliser
Phosphatidylethanolamine:	preference for hexagonal structure and hence mitigates against bilayer stability
Sphinogomyelin:	membrane bilayer stabiliser
Phosphatidylserine:	membrane bilayer stabiliser but interacts with Ca^{2+} to form non-bilayer structures
Cholesterol:	stabilises phosphatidylcholine bilayers by condensation but can disrupt unsaturated phosphatidylcholine bilayers unless sphingomyelin is present

Source: Cullis and DeKruijff (1979)

Electrical Properties of Membranes

The electrical properties of biological membranes need to be considered as work can be done on or by a membrane by electrical means. The charges of the major lipids of the erythrocyte membrane at physiological pH are given in Table 3. At this pH the net charge of the human erythrocyte membrane is negative and this charge is thought to result primarily from the sialic acid groups on the principal glycoprotein which is embedded on the outside leaflet and projects a carboxyl group (COO^-) into the extracellular solution (Finean *et al.*, 1978; Wintrobe, 1976).

Charged or polar groups at the surface of a biological membrane tend to attract counter-ions from the external electrolyte solution. This creates an electrical potential which is a function of distance away from the erythrocyte membrane. This potential decreases in magnitude from a value determined by the net negative charge at the membrane surface out into the electrically neutral external solution. Net surface potentials of biological cells, ψ_s, are typically -10 to -39 mV relative to the extracellular medium (Davies and Rideal, 1963; Finean *et al.*, 1978).

This surface charge, ψ_s, influences the surface pressure as well as the extent and rate of adsorption or desorption for soluble films (Davies and Rideal, 1963). When an ion

Table 3

Charges of the major phospholipids of the erythrocyte membrane at physiological pH

Phosphatidylcholine:	(+,-), (polar)
Phosphatidylethanolamine:	(-), (negative)
Sphingomyelin:	(+,-), (polar)
Phosphatidylserine:	(-,-), (negative)

Source: Finean et al. (1978)

desorbs from a charged surface it loses energy in the amount

$$z_j . \varepsilon . \psi_s$$

where z_j is the valency of the desorbing ion and ε is the electronic charge. Thus a negatively charged membrane component would have a lower desorption energy than an un-ionized or cationic component.

In thermodynamic terms the equilibrium distribution of membrane components between the membrane and its environment will be modified by this effect. Equilibrium of charged components must take into account the equality of electrochemical potentials, rather than the chemical potential alone, where the electrochemical potential is defined as:

$$\bar{\mu}_j = \mu_j + z_j . \varepsilon . \psi_s \qquad (7)$$

Thus equation (5) would be modified to read:

$$\left(\frac{X_j^e}{X_j^m}\right) = \exp \frac{(\mu_j^{o,m} - \mu_j^{o,e}) + z_j . \varepsilon . \psi_o}{RT} \qquad (8)$$

Whereas the unitary chemical potential difference $(\mu_j^{o,e} - \mu_j^{o,m})$ for an alkyl chain with 20 CH_2 groups might be approximately 17 Kcal/mole, the electrical energy associated with a univalent anion desorbing from a membrane potential of -40 mV to zero potential would require approximately 920 cal mol^{-1} which represents a 5 percent change in total free energy. This change in energy would change the ratio (X_j^e/X_j^m) by a factor of five.

Mechanical Properties of Membranes

Thermodynamic description of a membrane system must also
include the mechanical properties of the membrane since energy
can certainly appear in mechanical form in such a system.
For a thermodynamic model of membrane damage it is important
to appreciate the mechanical behaviour of membranes to interpret such reported phenomenon as mechanical resistance to
cellular volume reduction during dehydration and its relationship to injury. The elastic, visco-elastic, and plastic
behaviours of the cell membrane, primarily of human erythrocytes, have been reviewed (Evans and Skalak, 1980).

There are two major modes of elastic mechanical deformation
in a biological membrane: one is a shearing mode in which a
unit area of membrane is deformed at constant area, and the
other is a dilation mode in which unit area is extended or
contracted with a constant shape (Fig. 1). An arbitrary
elastic deformation can be expressed as some combination of
these two modes. The erythrocyte membrane shear modulus, μ,
is of the order 10^{-2} dyne cm^{-1}, whereas the surface elastic
modulus for area extension, K, is approximately 10^{2}-10^{3}
dyne cm^{-1}. Hence it takes little energy to stretch the red
cell membrane at constant area by shearing whereas the membrane is very resistant to changes in area. Thus during
freezing erythrocytes would be likely to deform preferentially
by large shearing at constant area rather than a concomitant
area and volume reduction during dehydration. In fact, an
assumption of constant area has often been made when modelling
freezing injury (Levin *et al*., 1976b; Mazur, 1963).

Visco-elastic behaviour of the biological membrane

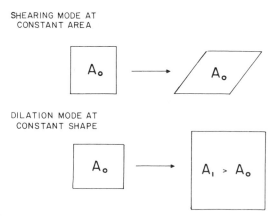

SHEARING MODE AT
CONSTANT AREA

DILATION MODE AT
CONSTANT SHAPE

Figure 1. *Elastic deformation modes in a biological membrane*

represents a non-conservative means of exchanging energy between the environment and the membrane. This type of membrane
response will occur during finite deformation rates when
membrane mechanical energy is irreversibly dissipated in the
form of a heat transfer to the environment. This characteristic of cell membranes may be significant during such
processes as rapid cellular dehydration or rehydration involving membrane deformations. This possibility and its
implications have not yet been explored with respect to
freezing injury.

Since the elastic shear modulus is 10^4-10^5 times smaller
than the area compressibility modulus, the expected response
of erythrocytes and cells with similar mechanical properties
for conservative or non-conservative processes would be a
shearing mode of deformation. For non-conservative processes
surface shear viscosity has been measured (η_e ~6.6 x 10^{-4}
Poise), but dilational viscosity has been more refractory to
examination (Evans and Skalak, 1980).

It should be mentioned that large shearing stresses can
cause the erythrocyte membrane to fail. That is, the membrane will no longer behave elastically. Instead it will
yield and exhibit plastic behaviour, and removal of the
shearing stress will not return the membrane to its original
configuration. This permanent deformation has resulted from
a reorganization of the membrane structure - specifically
this reorganization is thought to involve the spectrin molecules of the red cell and their associated proteins (Evans
and Skalak, 1980).

Estimates of the stresses necessary to initiate plastic
flow of the erythrocyte membrane by shearing range from
2×10^{-2} to 8×10^{-2} dyne cm^{-1}. Viscous dissipation of
mechanical energy will also occur for plastic flow and the
characteristic surface viscosity associated with this process
is approximately an order of magnitude larger than for
elastic dissipation: (η_p ~ 10^{-2} dyne-sec cm^{-1}).

Finally, bending of membranes in response to freezing or
other stress causes deformations that are resisted by membrane
rigidity. It is therefore possible to store energy in this
bending process. The bending modulus, B, estimated for
erythrocyte membranes is 10^{-13} to 10^{-12} dyne cm^{-1}. Generally
this bending free energy contribution for erythrocytes will
be negligible with respect to the shear elastic energy if
the curvature of the membrane has a characteristic radius
>0.25μm (Evans and Skalak, 1980). Therefore, unless the
erythrocyte is rather severely crenated or wrinkled not much
energy would be stored in bending during freezing.

Membrane-Associated Water

A crucial issue, not yet discussed, is the position of the
dividing surface between what is considered membrane and what
is not. Thermodynamically speaking, what will be considered
part of the system and what will be defined as environment?
Answering this question reveals the unique role played by
water in the membrane system.

   The motion and structure of water in biological systems
is likely to differ from that of bulk water (Tait and Franks,
1971). Analyses with spectroscopic probes reveal 10-12
bound water molecules and another 11 water molecules trapped
by each lecithin in lecithin/water systems (Chapman *et al.*,
1967; Le Neveu *et al.*, 1976). A variety of techniques have
estimated total amounts of membrane-bound water in the range
25-70 gm H_2O per 100 gm dry membranes or 20 to 40 percent by
weight of the total hydrated membrane (Finch and Schneider,
1975; Schneider and Schneider, 1972; Schneider *et al.*, 1979).
X-ray analysis indicates that approximately 20 to 30 percent
by weight hydration is necessary in order to avoid disruption
of the ordered bilayer pattern in erythrocyte and myelin
membranes (Finean, 1969). Water content lower than this level
may result in hexagonal and cubic structures of phospholipids
(Deamer *et al.*, 1970; Tinker and Pinteric, 1971). The binding
of water vapour to erythrocyte ghost membranes indicates that
the highest values obtained are similar to and greater than
binding energies of water in ice (Schneider and Schneider,
1972). These energies are assumed to be hydrogen bond energies for the tightly bound membrane water judging from the
results of infra red spectroscopy (Schneider *et al.*, 1979).

   This picture of the membrane suggests that a bilayer
membrane system which normally might be considered to consist
of the hydrocarbon region only, must necessarily be redefined
to include water. Hydration levels of 20 to 30 gm H_2O per
100 gm dry membrane are generally necessary for bilayer integrity (Finch and Schneider, 1975), and yet a tightly bound
monolayer of water ($\sim$5-6 kcal mol^{-1}) adheres to red cell
ghost membranes at 4 gm H_2O per 100 gm dry weight. The binding energy of membrane water decreases as additional hydration
layers are added (Schneider and Schneider, 1972) and bulk
water properties would be expected at some distance from the
membrane interface (Wilkinson *et al.*, 1977). These data
produce a somewhat unexpected result. Because of the great
difference in the molecular weights of water and typical
membrane components such as lipid and protein, the membrane
system defined above as a hydrated bilayer is predominantly
water on a mole basis, even in the "dehydrated" state (Table
4).

Table 4

Isotonic composition of the simplified model for the eryth-rocyte "half membrane"

Composition by weight (g)		(%)
Lipid	2.00×10^{-13}	35.0
Protein	2.00×10^{-13}	35.0
Water	1.71×10^{-13}	30.0
Composition by volume		
Lipid	$0.278 \ \mu^3$	39.4
Protein	$0.257 \ \mu^3$	36.4
Water	$0.171 \ \mu^3$	24.2
Molar composition		
Lipid	2.65×10^{-16}	2.71
Protein	3.11×10^{-18}	0.03
Water	9.52×10^{-15}	97.26

MEMBRANE THERMODYNAMICS

The value of a thermodynamic approach to membrane damage is not so much that it will immediately solve the unanswered problems of freezing injury, but rather, that it provides a conceptual framework and the potential for quantifying various changes of state within the membrane.

Thermodynamic Terminology

The first step in a general thermodynamic analysis begins with the definition of the system, which is any given mass or volume of interest. Everything not included in the system is defined as the environment. This requires the analyst to specify whether or not the analysis will be based upon a fixed mass system which is free to exchange energy and entropy with the environment but which is impermeable to mass transfer. This type of system is considered a closed system since no exchange of matter between the system and the environment is allowed. Alternatively a given volume of interest, the so-called control volume, may be defined as the system. This technique is applied to open systems which allow for exchange of mass as well as energy and entropy across the system boundaries. The most general open system analysis includes deformable control volumes as well as fixed control volumes.

Membrane components such as phospholipids have very low solubilities in the aqueous phase suggesting that a closed system analysis might be appropriate for a biological membrane. Whilst this may be a good assumption for many situations, there is considerable evidence suggesting that the membrane and its environment are in dynamic equilibrium. In some cases it may be that this equilibrium can be shifted dramatically by a change in the environment. Thus the membrane is best described for the purpose of modelling as an open system, which is defined by establishing a control surface at the inner boundary of the membrane and another at the outer boundary of the membrane (Fig. 2).

The second step in the thermodynamic analysis is the definition of the nature of the system. This is done by defining which properties are sufficient to determine completely the state of the system and then defining the relationship between these properties. This relationship is often called an equation of state which may be an analytical expression or it may be presented in other forms such as tables, charts, or figures. A familiar example of an equation state for gases at low pressures is the perfect gas law:

$$PV = mRT \qquad (9)$$

If the membrane system is made of one component then the equation of state would be for a pure substance. It is unlikely that any biological membrane system could be considered a pure substance due to the many membrane constituents and due to the intimately associated membrane water. For systems consisting of more than one component, various mixture models

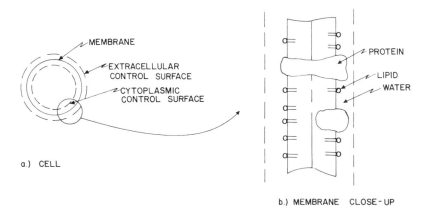

Figure 2. *Spherical membrane control volume system.*

must be considered. The analyst must assume or prove that
such a mixture is an ideal mixture and if it is not ideal
then non-ideal mixture models must be applied. The equation
of state including any mixture models place constraints on
the possible changes which can occur between equilibrium
states as the membrane system undergoes a quasi-static
process.

The quasi-static process is an assumed process model.
Since any real process occurs at a finite rate and is irreversible, it is not possible to apply classical equilibrium
thermodynamics to these processes. However, for quasi-
static processes the system is assumed to undergo changes
which are internally very close to equilibrium such that
equilibrium properties and relationships can be defined for
the system as it undergoes such a process.

Since the membrane system is composed of several different
components and since damage is assumed to involve some change
in the composition of the membrane, it is necessary to consider additional chemical equilibrium factors which would not
be necessary if the chemical composition of the system remained
unchanged.

Constraints Imposed by Chemical Equilibrium

The requirement of chemical equilibrium in a single phase
system is given by the Gibbs-Duhem equation:

$$0 = VdP - SdT - \Sigma n_j \, d\mu_j \qquad (10)$$

This describes the chemical potential of all components, n_j,
included in the system. Its importance is that it places
constraints on the changes which can actually occur between
equilibrium states.

If electrical effects are important then the chemical
potential must be generalized and expressed as the electrochemical potential, $\overline{\mu}_j$, where:

$$\overline{\mu}_j = \mu_j + z_j \varepsilon \psi_o \qquad (11)$$

Another factor which must be considered is surface tension,
γ. For bulk phases surface energy effects are negligible,
but for the cell membrane surface effects play a significant
role. Therefore equation (10) must be written to include
electrical properties and surface tension:

$$0 = VdP - SdT - \sum_{j=1}^{m} n_j \, d\bar{\mu}_j - Ad\gamma \qquad (12)$$

Thermodynamic representation of the membrane in this general form is given in several texts (Castellan, 1971; Davies and Rideal, 1963). This equation will now be applied to the erythrocyte membrane as a means of describing changes of state within the membrane which occur during damage

A final manipulation of the equation above will serve to illustrate the potential value of this expression. Assuming for simplicity that a single hypothetical membrane component species, LPS, was known to be a molecule lost from the membane during damage, then this component of interest could be isolated from the summation term on the right hand side of the equation above to yield:

$$n_{LPS} \, d\bar{\mu}_{LPS} = VdP - SdT - \sum_{j} n_j \, d\bar{\mu}_j - Ad\gamma \qquad (13)$$

This expression demonstrates an important point: it is possible to change the electrochemical potential of a "labile" membrane component, LPS, in several ways. If the imposed change increases the potential of the LPS component in the membrane system relative to the environment then a driving force would be established for a mass flux of LPS away from the membrane, possibly as desorption. Alternatively this driving force may provide the impetus for another form of reaction such as phase separation, microvesicle formation or membrane collapse. Note that the use of equation (13) requires assuming an equilibrium condition internal to the membrane. This is in contrast to the fact that the membrane system is not in equilibrium with its environment during the damage reaction.

Equation (13) also clearly shows that mechanical (dP, $d\gamma$), thermal (dT), as well as electrical and chemical effects ($d\bar{\mu}_j$) can establish the potential for alterations resulting in damage. The appearance of the thermal term is especially interesting in light of the lack of an understanding of the phenomenon of thermal shock (Farrant and Morris, 1973). Also it should be noticed that the application of one differential stress such as $d\bar{\mu}_j$ arising from changes in the jth component may not immediately result in the deleterious increase of the chemical potential of a labile component $d\bar{\mu}_{LPS}$ since that stress may be compensated for by an alteration of, say, dP. These considerations would be dictated by inter-

actions with the environment as well as the specific equation
of state for the membrane system in question where the
equation of state would define the relationship between the
various properties internal to the system.

The Membrane Equation of State

The interfacial membrane tension, $\overline{T}$, at the membrane interface may be resolved into two components, a surface tension,
γ, and a surface pressure, π (Davies and Rideal, 1963; Evans
and Skalak, 1980):

$$\overline{T} = \gamma - \pi \qquad (14)$$

The surface tension term, γ, can be thought of as a tension in the membrane caused by the hydrophobic effect. This
term is therefore related to $\Delta\mu_j^o$ in equation (5) and is
relatively insensitive to alterations in membrane area when
these changes are small (Evans and Skalak, 1980. On the other
hand there is a certain amount of repulsion between membrane
components arising from the thermal kinetic energy which results in a net two-dimensional momentum exchange which will
be defined in this analysis as the membrane pressure, π.
Israelachvili *et al.* (1980) present a similar view of the
pressure balance expected in a membrane system. These
authors discuss the nature of the effective momentum exchange
defined by Evans and Skalak (1980).
    In terms of membrane free energy changes, T is associated
with the total membrane free energy whereas γ is associated
with a free energy contribution due to the hydrophobic
interaction between the hydrocarbon and water at local interfacial regions. The surface pressure, π, can be associated,
in part, with the lateral momentum exchange of the non-polar
part of the amphiphiles constituting the membrane and in part
to the head group pressure. This total surface pressure can
also be equated to a surface osmotic pressure (Adamson, 1976).
Thus the total membrane free energy per molecule associated
with tension at the interface, F, is the sum of the free
energy identified with interfacial hydrocarbon/water interactions, F_w, and that identified with the non-polar and head
group portion of the membrane amphiphiles above, F_p (Evans
and Skalak, 1980):

$$F = F_w + F_p \qquad (15)$$

$$\left(\frac{\partial F}{\partial \overline{A}}\right)_{T,P} = \left(\frac{\partial F_w}{\partial \overline{A}}\right)_{T,P} + \left(\frac{\partial F_p}{\partial \overline{A}}\right)_{T,P} \tag{16}$$

where $\overline{A}$ is the area per molecule and:

$$\left(\frac{\partial F}{\partial \overline{A}}\right)_{T,P} = \overline{T} \tag{17}$$

$$\left(\frac{\partial F_w}{\partial \overline{A}}\right)_{T,P} = \gamma \tag{18}$$

$$\left(\frac{\partial F_p}{\partial \overline{A}}\right)_{T,P} = -\pi \tag{19}$$

To a first approximation the surface tension is independent of the area per molecule available to the membrane constituents since this contribution is by definition associated with the hydrophobic repulsion of the hydrocarbon chain out of the water phase. In other words, the surface tension is measuring the tendency for the water to squeeze hydrocarbons out of the aqueous phase independent of whether the amphiphiles are loosely or tightly packed together once they are in the membrane and out of the water.

The surface pressure, π, is the term which expresses how the amphiphiles interact with one another within the membrane. Surface packing affects the magnitude of the surface pressure and the momentum exchange between membrane molecules is also affected by temperature. Therefore the surface pressure is expected to be a function of area per molecule, $\overline{A}$, and temperature T. The surface pressure will also be affected by electrostatic effects or charged surfaces (Davies and Rideal, 1963; Israelachvili *et al.*, 1980). The surface tension is expected to be primarily a function of temperature but it is recognized that factors such as surface hydration levels will play important roles in certain cases. Thus:

$$\pi = \pi(\overline{A}, T) \tag{20}$$

$$\gamma = \gamma(T) \tag{21}$$

Since the surface pressure, π, measures interactions internal to the closed membrane system it is often called the membrane equation of state.

Biological membranes are generally modelled as a bilayer, the behaviour of which can be inferred from treating it as two monolayers. Monolayer systems have been studied extensively (Adamson, 1976; Davies and Rideal, 1963) and the assumption that bilayer behaviour can be approximated by the superimposition of two monolayers appears to be valid (Evans and Skalak, 1980).

At some temperatures the molecules constituting a monolayer may behave as a two-dimensional "gaseous film" yielding a curve such as curve A in Figure 3. The simplest equation of state for such a system would be:

$$\pi\overline{A} = kT \tag{22}$$

an equation similar to the perfect gas law (PV = RT). Equations of state in the gaseous region are developed for more complex cases by accounting for factors such as the "limiting area" of the amphiphile or the effects of mobile membrane molecules.

At other temperatures or for other membrane constituents significant van der Waals forces of attraction may result in a more cohesive nature to the membrane compared with the gaseous state. This is often called a "liquid expanded" state.

If the temperature and molecular structure of the membrane components are such that very strong cohesion between molecules exists then a condensed liquid or solid phase may exist. This is shown by curve B in Figure 3 (region 3-4).

The molecular nature of the membrane components would be expected to play a key role in the equation of state. Thus length of the hydrocarbon chain, charge, degree of saturation, molecular size, *etc.* are known to determine monolayer and membrane behaviour.

An added degree of complexity arises when modelling the equation of state for a membrane composed of many molecules rather than a single component. That is mixture models must be developed. Ideal mixing is the simplest case (Jones, 1975) but a membrane will exhibit non-ideal behaviour and this would have to be described for a more accurate model.

Returning to the question of the membrane tension and its constituent parts as expressed in equation (14) it is obvious that the total membrane tension is only partially described by the surface pressure since $\overline{T} = \gamma - \pi$. The cell membrane equation of state $\pi = \pi(\overline{A}, T)$ can be obtained directly from

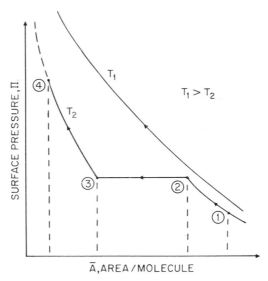

Figure 3. *Graphical equation of state data for typical amphiphilic monolayer system.*

cells by performing mechanical measurements (Mitchison and Swann, 1954) if these are performed reversibly at constant temperature (Evans and Skalak, 1980). In this case the change of $\overline{T}$ with area will be directly related to the change of π with area:

$$(\frac{\partial \overline{T}}{\partial \overline{A}})_{T,P} = (\frac{\partial \pi}{\partial \overline{A}})_{T,P} \tag{23}$$

since by definition $(\frac{\partial \gamma}{\partial \overline{A}})_{T,P} = 0$. These coefficients are related to the isothermal area compressibility modulus, K where:

$$K = \overline{A}(\frac{\partial \overline{T}}{\partial \overline{A}})_{T,P} \tag{24}$$

and K has a measured value of 450 dyne cm^{-1} at $25°C$ for erythrocytes (Evans and Waugh, 1977). In general, the membrane tension is also affected by temperature changes:

$$d\bar{T} = \frac{K}{\bar{A}} \left[d\bar{A} - (\frac{\partial \bar{A}}{\partial T})_{\bar{T}} \, dT \right] \tag{25}$$

where the quantity

$$\frac{1}{\bar{A}} (\frac{\partial \bar{A}}{\partial T})_{\bar{T}}$$

is the thermal area expansivity at constant membrane tension. All of the terms in the equation above are directly accessible by mechanical experiments performed at constant temperature. The first term is related solely to the surface pressure and therefore the internal equation of state, whereas the second term for thermal area expansivity includes the effects of changing surface tension with temperature. Thus constant temperature experiments mean that only the first term is retained and $d\bar{T}$ is related to $d\pi$ only.

Hypertonic exposure

To summarize this discussion and its relevance to freezing injury an example of cellular dehydration is considered. In the absence of external forces, the cell membrane will be in a stress-free state ($\bar{T} = 0$) or close to it (Evans and Skalak, 1980). Experimental support for this assumption is available for erythrocytes (Evans and Waugh, 1977), sea urchin eggs (Hiramoto, 1963), spinach protoplasts (Steponkus *et al.*, this volume), and algal cells (Baker, 1972). Thus in the isotonic state the membrane tension can be essentially zero with the surface tension being balanced by the surface pressure:

$$\bar{T} = 0 = \gamma_o - \pi_o \tag{26}$$

or

$$\gamma_o = \pi_o \tag{27}$$

As a cell is exposed to increasingly concentrated solution its volume is reduced in order to achieve osmotic equilibrium across the membrane. If the cell shrinks as a sphere then the volume reduction necessary for osmotic equilibration will be coupled with a surface area reduction, $V/A = r/3$, (Weist and Steponkus, 1978). This area reduction will not affect γ directly, according to the first order approximation made above, $\gamma \neq \gamma(\bar{A})$, but it will generally tend to increase the surface pressure as $\bar{A}$ decreases (Fig. 3). Equation (12) is now rewritten with the simplifying assumption that the membrane thickness is constant:

$$0 = -SdT - \Sigma n_i \, d\bar{\mu}_i - A d\bar{T} \qquad (28)$$

Hence the surface tension term has been combined with the surface pressure term to eliminate the hydrostatic pressure and yield the total membrane tension term.

Assuming that the cell is to be exposed to an isothermal, hypertonic exposure and that electrical effects can be ignored for the time being then:

$$0 = - \Sigma n_i \, d\mu_i - A d\bar{T} \qquad (29)$$

With spherical shrinkage the total extensive area A is reduced and if no membrane material is lost $\bar{A}$ is reduced. It would therefore be expected that membrane tension, $\bar{T}$, becomes negative since the surface tension is unchanged $\gamma = \gamma_0$ and the surface pressure increases $\pi > \pi_0$. There is experimental evidence showing that this occurs (Baker, 1972; Rand and Burton, 1964; Williams and Takahashi, 1978). Under these conditions it would be expected that the summation term $\Sigma n_i d\mu_i$ would increase. This means that the average chemical free energy in the membrane has increased, which would provide the driving force for a reaction of one or more components n_i. If one of the components were preferentially more labile than the others, this component might be driven from the membrane and be desorbed as suggested by Lovelock (1954a). Desorption could occur at constant total area A, for example the area corresponding to "minimum volume". This desorption process would increase the area per molecule $\bar{A}$ from the value assumed in the shrunken state, decrease π and tend to increase the membrane surface tension back to a stress-free state ($\bar{T} = 0$). The release of micelles or microvescicles could be viewed in much the same manner.

If this were to occur the shrunken cell would be limited in its ability to return to the isotonic state unless material could reversibly return to the membrane during deplasmolysis (Steponkus, 1978a; Williams, 1981). Instead of the desorption of a single component another possibility would be the increase of total surface area (and reduction of membrane tension) by the formation of membrane microvesicles or blebs. There is experimental evidence for this event (Araki, 1979; Fujikawa, this volume) and a mechanism for the necessary membrane alterations for the formation of blebs has been described (Cullis and DeKruijff, 1979). This behaviour is similar to the formation of emulsions which occurs when $\bar{T} < 0$ (Davies and Rideal, 1963).

It is important to realize that the above sequence represents only one possible response of a cell to hypertonic exposure. The thermodynamic analysis described here is therefore not meant to predict precisely how any given cell-type might respond. The ability to predict quantitatively such a response will depend upon further experimental work. It should be clear however, that the approach proposed here has potential value in interpreting existing data and revealing where gaps exist in our knowledge.

Consider two other possible responses of a cell exposed to a hypertonic solution. In the first case the cell reaches osmotic equilibrium by losing water but undergoes a negligible surface area change. For example, a cell might flatten in a "pancake" mode or it might shrink with projections on its surface such as with a crenated red cell. This response could conceivably occur at constant membrane surface area and thus the surface tension term $(-Ad\bar{T})$ would not be expected to contribute to a change in the total chemical free energy within the membrane ($\Sigma n_i.d\mu_i$). Thus the geometry is expected to play a role in the damage response during hypertonic exposure or freezing. This point has been recognized by Weist and Steponkus (1978a) in work with spinach protoplasts and by Israelachvili *et al.* (1976, 1980) in a theoretical study of micelle formation.

As a final example, the effect of multiple membrane constituents and their combined properties as a mixture may be discussed. If the unstressed, isotonic cell exists in state 1 on Figure 3 then an isothermal compression of the membrane by membrane shrinkage to state 2 may result in a rather small contribution from the surface tension term $(-Ad\bar{T})$. This change may lead to an insignificant or immeasurable alteration of the membrane or release of membrane component(s). If the isotonic membrane state is at state 2 on Figure 3 then an isothermal membrane compression to state 3 may result in a membrane phase transition with a lateral separation of membrane components. There is experimental evidence to show that such events occur in biological membranes (Höchli and Hackenbrock, 1976; Verma and Wallach, 1976; Pringle and Chapman, this volume). In this case $d\bar{T} = 0$ and a reversible phase separation would be expected to occur with a change in membrane free energy level at constant surface pressure π. It might also be the case that the membrane system could exist in state 3 in Figure 3 such that compression over a small area increment would result in dramatic changes in $d\bar{T}$ (or $d\pi$) causing collapse of this condensed state. There is evidence indicating the collapse of monolayer systems (Ries, 1979; Williams, 1981).

While cells may not have the capacity to alter membrane composition at a rate comparable to that at which an external disturbance is applied, some cells apparently have the genetic means to alter membrane composition in "anticipation of" external disturbance. Such adaptations with respect to plant cold hardening have been reviewed (Williams, 1981).

In summary, a thermodynamic interpretation of membrane tension indicates that an increased osmotic pressure in the bulk solution would be expected to increase the membrane surface pressure (decreasing the membrane surface tension). It appears that this increased pressure may result in a number of responses depending upon the initial state of the membrane, the degree of perturbation, the internal equation of state, and the solubility of the membrane components in the extramembranous solution. It is not clear that an osmotic pressure gradient *per se* is responsible for membrane damage.

A THERMODYNAMIC/KINETIC MODEL FOR MEMBRANE DAMAGE

In this section an example of the thermodynamic methodology presented above is applied to the case of human erythrocytes exposed to hypertonic solutions of sodium chloride. This example shows in detail the extent to which information is known and how much must be assumed in order to provide quantitative predictions of cellular damage.

Definition of the System

The membrane is defined as an open system to allow for the exchange of membrane components. The control volume of interest is enclosed between two control surfaces, one immediately on the inside of the bilayer and one immediately on the outside of the bilayer. The control volume is thus a shell (Fig. 2). The control volume is taken to be non-deformable at this stage of the analysis. The contents of the control volume are described as follows:

1) The constituents of the erythrocyte are assumed to consist of lipid, protein, and water - no distinction is made at this point between different types of lipid or protein. Carbohydrates are neglected.
2) The membrane is assumed to consist of 50 percent dry weight protein, 50 percent dry weight lipid.
3) The membrane is assumed to be sufficiently hydrated in the isotonic state such that the weight of membrane water

represents 30 percent of the total membrane weight (Finch and Schneider, 1975).

4) The total lipid content by weight of the human erythrocyte membrane is taken to be 4.0×10^{-13} gm cell^{-1} (Wintrobe, 1976).

5) The molar volume of membrane water was taken to be constant and equal to that of bulk water 18 cm^3-mole^{-1} (Evans and Skalak, 1980; Robinson and Stokes, 1959).

6) The molar volume of the lipid was taken as a constant due to the lack of detailed equation of state information. This volume was based upon a cylinder of length 25Å and an effective diameter of 70Å^2 (Evans and Skalak, 1980; Israelachvili *et al.*, 1976).

7) The molar volume of the protein was calculated on the basis of a spherical molecule of diameter 64Å. This was taken as a mean value, a more detailed analysis of typical erythrocyte membrane proteins is given by Bretscher (1973).

8) The molecular weight of water was assumed to be 18.0; that of the average lipid 750; and that of the protein 64,000.

9) The typical isotonic erythrocyte surface area, A_m, was taken to be 135 μ^2 and its volume, V_m, as 100 μ^3 (Canham and Burton, 1968).

10) The membrane is assumed to be symmetrical.

11) Electrical effects are neglected.

12) Ideal mixing is assumed such that the total membrane volume is simply the sum of the partial volumes of each component.

13) Na$^+$ and Cl$^-$ in the membrane phase can be taken as insignificant to a first order approximation.

The information above is sufficient to characterize the membrane on a weight, volume, and a mole basis (Table 4). Even though water constitutes 24 percent by volume or 30 percent by weight it is obvious that on a molar basis the membrane is dominated by water. It is significant that the chemical free energy of any system is related to the mole fractions and not to the weights or volumes of membrane constituents. The importance of not neglecting water in the interfacial phase has been emphasezed elsewhere (Adamson, 1976).

The erythrocyte membrane is typically taken to be approximately 75-90Å thick (Singer and Nicolson, 1972; Wintrobe, 1976). For erythrocyte dimensions used here the volume of the membrane is well-approximated, to better than 99 percent accuracy, by the product of membrane surface area and membrane thickness:

$$V_m = A_m \cdot t_m \qquad\qquad (30)$$

Since V_m is available from the summation of partial vol-
umes, V_j, above $\quad (V_m = \sum_{j=1}^{3} n_j \overline{V}_j) \quad$ and A_m is known from the
literature, (35 μ^2) the membrane thickness is a derived
quantity. For the data stated above the derived membrane
thickness under isotonic conditions would be 80 $\overset{o}{A}$, for non-
aqueous material which would stain for electron microscopy.
The agreement between experimental measurements and the present model is thus encouraging.

Description of the Damage Process

There are a number of ways in which the erythrocyte membrane
may be damaged. Unfortunately the exact mechanism(s) remain(s)
to be defined; Lovelock proposed a dissolution of the red
cell membrane resulting from freezing (Lovelock, 1955a).
Evidence that suggests that microvesicles may be lost from
the membrane has been presented (Araki, 1979; Hanahan and
Mitchell, 1966; Fujikawa, this volume). Recent findings of
membrane lipid polymorphism may provide a basis for the
formation of such microvesicles (Cullis and DeKruijff, 1979).
    In this section a hypothetical case of damage provides
a specific illustration of the modelling procedure. Damage
occurs when the stabilizing influence of the hydrophobic
effect is affected by external perturbations. Such perturbations provide the free energy changes necessary to reduce the
activation energy barrier to damage. The particular damage
might be a re-structuring within the membrane system which
could render the cell permeable to solutes (small and large).
In this case the membrane as a whole may remain intact but
the global membrane hydrophobic interaction may be altered
such that the membrane is now more porous.
    On the other hand the membrane may "break down" more
dramatically by collapsing or by the release of molecules,
micelles, or microvesicles. Kinetically these latter reactions
may turn out to have important similarities. In particular
the hydrophobic interaction between water and hydrocarbon
may play the key role not only in determining equilibrium
considerations but also for determining the kinetic behaviour
of the system. In the specific example to follow it is
hypothesized that the key activation step required to undergo
the destructive process involves exposing hydrocarbon chains

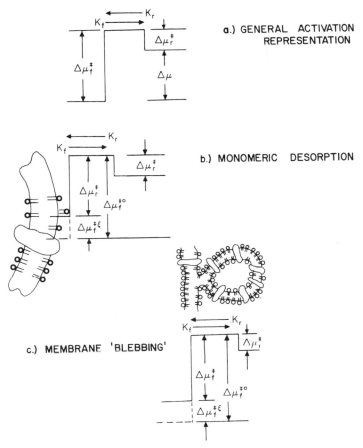

Figure 4. *The generalized activation reaction illustrated for two specific damage mechanisms.*

to water (Figure 4).

The process of membrane damage is modelled kinetically as being rate-limited by an effective unimolecular reaction at the membrane surface which is a result of the hypertonic exposure. The details of this approach have been discussed previously (McGrath *et al.*, 1978, 1980, 1981). The major assumptions are:

1) The equilibrium state, which occurs when the chemical potential of the labile species is spatially uniform is governed by a difference in the standard state chemical potentials, $\Delta\mu^{\circ}$, (Fig. 4a).

2) The kinetics at the membrane are determined by a forward

reaction activation energy difference $\Delta\mu_f^{\neq}$ and a reverse

reaction activation energy difference $\Delta\mu_f^{\neq}$ where $\Delta\mu_f^{\neq}$ =
$\Delta\mu_r^{\neq} + \Delta\mu^{\circ}$.

3) Experimental evidence for human erythrocyte destruction suggests that exposure to hypertonic sodium chloride reduces $\Delta\mu^{\circ}$ and $\Delta\mu_f^{\neq}$ because increasing amounts of membrane components are observed in the extracellular space (Araki, 1979; Morris, 1975; Fujikawa, this volume, Foote *et al.*, unpublished results) and the haemolysis rate increases (McGrath, 1977). These data would be consistent with a simple increase in the standard free energy of the labile component, LPS, resulting from the initial hypertonic exposure as is shown schematically in Figure 4. A change in $\Delta\mu_r^{\neq}$ is not necessary to explain existing data and with no evidence to the contrary it is left as invariant with respect to hypertonic exposure. Such external perturbations change standard state free energies and drive reaction kinetics (Glasstone *et al.*, 1941).

4) The reduction in $\Delta\mu_f^{\neq}$ at constant $\Delta\mu_r^{\neq}$ results in a net removal of membrane components and it is assumed that the change in $\Delta\mu_f^{\neq}$ due to perturbation, $\Delta\mu_f^{\neq\xi}$ is available from the Gibbs-Duhem relationship within the membrane phase. Note that $\Delta\mu_f^{\neq}$ would be variable dependent upon the local environment in any state.

In this model a distinction is made between two types of standard state free energies such as those given in Figure 4. One type of standard state is $\Delta\mu_j^{\circ}$ which has been used in this model to denote the standard free energy difference between component j in the membrane and the environment. This free energy difference yields a partition coefficient which describes the difference in concentrations of component j in the membrane and the environment. This $\Delta\mu_j^{\circ}$ is a function of the environment in the general case and therefore differs from a "true" standard free energy difference $\Delta\mu_j^{*}$ where $\Delta\mu_j^{*} = \Delta\mu_j^{*}(T,P)$, independent of the concentration. This effect of an apparently variable standard state free energy resulting from environmental changes is discussed with respect to rate processes by Glasstone *et al.* (1941).

The erythrocyte exposed to hypertonic sodium chloride is taken as an example. In the isotonic case there would be a

given $\Delta\mu_{j1}^{o}$. After exposure to a hypertonic sodium chloride
solution all experimental evidence points to a redistribution
of membrane material. Therefore a new standard state free
energy $\Delta\mu_{j2}^{o}$ would be measured in this new equilibrium state.
In this model the Gibbs–Duhem relationship is used to calculate the instantaneous standard state free energy $(\Delta\mu_{j}^{o})$
where the external perturbations cause no alteration of the
"true" standard state free energy $(d\mu_{j}^{*} = 0)$

In the isotonic state a dynamic equilibrium is established
such that the net rate of exchange of a labile component, LPS,
is zero. In the general case the net rate of exchange away
from the membrane (in whatever form) is expressible in terms
of specific reaction rates and concentrations per unit area
at the membrane surface. Therefore the general net forward
reaction for the release of membrane material is:

$$\frac{d\Gamma_m}{dt} = K_r\Gamma_e - K_f\Gamma_m \qquad (31)$$

where the mole concentration per unit area at the interface is
denoted by Γ and the specific reaction rates in the reverse
and forward directions are given as K_r and K_f, respectively.
Note that Γ has a straightforward interpretation when each
half of the membrane is taken to be a monolayer rather than
a multilayered system.

It has been assumed that this damage reaction is rate-
limited by the reaction at the surface rather than by diffusion away from the surface. (A kinetic model somewhat similar
to the model presented here is available for a diffusion rate-
limited model of desorption from monolayers.) Consequently,
for the surface reaction limited model, a two-compartment
analysis is pursued wherein the spatial gradient of the membrane component is not important – the component is either in
the membrane phase or out of it $(\Gamma_m$ or $\Gamma_e)$. Such a two-
compartment analysis allows conversion of equation (31) above
to the form:

$$\frac{dC_m}{dt} = K_rC_e\left(\frac{V_e}{V_m}\right) - K_fC_m \qquad (32)$$

where the respective concentrations, are given as C (moles
volume^{-1}) (Fig. 5).
The specific reaction rates are written in the form:

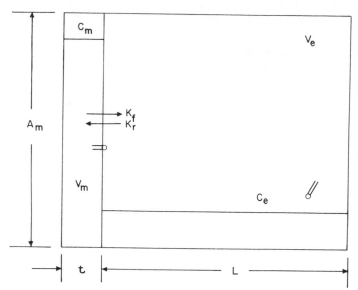

Figure 5. *Lumped, two-compartment model for the exchange of membrane material.*

$$K_i = \left(\frac{KT}{h}\right) \cdot \exp\left[-\Delta\mu_i^{\neq}/RT\right] \qquad (33)$$

where $i = f$ for the forward reaction and $i = r$ for the reverse reaction. It is interesting to note that similar kinetic methodology has been applied in the study of resonance times of lipids in bilayers as these lipids exchange with aqueous environments (Israelachvili *et al.*, 1980).

In the isotonic equilibrium state there is no net exchange of membrane material:

$$\frac{dC_m}{dt} = 0 = K_r C_e \left(\frac{V_e}{V_m}\right) - K_f C_m \qquad (34)$$

The initial external and membrane concentrations C_e^{iso}, C_m^{iso}, as well as values for V_e and V_m may be calculated from data in the literature (Lovelock, 1955a) or from assumptions made above. The forward specific reaction rate, K_f, is obtained from haemolysis kinetics experiments performed on a stop flow system (McGrath, 1977). Equation (34) is then solved for the only remaining unknown, the invariant, K_r.

Exposure to a hypertonic medium causes a change in $\Delta\mu_f^{\neq}$

such that it is instantaneously reduced below its isotonic equilibrium value $\Delta\mu_f^{\neq iso}$. The instantaneous $\Delta\mu_f^{\neq}$ would be expressed as:

$$\Delta\mu_f^{\neq} = \Delta\mu_f^{\neq iso} - \Delta\mu_f^{\neq\xi} \tag{35}$$

where $\Delta\mu_f^{\neq\xi}$ is obtained from the Gibbs-Duhem relationship.

For the case of an isothermal, hypertonic exposure with constant membrane surface area, it can be shown that water plays a dominant role in the Gibbs-Duhem relationship for the present model such that the general form given as equation (13) can be simplified to:

$$d\mu_{LPS} = -(n_w/n_{LPS})\ d\mu_w \tag{36}$$

The hypertonic exposure represents a perturbation in the extramembranous phase which causes a change in the chemical potential of water outside the membrane volume:

$$\Delta\mu_w^{\xi} = \int_{\mu_w^{iso}}^{\mu_w^{\xi}} d\mu_w \tag{37}$$

However, it has been assumed that with respect to the damage reaction of interest here, (desorption, microvesicle formation or otherwise) that the equilibrium of membrane water with water outside the membrane system is instantaneous (Sidel and Solomon, 1957). Therefore any change in external water chemical potential is instantly reflected in a like change within the membrane phase. Thus the differential change of water chemical potential imposed externally would be the same differential change which would be appropriate within the membrane as expressed in the simplified Gibbs-Duhem relationship (equation 36).

It is assumed in this model that the change in forward activation energy due to perturbation, $\Delta\mu_f^{\neq\xi}$, is equivalent to the integration of equation (36).

$$\Delta\mu_f^{\neq\xi} = \int_{\mu_f^{\neq iso}}^{\mu_f^{\neq\xi}} d\mu_{LPS} \tag{38}$$

Assuming that during this change from the isotonic state to the instantaneous perturbed state the ratio of the number of moles of water to lipid involved in the damage process remains constant:

$$\beta = (n_w/n_{LPS}) \qquad (39)$$

then

$$\Delta\mu_f^{\neq\xi} = -\beta RT \ln (a_w^{\xi}/a_w^{iso}) \qquad (40)$$

As outlined by McGrath et $al.$ (1978) this result combined with equation (35) implies that a plot of $\Delta\mu_f^{\neq}$ as a function of $\ln (a_w^{\xi}/a_w^{iso})$ would have an intercept value of $\Delta\mu_f^{\neq iso}$ in the isotonic state and a slope of $-\beta RT$ (Fig. 6).

Haemolysis kinetic data taken from a stop flow system at $25°C$ for various hypertonic sodium chloride concentrations have been plotted in the theoretical form suggested above with a coefficient of determination, r^2, of 0.997 (McGrath, 1977; McGrath et $al.$, 1978) (Fig. 6).

The quantitative relationship between the change of water activity and the driving potential for the desorption reaction as determined from the stop flow haemolysis kinetics ($i.e.$ values for $\Delta\mu_f^{\neq iso}$ and β) is used in the computer simulations given in the next section.

It may be worth speculating on the derived value of β (approximately 40) for the haemolysis reaction. This suggests that for the perturbation step which activates the reacting molecule, LPS, forty molecules of water are involved. If the reaction were a desorption reaction and the reacting LPS molecule were a diacyl phospholipid, an activation step inserting 20 CH_2 groups into water would involve approximately 40 water molecules. Insertion of 20 CH_2 groups of a typical erythrocyte phospholipid represents an activation step which involves exposing approximately one half of the hydrocarbon portion of amphiphile to the aqueous phase. Equilibrium measurements show that it requires approximately 850 calories per mole of CH_2 to remove CH_2 groups from an organic phase to an aqueous phase (Davies and Rideal, 1963; Israelachvili et $al.$, 1976; Tanford, 1978, 1980). A finite reaction rate would involve a different free energy change but this estimate shows that the proposed "half exposure" activation step required for the desorption reaction would require a minimum free energy change of approximately 17 kcal mol^{-1}. This value of free energy is typical of those used to calculate lipid exchange rates between the aqueous medium and aggregate structures such as bilayers (Israelachvili et $al.$, 1980). This is quite reasonable when compared with the range

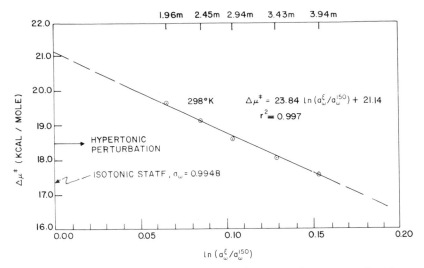

Figure 6. *Free energy as a function of the perturbation of water activity for human erythrocyte haemolysis.*

of 17 to 21 kcal mol^{-1} derived from haemolysis kinetic experiments (McGrath, 1977; McGrath *et al.*, 1978).

COMPUTER SIMULATIONS OF MEMBRANE DAMAGE

The membrane model and the hypothetical process of damage described above have been programmed on a digital computer to solve the differential equation describing the rate of release of membrane component:

$$\frac{dC_m}{dt} = K_r C_e \left(\frac{V_e}{V_m}\right) - K_f C_m \tag{41}$$

These results have been published recently (McGrath *et al.*, 1980, 1981). Equation (41) is a first-order, ordinary differential equation which is non-linear since the forward specific reaction rate, K_f is a function of the LPS membrane concentration C_m. The equation is solved subject to the initial conditions:

$$C_e = C_e^o \tag{42}$$

and

$$C_m = C_m^o \tag{43}$$

where K_r is determined from the isotonic equilibrium conditions and

$$K_f = (\frac{KT}{h}) \exp \left[\frac{-(\Delta\mu_f^{\neq iso} - \Delta\mu_f^{\neq\xi})}{RT} \right] \tag{44}$$

The forward specific reaction rate is a quantity which has an instantaneous value determined by the value of $\Delta\mu_f^{\neq\xi}$:

$$\Delta\mu_f^{\neq} = -\beta RT \ln(a_w^\xi/a_w^{iso}) \tag{45}$$

where β, $\Delta\mu_f^{\neq iso}$, and a_w^{iso} are known (McGrath, 1977; McGrath *et al.*, 1978).

Simulation of a particular damage reaction requires that a haematocrit be chosen such that the ratio (V_e/V_m) may be determined. Once the level of hypertonic exposure is chosen a_w^ξ would be known. In all cases ideal solution behaviour is assumed which is a common simplifying assumption, and $x_w^m = x_w^e$ implying that $\mu_w^{o,m} = \mu_w^{o,e}$ for $\mu_w^m = \mu_w^e$ to be true at all times. This assumption means that all water involved in exchange is identical to bulk water. Equation (41) is then integrated numerically with a simple explicit, forward-difference technique to yield the new membrane concentration of LPS, C_m at each time step.

In the absence of specific equation of state information, water is assumed to replace the volume vacated by the desorbed membrane component, LPS, with no protein flux. As water fills the membrane control volume in this process the instantaneous membrane water activity, a_w^ξ, in equation (45) is adjusted to allow further integration of equation (41). The dissolution of the membrane by this process (assumed to occur at constant total membrane surface area) eventually would leave the membrane leaky to macromolecules such as haemoglobin. It is not known what critically low value of C_m would allow the loss of haemoglobin in this model.

To be acceptable the computer model must successfully predict two basic characteristics of the desorption reaction which match characteristics of the experimentally observed haemolysis reactions. The first criterion is that a larger hypertonic perturbation must result in a larger release of membrane material and thus a larger amount of haemolysis in the population. The second requirement is that the character-

istic times associated with haemolysis and the predicted
desorption reaction times must be comparable. Both requirements are met (Fig. 7). In virtually all cases simulated
thus far the characteristic times predicted for the desorption
reaction are approximately 1-3 minutes, which match the times
observed for haemolysis (McGrath, 1977).

Unless otherwise stated the standard conditions for these
computer simulations were: 298°K; 3M NaCl exposure; 10
percent haematocrit; membrane hydration - 30 percent total
membrane weight; membrane dry weight content 50 percent lipid,
50 percent protein. The comparisons of levels of response to
various perturbations and the comparisons of characteristic
transient times is shown in Figure 7.

The predicted effect of haematocrit is shown in Figure 8.
The simulations predict that increased haematocrits are
beneficial in that less membrane material is expected to be
lost at a higher cell density.

The effects of shrinkage morphology have already been
discussed. The computer simulations predict that cell shrinkage at constant surface area would be less damaging than

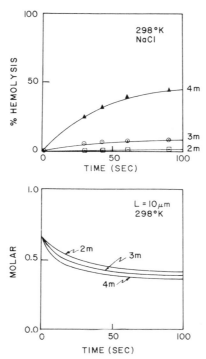

Figure 7. *Comparison of characteristic times for haemolysis
data and computer predicted membrane component loss.*

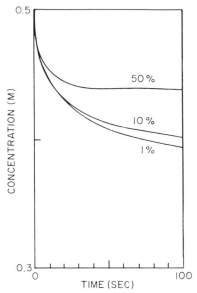

Figure 8. *Predicted effect of cell shrinkage morphology on the exchange of membrane material.*

shrinkage as a sphere which is shown in Figure 9. This assumes that equation (36) is valid so that membrane tension contributions are negligible as well as contributions from electrical effects and "trace" levels of membrane components.

CONCLUSIONS

The contributions of the thermodynamic modelling presented above are not explicitly related to an improvement in an understanding of the precise mechanism of freezing injury. This model must rely instead upon sound data from the field of low temperature science which provide the basis for mechanistic detail. However the model can perform a useful function in the future assuming that general validity can be established in that postulated mechanisms of injury may be used as input to the model and the consequences of these hypotheses may be studies.

One of the important contributions of this model is that it is a non-equilibrium description of membrane alterations. Consequently the major feature of the model is the thermo-dynamic formulation of the problem which provides the necessary ground rules for interpreting existing data in a unified manner. The Gibbs-Duhem equation applied to the membrane system is an important example of the type of result which

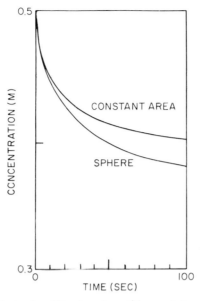

Figure 9. *Predicted effect of cell packing (haematocrit) on the exchange of membrane material.*

can be expected from this approach. This equation places constraints on the possible changes between equilibrium states in a given process. Furthermore the individual free energy contributions in this expression explicitly denote thermal, mechanical, and electrochemical changes. Thus thermodynamic analysis reveals an expression which relates all of the necessary forms of changes of state within the membrane which are known to cause damage.

This model implicitly assumes that the fundamental cause of freezing injury is an alteration of the membrane. It assumes that the various perturbations given above can cause free energy changes which in turn alter the activation free energy barriers for particular reactions.

The system of interest in this analysis has been defined as an open system capable of exchanging mass with its environment in order to account for the apparent loss of membrane material in many cases of stress.

It appears that with respect to the constituents of a typical biological membrane, there are many more components then can be analysed conveniently. However, the Gibbs-Duhem equation indicates that for components present only in trace amounts, the free energy contribution of these terms may be negligible so that a simplified model of the membrane may be a good approximation to the actual membrane in some cases.

A review of membrane-associated water suggests that the membrane system must include water. In fact, on a mole basis water is easily the dominant component of the membrane and would therefore be expected to dominate overall free energy changes at the interface.

The stability of the membrane system appears to be due to a balance between the hydrophobic effect and an internal surface pressure. The hydrophobic effect describes the strong cohesive nature of water which results in amphiphilic molecules being "squeezed" into structures such as bilayers or micelles. These structures produce a lower free energy state than randomly distributed amphiphiles. The ordering of water molecules in the vicinity of amphiphiles in solution represents an unfavourably low entropy state and the surface tension at a membrane interface has been identified with this hydrophobic effect. Since the overall membrane tension in a biological cell appears to be zero or negligible in the normal or isotonic case, a repulsive surface pressure is defined which offsets the surface tension in the membrane stress-free state. The surface pressure is an affect internal to the membrane system arising from a net exchange of molecular momentum.

The thermo-mechanical coupling within the membrane system is typically described by an equation of state which relates the surface pressure, temperature, and molecular area. The electro-chemical effects occurring within the membrane are described by the electrochemical potential. The definition of the electrochemical potential shows that superpositioning of the electrical free energy and the chemical potential free energy is assumed. Temperature plays a role in both the mechanical aspects of the membrane (equation of state) and the eletrochemical aspects of the membrane, but the explicit expression for the effects of a temperature change is in the "-SdT" term of the Gibbs-Duhem equation. Thus, the Gibbs-Duhem equation completes the description of the total coupling within the system since it constrains the allowable changes in thermal, electrochemical, and mechanical free energy forms:

$$0 = -SdT - \Sigma_i n_i \, d\overline{\mu}_i - Ad\overline{T} \qquad (28)$$

Several interesting features arise from this interpretation. For instance, it is apparent that the membrane surface area, A, plays a role in the total free energy state of the membrane. This term explains why cell geometry during shrinkage would be expected to affect cell recovery.

Another interesting point concerning this equation is that since membrane entropy will always be a positive quantity,

a temperature reduction will always yield an increase in
average membrane component free energy from the "-SdT" term.
This tendency could provide the increased free energy necessary to drive a particular reaction leading to membrane
failure. This result indicates that there may be a theoretical basis for the phenomenon of thermal shock.

The future potential of the model is related to the fact
that all forms of free energy appear in the Gibbs-Duhem
equation. Since changes in the free energy are used to calculate activation free energies, there is the possibility of
defining instantaneous damage rate equations when several
perturbations occur simultaneously, such as during freezing.

To summarize the membrane as a thermodynamic system it
could be said that the overall state of the membrane and its
stability is related to both the environment and conditions
within the membrane. The solubility of membrane components
is very low in the aqueous phase under normal conditions.
Depending upon the nature of the amphiphilic molecules comprising the system the internal condition of the membrane
may be best described as "gaseous", or as an "expanded
liquid", etc. External disturbances will cause a complex
response which in some cases may be resolved primarily by
internal changes (such as a phase separation) whereas in
other cases free energy increases may act to drive a reaction
favouring substantial exchange of material between the membrane and its environment.

It is not surprising that cells may have evolved membranes
which will respond differently to various stresses. Cells
containing membrane molecules which are highly insoluble in
aqueous phases may respond to freezing by storing substantial
levels of energy in an elastic mechanical work mode until
failure occurs by collapse. On the other hand other cells
may have membranes composed of relatively soluble molecules.
Environmental stress imposed upon these cells may lead to an
inconsequential equilibrium increase of mechanical tension
but a rather large exchange of mass between the membrane and
its surroundings.

This analysis demonstrates the importance of defining the
initial state of the membrane and the environment. Stresses
imposed upon cells in different initial states generally will
result in different responses and end states.

A key area of further research is the derivation of suitable equations of state for biological cells of interest.
An immediate need is the problem of dealing with membrane
component mixtures. In this respect, recent cryobiology
research using liposomes which have membranes comprised of a
small number of molecules (two or three) may be useful as
systems simple enough to be amenable to analysis (Morris and
McGrath, 1981a).

ACKNOWLEDGEMENTS

The author would like to express his gratitude to the Whitaker
Foundation for providing initiation support to aid in establishing the Bioengineering Transport Processes Laboratory and
for continuing support of the research activities of the
laboratory, including research described in this chapter.

NOMENCLATURE

a = activity
A = area
$\overline{A}$ = area per molecule
B = elastic bending modulus
C = concentration
d = differential amount
F = free energy per molecule
k = Boltzmann constant or specific reaction rate
K = isothermal area compressibility modulus
M = molar
n = number of moles
P = absolute pressure
R = universal gas constant
S = entropy
T = absolute temperature
$\overline{T}$ = membrane interfacial tension
v = specific volume
V = volume
X = mole fraction
Z = valency

GREEK SYMBOLS

β = ratio of moles of water to moles of lipid
Γ = moles per unit membrane surface area
γ = membrane surface tension
ε = electronic charge
η = shear viscosity
μ = chemical potential
$\overline{\mu}$ = electrochemical potential
π = membrane surface pressure
ψ = electrical potential

SUBSCRIPTS

e   = elastic
f   = forward

SUBSCRIPTS (Cont.)

i	=	chemical species i
j	=	chemical species j
LPS	=	hypothetical membrane component species
m	=	membrane
o	=	stress-free reference state
p	=	non-polar, hydrocarbon region
P	=	plastic
r	=	reverse
s	=	surface
w	=	water or polar, aqueous region

SUPERSCRIPTS

e	=	environment
iso	=	isotonic reference state
m	=	membrane
o	=	reference state
$\neq$	=	related to the activation step
ξ	=	due to the perturbation process
*	=	absolute reference state
1	=	state 1
2	=	state 2

REFERENCES

Abbas, C.A. and Card, G.L. (1980). *Biochim. biophys. Acta* 602, 469-476.

Abbott, B.C., Howarth, J.V. and Ritchie, J.M. (1965). *J. Physiol.* 178, 368-383.

Abernethy, J.L. (1967a). *J. Chem. Educ.* 33, 177-180.

Abernethy, J.L. (1967b). *J. Chem. Educ.* 44, 364-370.

Adamson, A.W. (1976). "Physical Chemistry of Surfaces", Third edition, John Wiley and Sons, New York.

Ahkong, Q.F., Fisher, D., Tampion, W. and Lucy, J.A. (1975). *Nature* 253, 194-195.

Ahluwalia, B. and Holman, R.T. (1969). *J. Reprod. Fert.* 18, 431-437.

Aloia, R.C. (1980). *Fed. Proc.* 39, 2974-2979.

Amelunxen, R.E. and Murdock, A.L. (1978). *CRC Critical reviews in microbiology* 6, 343-390.

Angell, C.A. (1981). *In* "Water - A Comprehensive Treatise", Vol. 7., (F. Franks, ed.) Plenum Press, New York, (In press).

Anner, B.C. and Jørgensen, P.L. (1979). *In* "(Na$^+$+K$^+$)-ATPase, Structure and Kinetics, (J.C. Skou and J.G. Nørby, eds.) pp. 87-97, Academic Press, London.

Ansell, G.B. and Hawthorne, J.N. (1964). "Phospholipids: Chemistry, Metabolism and Function". Elsevier, Amsterdam.

Apeland, J. (1966). *Bull. Int. Inst. Refrig.* 46, 325-335.

Araki, T. (1977). *Biochim. biophys. Acta* 496, 532-546.

Araki, T. (1979). *FEBS Lett.* 97, 237-240.

Arias, I., Williams, P.M. and Bradbeer, J.W. (1976) *Planta* 131, 135-139.

Arrhenius, S.A. (1889). *Physik. Chem.* 4, 226-248.

Asano, Y., Matsui, H., Nakao, M. and Nagano, K. (1970). *Biochim. biophys Acta* 219, 169-178.

Augustus, J. (1976). *Biochim. biophys. Acta* 419, 63-75.

Babel, W., Rosenthal, H.A. and Rapaport, S. (1972). *Acta biol. med. germ.* 28, 565-576.

Backmann, L. and Schmitt, W.W. (1971). *Proc. Nat. Acad. Sci. USA.*, 68, 2149-2152.

Bagnall, D.J. (1979). *In* "Low temperature stress in crop
    plants: The role of the membrane" (J.M. Lyons, D. Graham
    and J.K. Raison, eds) pp. 67-80. Academic Press, New York
    and London.
Bagnall, D.J. and Wolfe, J. (1978). *J. Exp. Bot.* 29, 1231-
    1242.
Baker, H. (1972). *Cryobiology* 9, 283-288.
Baker, J.H. and Smith, D.G. (1972). *J. appl. Bact.* 35, 589-
    596.
Baker, P.F. and Willis, J.S. (1972). *J. Physiol.* 224, 441-462.
Bangham, A.D. and Hancock, J.L. (1955). *Nature* 176, 656.
Bangham, A.D., de Gier, J. and Greville, G.P. (1967). *Chem.
    Phys. Lipids* 1, 225-246.
Bangham, A.D., Hill, M.W. and Miller, N.G.A. (1974). *In*
    "Methods in membrane biology" (E.D. Korn, ed), 1, 1-68,
    Plenum Press.
Bank, H. (1973). *Cryobiology* 10, 157-170.
Bank, H. (1974). *Exp. Cell Res.* 85, 367-376.
Bank, H. and Mazur, P. (1973). *J. Cell Biol.* 57, 729-742.
Barnett, L.B., Adams, R.E. and Ramsey, J.A. (1974). *Life
    Sciences* 14, 653-658.
Barnett, R.E. and Palazotto, J. (1974). *Ann. N.Y. Acad. Sci.*
    242, 69-76.
Baxter, R.M. and Gibbons, N.E. (1962). *Can. J. Microbiol.*
    8, 511-517.
Behrisch, H. (1978). *In* "Strategies in Cold: natural torpidity
    and thermogenesis". (L.C.H. Wang and J.W. Hudson, eds.),
    pp. 461-482. Academic Press, New York.
Beljkevic, V.I., Kljucareva, Z.S., Rombe, S.M. and Filaretova,
    N.V. (1959). *Trudy vses. nauchno-issled. Inst. Zhivotn.*
    23, 211-231.
Bennoun, P. (1972). *C.r. hebd. Séanc. Acad. Sci.* Paris Series
    D, 275, 1777-1778.
Benson, R.W., Pickett, B.W., Komarek, R.J. and Lucas, J.J.
    (1967). *J. Anim. Sci.* 26, 1078-1081.
Berger, H., Jones, P. and Hanahan, D.J. (1972). *Biochim.
    biophys. Acta* 260, 617-629.
Berthon, B., Claret, M., Mazet, J.L. and Poggioli, J. (1980).
    J. Physiol. 305, 267-277.
Beyer, R.E. (1972). *In* "Hibernation and hypothermia, persp-
    ectives and challenges" (F.E. South, J.P. Hannon, J.R.
    Willis, E.T. Pengelley and N.R. Alpert, eds.) pp. 17-53.
    Elsevier, Amsterdam.
Bhakoo, M. and Herbert, R.A. (1979). *Arch. Microbiol.* 121,
    121-127.
Bhakoo, M. and Herbert, R.A. (1980). *Arch. Microbiol.* 126,
    51-55.

Biagi, B.A. and Giebisch, G. (1979). *Amer. J. Physiol.* 236, F302-310.

Bick, H. (1972). *Ciliated Protozoa* World Health Organisation, Geneva.

Birillo, I.M. and Puhaljskii, L.H. (1936). *Probl. Zhivotn.* 10, 24-40.

Bishop, D.G., Kenrick, J.R., Bayston, J.H., Macpherson, A.S., Johns, S.R. and Willing, R.I. (1979). *In* "Low temperature stress in crop plants: The role of the membrane" (J.M. Lyons, D. Graham and J.K. Raison, eds.) pp. 375-390. Academic Press, New York and London.

Blackshaw, A.W. (1954). *Aust. J. biol. Sci.* 7, 573-582.

Blackshaw, A.W. (1958). *Aust. J. biol. Sci.* 11, 581-588.

Blackshaw, A.W. and Salisbury, G.W. (1957). *J. Dairy Sci.* 40, 1099-1106.

Blaker, W.D. and Moscatelli, E.A. (1978). *J. Neurochem.* 31, 1513-1518.

Blaker, W.D. and Moscatelli, E.A. (1979). *Lipids.* 14, 1027-1031.

Blok, M.C., van Deenen, L.L.M. and de Gier, D. (1976). *Biochim. biophys. Acta* 433, 1-12.

Blok, M.C., van Deenen, L.L.M. and de Gier, D. (1977). *Biochim. biophys. Acta* 464, 509-518.

Blostein, R. (1970). *J. Biol. Chem.* 245, 270-275.

Blough, H.A. (1975). *Proc. 3rd Intl. Congress Virology,* W36, Madrid.

Bock, P.E. and Frieden, C. (1978). *Trends in Biochem. Science* 3, 100-103.

Boender, J. (1968). *VI Int. Congr. Anim. Reprod. Artif. Insem. (Paris)* 2, 1217-1219.

Bogart, R. and Mayer, D.T. (1950). *J. Anim. Sci.* 9, 143-152.

Borden, D., Whitt, G.S. and Nanney, D.L. (1973). *J. Protozool.* 20, 693-700.

Borochov, A. and Borochov, H. (1979). *Biochim. biophys. Acta* 550, 546-549.

Bottomley, J.M., Kramers, M.T.C. and Chapman, D. (1980) *FEBS Lett.* 119, 261-264.

Bowler, K. and Duncan, C.J. (1968). *Comp. Biochem. Physiol.* 24, 1043-1054.

Bowler, K. and Duncan, C.J. (1969). *Physiol. Zool.* 42, 211-219.

Bradley, M.P., Forrester, I.T. and Rayns, D.G. (1979). *Proc. Univ. Otago Med. school* 57, 3-5.

Bradley, M.P., Rayns, D.G. and Forrester, I.T. (1980). *Archs Androl.* 4, 195-204.

Bramlage, W.J., Leopold, A.C. and Parrish, D.J. (1978). *Plant Physiol.* 61, 525-529.

Brandts, J.F. (1967). *In* "Thermobiology" (A.H. Rose, ed.),
    pp. 25-72. Academic Press, New York and London
Brandts, J.F., Fu, J. and Nordin, J.H. (1970). *In* "The
    frozen cell", (G.E.W. Wolstenholme and M. O'Connor, eds.),
    Ciba Foundation Symp. pp. 189-208
Branton, D. (1966). *Proc. Nat. Acad. Sci.* USA 55, 1048-1056.
Branton, D. (1969). *Ann. Rev. Plant Physiol.* 20, 209-238.
Brasitus, T.A., Tall, A.R. and Schachter, D. (1980). *Bio-
    chemistry* 19, 1256-1261.
Breathnach, A.S., Gross, M., Martin, B. and Stolinski, C.
    (1976). *J. Cell Sci.* 21, 437-448.
Bretscher, M.S. (1973). *Science* 181, 622-629.
Brock, T.D. (1975). *J. gen. Microbiol.* 89, 285-292.
Brock, T.D. (1978). "Thermophilic microorganisms and life at
    high temperatures". Springer, New York.
Brotherus, J.R., Jost, P.C., Griffith, O.H., Keana, J.F.W.
    and Hokin, L.E. (1980). *Proc. Nat. Acad. Sci.* 77, 272-276.
Brown, A.D. and Borowitzka, L.J. (1979). *In* "Biochemistry
    and Physiology of Protozoa", 2nd edition 1, (M. Levandowsky
    and S.H. Hunter, eds.) pp. 139-190. Academic Press, New York.
Brown, C.M. and Minnikin, D.E. (1973). *J. gen. Microbiol.*
    75, IX.
Brown, C.M. and Rose, A.H. (1969). *J. Bact.* 99, 371-378.
Brownlie, L.E. (1966). *J. appl. Bact.* 29, 447-454.
Burke, M.J., Gusta, L.V., Quamme, H.A., Weiser, C.J. and
    Li, P.H. (1976). *Ann. Rev. Plant Physiol.* 27, 507-528.
Butcher, R.W. (1959). *Minist. Acric. Fish Food (GB)* Ser. IV
    part 1, 1-74.
Butler, K.W., Tattrie, N.H. and Smith, I.C.P. (1974).
Biochim. biophys. Acta 363, 351-360.
Butler, W.J. and Roberts, T.K. (1975). *J. Reprod. Fert.* 43,
    183-187.
Byrne, P. and Chapman, D. (1964). *Nature* 202, 987-988.
Cadenhead, D.A., Kellner, B.M.J. and Müller-Landau, F. (1975).
    Biochim. biophys. Acta 382, 253-259.
Cameron, I.C. (1967). *J. Protozool.* 14, Suppl. 7.
Canham, P.B. and Burton, A.C. (1963). *J. Cell comp. Physiol.*
    61, 245-253.
Cannon, B. and Polnaszek, C.F. (1976). *In* "Regulation of
    Depressed Metabolism and Thermogenesis", (L. Jansky and
    X.J. Musacchia, eds.) pp. 93-116. Charles C. Thomas,
    Springfield, IL, USA.
Cannon, B., Polnasek, C.F., Butler, K.W., Eriksson, L.E.G.
    and Smith, I.C.P. (1975). *Archs. Biochem. Biophys.* 167,
    505-518.
Careri, G., Fasella, P. and Gratton, E. (1979). *Ann. Rev.
    Biophys. Bioeng.* 8, 69-97.

Carpenter, D.O. and Alving, B.O. (1968). *J. gen. Physiol.* 52, 1-21.

Carr, M.K.V. and Hough, M.N. (1978). *In* "Forage Maize" (E.S. Bunting, B.F. Pain, R.H. Phipps, J.M. Wilkinson and R.E. Gunn, eds.) p. 15. Agricultural Research Council, London.

Castellan, G. (1971). *Physical Chemistry* Second edition, Addison-Wesley.

Chandler, D.E. and Heuser, J. (1979). *J. Cell Biol.* 83, 91-108.

Chang, M.C. and Walton, A. (1940). *Proc. Roy. Soc. B* 129, 517-527.

Chapman, D. (1966). *Ann. N.Y. Acad. Sci.* 137, 745-754.

Chapman, D. (1967). *In* "Thermobiology" (A.H. Rose, ed.) p. 123 Academic Press, London and New York.

Chapman, D. (1969). *Lipids* 4, 251-260.

Chapman, D. (1975). *Q. Rev. Biophys.* 8, 185-235.

Chapman, D. and Urbina, J. (1971). *FEBS Lett.* 12, 169-172.

Chapman, D. and Wallach, D.F.H. (1968). *In* "Biological Membranes: Physical Fact and Function". (D. Chapman, ed.), p.125-202. Academic Press, New York.

Chapman, D., Williams, R.M. and Ladbrooke, B.D. (1967). *Chem. Phys. Lipids* 1, 445-475.

Chapman, D., Urbina, J. and Keough, K.M. (1974). *J. Biol. Chem.* 249, 2512-2521.

Chapman, D., Cornell, B.A. and Quinn, P.J. (1977a) *FEBS Symposium* 42, (G. Semenaz and E. Carafoli, eds.) pp. 72-85. Springer Verlag, Berlin.

Chapman, D., Cornell, B.A., Eliasz, A.W. and Perry, A. (1977b). *J. Mol. Biol.* 113, 517-538.

Chapman, D., Peel, W.E. and Quinn, P.J. (1978). *Ann. N.Y. Acad. Sci.* 308, 67-82.

Chapman, D., Gomez-Fernandez, J.C. and Goni, F.M. (1979). *FEBS Lett.* 98, 211-223.

Charnock, J.S. (1978). *In* "Strategies in Cold: Natural Torpidity and Thermogenesis" (L.C.H. Wang and J.W. Hudson, eds.) pp. 417-460. Academic Press, New York.

Charnock, J.S. and Simonson, L.P. (1978a). *Comp. Biochem. Physiol.* 593, 223-229.

Charnock, J.S. and Simonson, L.P. (1978b). *Comp. Biochem. Physiol.* 60B, 433-439.

Charnock, J.S., Gibson, R.A., McMurchie, E.J. and Raison, J.K. (1980). *Molec. Pharm.* 18, 476-482.

Chen, S.S.C. and Varner, J.E. (1969). *Plant Physiol.* 44, 770-774.

Cherry, R.J. (1979). *Biochim. biophys. Acta* 559, 289-327.

Cherry, R.J., Bürkli, A., Busslinger, M., Schneider, G. and Parish, G.R. (1976). *Nature* 263, 389-393.

Chien, S., Sung, K.P., Skalak, R., Usami, S. and Tozeren, A.
    (1978). *Biophys. J.* 24, 463-487.
Chin, J.H. and Goldstein, D.B. (1977). *Science* 196, 684-685.
Chin, J.H., Parsons, L.M. and Goldstein, D.B. (1978).
    Biochim. biophys. Acta 513, 358-363.
Choong, C.H. and Wales, R.G. (1962). *Aust. J. biol. Sci.* 15,
    543-551.
Christiansen, M.N. (1963). *Plant Physiol.* 38, 520-522.
Christiansen, M.N. (1967). *Plant Physiol.* 42, 431-433.
Christiansen, M.N. (1968). *Plant Physiol.* 43, 743-746.
Christiansen, M.N., Carns, H.R. and Slyter, D.J. (1970).
    Plant Physiol. 46, 53-56.
Cillis, G., Peterson, R., Russell, L., Hook, L. and Freund, M.
    (1978). *Prep. Biochem.* 8, 363-378.
Claret, M., Garay, R. and Giraud, F. (1978). *J. Physiol.*
    274, 247-264.
Clarke, A., Coulson, G. and Morris, G.J. (submitted). *Plant*
    Physiol.
Clarke, E.C.W. and Glew, D.N. (1980). *J. Chem. Soc. Faraday*
    Trans. I 76, 1911-1916.
Clarkson, D.T. and Gerloff, G.C. (1979). ARC Letcombe Labor-
    atory Annual Report for 1978. pp. 53-54. Agricultural Research
    Council, London.
Clowes, A., Cherry, R.J. and Chapman, D. (1971). *Biochim.*
    biophys. Acta 249, 301-307.
Cochrane, V.W. (1958). "Physiology of fungi". Wiley, New York.
Cogan, U., Shinitsky, M., Weber, G. and Nishida, T. (1973).
    Biochemistry 12, 521-528.
Cole, K.S. (1932). *J. Cell Comp. Physiol.* 4, 421-433.
Cone, R.A. (1972). *Nature* 236, 39-43.
Connor, J.A. and Stevens, C.F. (1971). *J. Physiol.* 213, 1-20.
Conner, R.L. and Stewart, B.Y. (1976). *J. Protozool.* 23,
    193-196.
Cook, G.M.W. and Stoddart, R.W. (1973). "Surface Carbohydrates
    of the Eukaryotic Cell", Academic Press, New York.
Cooney, G.C. and Emerson, R. (1964). "Thermophilic fungi. An
    account of their biology, activities, and classification".
    Freeman, San Francisco.
Cossins, A.R. (1976). *Lipids* 11, 306-316.
Cossins, A.R. (1977). *Biochim. biophys. Acta* 470, 395-411.
Cossins, A.R. (1981). *In* "Fluorescent Probes in Proteins and
    Membranes" (G. Beddard and M.A. West, eds.). Academic
    Press, London. (In press).
Cossins, A.R. and Prosser, C.L. (1978). *Proc. Natl. Acad.*
    Sci. USA. 75, 2040-2043.
Cossins, A.R. and Wilkinson, H.L. (1982). *J. Therm. Biol.*
    In press.

Cossins, A.R., Friedlander, M. and Prosser, C.L. (1977).
  J. Comp. Physiol. 120, 109-121.
Cossins, A.R., Christiansen, J.A. and Prosser, C.L. (1978).
  Biochim. biophys. Acta 511, 442-454.
Cossins, A.R., Kent, J. and Prosser, C.L. (1980). *Biochim.
  biophys. Acta* 599, 341-358
Costello, M.J. and Gulik-Krzywicki, T. (1976). *Biochim.
  biophys. Acta* 455, 412-432
Costerton, J.W.L., Ingram, J.M. and Cheng, K.J. (1974).
  Bact. Rev. 38, 87-115.
Creencia, R.P. and Bramlage, W.J. (1971). *Plant Physiol.*
  47, 389-394.
Cremel, G., Rebel, G., Canguilhem, B., Rendon, A. and
  Waksman, A. (1979). *Comp. Bioch. Physiol.* 63A, 159-168.
Cronan, J.E. Jr. (1975). *J. Biol. Chem.* 250, 7074-7077.
Cronan, J.E. Jr. (1978). *Ann. Rev. Biochem.* 47, 163-189.
Cronan, J.E. and Gelman, E.P. (1973). *J. Biol. Chem.* 248,
  1188-1195.
Cross, J. and McMahon, D. (1976). *Molec. gen. Genetics* 147,
  169-178.
Crowe, J.H., Crowe, L.M. and Deamer, D.W. *In* Biophysics of
  Water. (F. Franks, ed.) John Wiley and Sons, London,
  (In press).
Crowley, J., Ways, P. and Jones, J.W. (1965). *J. Clin.
  Invest.* 44, 989-998.
Crozier, W.J. (1924). *Proc. Natl. Acad. Sci.* USA, 10,
  461-464.
Cullen, J., Phillips, M.C. and Shipley, G.G. (1971).
  Biochem. J. 125, 733-742.
Cullis, P.R. and de Kruijff, B. (1979). *Biochim. biophys.
  Acta* 559, 399-420.
Curatolo, E., Verma, S.P., Sakura, J.D., Small, D.M., Shipley,
  G.G. and Wallach, D.F.H. (1978). *Biochemistry* 17, 1802-
  1807.
Czygan, F.C. (1970). *Arch. Mikrobiol.* 74, 69-76.
Dalziel, A.E. and Breidenbach, R.W. (1979). *In* "Low
  temperature stress in crop plants: The role of the mem-
  brane", (J.M. Lyons, D. Graham and J.K. Raison, eds.),
  pp. 319-326. Academic Press, New York and London.
Darin-Bennett, A. and White, I.G. (1977). *Cryobiology* 14,
  466-470.
Darin-Bennett, A., Poulos, A. and White, I.G. (1973). *Aust.
  J. biol. Sci.* 26, 1409-1420.
Darin-Bennett, A., Poulos, A. and White, I.G. (1974). *J.
  Reprod. Fert.* 41, 471-474.
Darin-Bennett, A., Poulos, A. and White, I.G. (1976).
  Andrologia 8, 37-45.

Darin-Bennett, A., White, I.G. and Hoskins, D.D. (1977).
  J. Reprod. Fert. 49, 119-122.
Darke, A., Finer, E.G., Flook, A.G. and Phillips, M.C. (1972).
  J. Mol. Biol. 63, 265-278.
Davies, H.V. and Pinfield, N.J. (1980). *Z. Pflanzenphys.*
  96, 59-65.
Davies, J. and Rideal, E. (1963). *Interfacial Phenomena*
  Academic Press, New York.
Dawson, R.M.C., Hemington, N. and Davenport, J.B. (1962).
  Biochem. J. 84, 497-501.
Deamer, D.W. and Branton, D. (1967). *Science* 158, 655-657.
Deamer, D.W., Leonard, R., Tardieu, A. and Branton, D. (1970).
  Biochim. biophys. Acta 219, 47-60.
Dean, W.L. and Tanford, C. (1978). *Biochemistry* 17, 1683-1690.
Deibel, R.H. and Seeley, H.W. *In* "Bergey's manual of deter-
  minative bacteriology" 8th Edn. (S.T. Cowan, J.G. Holt,
  J. Liston, R.G.E. Murray, C.F. Niven, A.W. Ravin and
  R.Y. Stanier, eds.) pp. 490-517. Wilkins and Wilkins,
  Baltimore.
Demel, R.A., Geurts van Keroel, W.S.M. and van Deenan, L.L.M.
  (1972). *Biochem. biophys. Acta* 266, 26-40.
Deuticke, B. and Ruska, C. (1976). *Biochim. biophys. Acta*
  433, 638-653.
DeVries, A.L. (1980). *In* "Animals and Environmental Fitness",
  (R. Gilles, ed.) pp. 583-607. Pergamon Press, Oxford.
Dierolf, B.M. and McDonald, H.S. (1969). *Z. Vergl. Physiol.*
  62, 284-290.
Diller, K.R. and Cravalho, E.G. (1970). *Cryobiology* 7,
  191-199.
Di Rienzo, J.M. and Inouye, M. (1979). *Cell* 17, 155-161.
Dixon, M. and Webb, E.C. (1964). "The Enzymes", Longmans, London
Dixon, W.L., Franks, F. and ap Rees, T. (1981). *Phytochemistry*
  (In press).
Dott, H.M. and Dingle, J.T. (1968). *Exp. Cell Res.* 52, 523-
  540.
Dott, H.M. and Foster, G.C. (1972). *J. Reprod. Fert.* 29,
  443-445.
Douce, R. and Joyard, J. (1979). *Adv. Bot. Res.* 7, 1-117.
Douzou, P. (1977). "Cryobiochemistry", p.141. Academic Press,
  London.
Dowgert, M.F. and Steponkus, P.L. (1979). *Plant Physiol.*
  63, S-76.
Dowgert, M.F. and Steponkus, P.L. (1981). *Plant Physiol.*
  (In press).
Downton, W.J.S. and Hawker, J.S. (1975). *Phytochem.* 14,
  1259-1263.
Drew, M. and McLaren, S. (1970). *Physiologia Pl.* 23, 544-
  560.

Dueling, H.J. and Helfrich, W. (1976). *Biophys. J.* 16, 861-868.

Dunham, P.B. and Sellers, D.A. (1980). *Fed. Proc.* 39, 1840.

Dunham, P.B., Stewart, G.W. and Ellory, J.C. (1980). *Proc. Nat. Acad. Sci.* USA. 77,1711-1715.

Eaks, I.L. and Morris, L.L. (1956). *Plant Physiol.* 31, 308-314.

Edwards, M. (1976). *Plant Physiol.* 58, 237-239.

Elford, B.C. and Solomon, A.K. (1974). *Nature* 248, 522-524.

Elgsaeter, A. and Branton, D. (1974). *J. Cell Biol.* 63, 1018-1030.

Ellory, J.C. and Willis, J.S. (1976). *Biochim. biophys. Acta* 443, 301-305.

Ellory, J.C. and Willis, J.S. (1978). *J. Physiol. Lond.* 275, 62P.

Ellory, J.C. and Willis, J.S. (1981). *J. Physiol. Lond.* (In press).

Engelman, D.M. and Rothman, J.E. (1972). *J. Biol. Chem.* 247, 3694-3695.

Erwin, J. and Bloch, K. (1963). *J. biol. Chem.* 238, 1618-1624.

Esfahan, M., Limbrick, A.R., Knutton, S., Oka, T. and Wakil, S.J. (1971). *Proc. Nat. Acad. Sci.* USA. 68, 3180-3184.

Esser, A.F. and Souza, K.A. (1974). *Proc. Nat. Acad. Sci.* USA. 71, 4111-4115.

Estep, T.N., Mountcastle, D.B., Biltonen, R.L. and Thompson, T.E. (1978). *Biochemistry* 17, 1984-1989.

Etemadi, H.H. (1980). *Biochim. biophys. Acta* 604, 347-472.

Evans, E.A. and Skalak, R. (1980). "Mechanics and thermodynamics of biomembranes". CRC Press, Florida.

Evans, E.A. and Waugh, R. (1977). *Biophys. J.* 20, 307-313.

Fang, L.S.T. (1971). "Enzymatic Basis of Cold Adaptation of Ion Transport in Hibernators". Ph.D. Thesis, University of Illinois, Urbana.

Fang, L.S.T. and Willis, J.S. (1974). *Comp. Biochem. Physiol.* 48A, 687-698.

Farrant, J. (1977). *Phil. Trans. Roy. Soc.* B278, 191-203.

Farrant, J. and Morris, G.J. (1973). *Cryobiology* 10, 134-140.

Farrant, J. and Woolgar, A.E. (1972). *Cryobiology* 9, 9-15.

Farrell, J. and Rose, A.H. (1967). *In* "Thermobiology". (A.H. Rose, ed.) Academic Press, London and New York.

Farrell, J. and Rose, A.H. (1968). *J. gen. Microbiol.* 50, 429-439.

Fasman, G.D. (1975). "CRC Handbook of Biochemistry and Molecular Biology. Lipids, Carbohydrates, Steroids". 3rd Edition. CRC Press, Cleveland, Ohio, USA.

Fennema, O. (1975). *In* "Water Relations of Foods", (R.B. Duckworth, ed.) pp. 397-416. Academic Press, London.

Ferber, E., Resch, K., Wallach, D.F.H. and Imm, W. (1972).
  Biochim. biophys. Acta 266, 494-504.
Ferguson, K.A., Conner, R.L., Mallory F.B. and Mallory, C.W.
  (1972). *Biochim. biophys. Acta* 270, 111-116.
Fettiplace, R. and Haydon, D.A. (1980). *Physiological Reviews*
  60, 510-550.
Finch, E.D. and Schneider, A.S. (1975). *Biochim. biophys.
  Acta* 406, 146-154.
Finean, J.B. (1969). *Q. Rev. Biophys.* 2, 1-23.
Finean, J.B., Coleman, R. and Michell, R.H. (1978).
  "Membranes and their cellular functions", Second Edition,
  Halsted Press, John Wiley and Sons, New York.
Finnes, G. and Matches, J.P. (1974). *Can. J. Microbiol.* 20,
  1639-1645.
Fishbarg, J. (1971). *J. Physiol.* 224, 149-171.
Fletcher, N.H. (1970). "Chemical Physics of Ice", Cambridge
  University Press, Cambridge.
Fogg, G.E. (1967). *Phil. Trans. Roy. Soc.* B252, 279-287.
Foote, E., McGrath, J. and Pegg, D. Unpublished data.
Forster, J. (1887). *Centr. Bakteriel Parasitenk.* 2, 337-340.
Foulkes, J.A. (1977). *J. Reprod. Fert.* 49, 277-284.
Fourcans, B. and Jain, M.K. (1974). *Adv. Lipid Res.* 12,
  147-226.
Fowler, V. and Branton, D. (1977). *Nature* 268, 23-26.
Fox, C.F. and Tsukagoshi, T. (1972). "Membrane Research",
  pp.145-153. Academic Press, New York.
Franks, F. (1977). *Phil. Trans. Roy. Soc.* B278, 33-56.
Franks, F. (1981). *In* "Water - A Comprehensive treatise",
  Vol. 7, (F. Franks, ed.) Plenum Press, New York. (In press).
Franks, F. and Bray, M. (1980). *Cryo-Letters* 1, 221-226.
Friedman, H., Lu, P. and Rich, A. (1969). *Nature* 223, 909-
  913.
Friedman, S.M. (1978). "Biochemistry of thermophily",
  Academic Press, New York.
Friedman, S.M., Nakashima, M. and McIndoe, R.A. (1977).
  Can. J. Physiol. Pharmacol. 55, 1301-1310.
Fujikawa, S. (1980). *Cryobiology* 17, 351-362.
Fujikawa, S. (1981). *J. Cell Sci.* (In press).
Fulco, A.J. (1970). *J. Biol. Chem.* 244, 889-896.
Fulco, A.J. (1972). *J. Biol. Chem.* 247, 3511-3519.
Fulco, A.J. (1974). *Ann. Rev. Biochem.* 43, 215-241.
Fulco, A.J. and Fujii, D.K. (1980). *In* "Membrane Fluidity",
  M. Kates and A. Kuksis, eds.),pp. 77-98, Humana, Clifton,
  New Jersey.
Fulford, A.J.C. and Peel, W.E. (1980). *Biochim. biophys.
  Acta* 598, 237-247.

Fukushima, H., Watanabe, T. and Nozawa, Y. (1976). *Biochim. biophys. Acta* 436, 249-259.

Fukushima, H., Nagao, S. and Nozawa, Y. (1979). *Biochim. biophys. Acta* 572, 178-182.

Funder, J. and Wieth, J.O. (1980). *In* "Membrane Transport in Erythrocytes: Relations between function and molecular structure", (U.V. Lassen, H.H. Ussing and J.O. Weith, eds.), pp. 520-527. Muncksgaard.

Garber, M.P. and Steponkus, P.L. (1976). *Plant Physiol.* 57, 673-680.

Garrahan, P.J. and Glynn, I.M. (1967). *J. Physiol.* 192, 237-256.

Gaughran, E.R.L. (1947). *Bact. Rev.* 11, 189-225.

Gavish, B. and Werber, M.M. (1979). *Biochemistry* 18, 1269-1275.

Gebauer, M.R., Pickett, B.W., Komarek, R.J. and Gaunya, W.S. (1970). *J. Dairy Sci.* 53, 817-823.

Gebhardt, C., Gruler, H. and Sackmann, E. (1977). *Z. Naturforsch.* 32, 581-596.

Gerritsen, W.J., Verkleij, A.J., Zwall, R.F.A. and van Deenen, L.L.M. (1978). *Eur. J. Biochem.* 85, 255-261.

Ghosh, D., Lyman, R.L. and Tinoco, J. (1971). *Chem. Phys. Lipids* 7, 173-184.

Giaquinta, R.T. and Geiger, D.R. (1973). *Plant Physiol.* 51, 372-377.

de Gier, J., Mandersloot, J.G. and van Deenan, L.L.M. (1968). *Biochim. biophys. Acta* 150, 666-675.

Gill, C.O. (1975). *J. gen. Microbiol.* 89, 293-298.

Gill, C.O. and Suisted, J.R. (1978). *J. gen. Microbiol.* 104, 31-36.

Gimmler, H., Heilman, B., Demming, B. and Hartung, W. (1981). *Z. Naturforschg.* Submitted.

Gingell, D. (1976). *In* "Mammalian Cell Membranes", (G.A. Jamieson and D.M. Robinson, eds.) pp. 198-223. Butterworths, London.

Gladcinova, E.F. (1937). *Usp. Zooteh. Nauk.* 4,

Glasstone, S., Laidler, K. and Eyring, H. (1941). "The Theory of Rate Processes", McGraw-Hill, New York.

Glynn, I.M. (1957). *J. Physiol.* 136, 148-173.

Glynn, I.M. and Hoffman, J.F. (1971). *J. Physiol.* 218, 239-256.

Glynn, I.M. and Karlish, S.J.D. (1975). *Ann. Rev. Physiol.* 37, 13-55.

Goldfine, H. (1972). *Advances in Microbial Physiology* 8, 1-58.

Goldstein, D.B. (1976). *Life Sciences* 18, 553-562.

Goldman, S.S. (1975). *Am. J. Physiol.* 228, 834-838.

Goldman, S.S. and Albers, R.W. (1975). *Arch. Biochem. Biophys.* 169, 540-544.

Goldman, S.S. and Albers, R.W. (1979). *J. Neurochem.* 32, 1139-1142.

Goldman, S.S. and Willis, J.S. (1973). *Cryobiology* 10, 212-217.

Gomez-Fernandez, J.C., Goni, F.M., Bach, D., Restall, C.J. and Chapman, D. (1980). *Biochim. biophys. Acta* 598, 502-516.

Gomperts, B. (1977). "The plasma membrane: Models for structure and function", Academic Press, New York.

Gordon, L.M., Sauerheber, R.D. and Esgate, J.A. (1978). *J. Supramolec. Struct.* 9, 299-326.

Gorman, A.L.F. and Marmor, A.F. (1970). *J. Physiol.* 210, 897-931.

Gottfried, E.L. and Rapport, M.M. (1962). *J. biol. Chem.* 237, 329-333.

Gould, G.W. and Measures, J.C. (1977). *Phil. Trans. Roy. Soc.* B278, 151-165.

Graham, D., Hockley, D.G. and Patterson, B.D. (1979). *In* "Low temperature stress in crop plants: The role of the membrane", (J.M. Lyons, D. Graham and J.K. Raison, eds.), pp. 453-461. Academic Press, New York and London.

Grant, C.W.M. and McConnell, H.M. (1974). *Proc. Natl. Acad. Sci. USA.* 71, 4653-4657.

Gray, G.M. (1960). *Biochem. J.* 77, 82-91.

Griffith, O.H., Dehlinger, P.J. and Van, S.P. (1974). *J. Membrane Biol.* 15, 159-192.

Griffith, O.H., Jost, P.C., Capaldi, R.A. and Vanderkooi, G. (1973). *Ann. N.Y. Acad. Sci.* 222, 561-573.

Grisham, C.M. and Barnett, R.E. (1973). *Biochim. biophys. Acta* 311, 417-423.

Grogan, D.E., Mayer, D.T. and Sikes, J.D. (1966). *J. Reprod. Fert.* 12, 431-436.

Gross, E. and Packer, L. (1965). *Biochem. biophys. Res. Commun.* 20, 715-719.

Grossman, A. and Togasaki, R.K. (1979). *Plant Physiol.* 63, (Supplement), 145.

Grout, B.W.W., Morris, G.J. and Clarke, A. (1980). *Cryo-Letters* 1, 251-256.

Grout, B.W.W., Shelton, K., Coulson, G. and Morris, G.J. (1981). *Cryo-Letters* (In press).

Gruener, N. and Avidor, Y. (1966). *Biochem. J.* 100, 762-767.

Grunze, M. and Deuticke, B. (1974). *Biochim. biophys. Acta* 356, 125-130.

Guinn, G. (1971). *Crop Sci.* 11, 101-102.

Gurr, M.I. and James, A.T. (1971). "Lipid biochemistry: an Introduction", Chapman and Hall, London.

Hackenbrock, C.R., Höchli, M. and Chau, R.M. (1976). *Biochim. biophys. Acta* 455, 466-484.

Hammerstedt, R.H., Amann, R.P., Rucinsky, T., Morse II, P.D.
    Lepock, J., Snipes, W. and Keith, A.D. (1976). *Biol.
    Reprod.* 14, 381-397.
Hammerstedt, R.H., Keith, A.D., Snipes, W., Amann, R.P.
    Arruda, D. and Griel, L.C. (1978). *Biol. Reprod.* 18,
    686-696.
Hancock, J.L. (1951). *Nature* 167, 323-324.
Hancock, J.L. (1952). *J. exp. Biol.* 29, 445-453.
Hannon, J.P. (1958). *Circ. Res.* 6, 771-778.
Hanson, M.R. and Bogorod, L. (1978). *J. gen. Microbiol.* 105,
    253-262.
Harder, W. and Veldkamp, H. (1965). *Arch. Microbiol.* 59, 723-
    730.
Harder, W. and Veldkamp, H. (1971). *Antonie van Leeuwenhoek*
    37, 51-63.
Harris, J.E. (1941). *J. Biol. Chem.* 141, 579-595.
Harris, P. and James, A.T. (1969). *Biochem. J.* 112, 325-330.
Harrison, R. and Lunt, G.C. (1975). "Biological Membranes",
    Blackie, Glasgow and London.
Harrison, R.A.P. and White, I.G. (1972). *J. Reprod. Fert.*
    30, 105-115.
Hartree, E.F. and Mann, T. (1959). *Biochem. J.* 71, 423-434.
van Hasselt, Ph.R. and Strikwerda, J.T. (1976). *Physiol.
    Plant.* 37, 253-257.
Hasted, J.B. and Sahidi, M. (1976). *Nature* 262, 777.
Hatefi, Y. and Hanstein, W.G. (1974). *In* "Methods in
    Enzymology", Vol. XXXI, A, (S. Fleischer and L. Packer,
    eds.) pp. 770-790. Academic Press, New York.
Hauser, H. and Dawson, R.M.C. (1967). *Eur. J. Biochem.* 1,
    61-69.
Hazel, J. (1973). *In* "Effects of temperature on ectothermic
    organisms", (W. Weiser, ed.), Springer-Verlag, Berlin.
Hazel, J. and Prosser, C.L. (1974). *Physiol. Rev.* 54, 620-
    677.
Hearse, D.J., Humphrey, S.M. and Bullock, G.R. (1978). *J.
    Mol. Cell Cardiol.* 10, 641-668.
Heber, U. (1967). *Plant Physiol.* 42, 1343-1350.
Heber, U. (1968). *Cryobiology* 5, 188-201.
Heber, U. (1969). *Biochim. biophys. Acta* 180, 302-319.
Heber, U. (1970). *In* Ciba Found. Symp. "The Frozen Cell",
    (G.E.W. Wolstenholme and M. O'Connor, eds.) pp. 175-188,
    J. and A. Churchill, London.
Heber, U. and Ernst, R. (1967). *In* "Cellular Injury and
    Resistance in Freezing Organisms", (E. Asahina, ed.)
    Proc. Internat. Conf. Low Temp. Sci. Vol. II, pp. 63-77,
    Bunyeido Print. Co., Sapporo.
Heber, U. and Heldt, H.W. (1981) *Ann. Rev. Plant Physiol.*
    (In press).

Heber, U. and Kempfle, M. (1970). *Z. Naturforsch* <u>25b</u>, 834–842.

Heber, U. and Santarius, K.A. (1964). *Plant Physiol.* <u>39</u>, 712–719.

Heber, U. and Santarius, K.A. (1976). *In* "Water und Plant Life", (O.L. Lange, L. Kappen and E. -D. Schulze, eds.), Ecological Studies, Vol. 19, pp. 253–267, Springer, Berlin and Heidelberg.

Heber, U., Tyankova, L. and Santarius, K.A. (1971). *Biochim. biophys. Acta* <u>241</u>, 578–592.

Heber, U., Tyankova, L. and Santarius, K.A. (1973). *Biochim. biophys. Acta* <u>291</u>, 23–37.

Heber, U., Volger, H., Overbeck, V. and Santarius, K.A. (1979). *In* "Proteins at Low Temperatures", (O. Fennema, ed.), Adv. Chem. Ser., No. 180, pp. 159–189, American Chem. Soc., Washington, D.C.

Henderson, R. and Unwin, P.N.T. (1975). *Nature* <u>257</u>, 28–32.

Hepler, L.G. and Woolley, E.M. (1973). *In* "Water – A Comprehensive Treatise", Vol. 3, (F. Franks, ed.) pp. 145–172, Plenum Press, New York.

Herbert, R.A. and Bell, C.R. (1977). *Arch. Microbiol.* <u>113</u>, 215–220.

Herbert, R.A. and Bhakoo, M. (1979). *In* "Growth in Cold Environments", (A.D. Russell, ed.), Society of Applied Bacteriology Technical Series, <u>13</u>, 1–16.

Herring, F.H., Tatischeff, I. and Weeks, G. (1980). *Biochim. biophys. Acta* <u>602</u>, 1–9.

Hesketh, T.R., Smith, G.A., Houslay, M.D., McGill, K.A., Birdsall, N.J.M., Metcalfe, J.M. and Warren, G.B. (1976). *Biochemistry* <u>15</u>, 4145–4151.

Hibbitt, K.G. and Benians, M. (1971). *J. gen. Microbiol.* <u>68</u>, 123–128.

Hidalgo, C., Thomas, D.D. and Ikemoto, N. (1978). *J. Biol. Chem.* <u>78</u>, 6879–6887.

Hilbig, R. and Rahmann, H. (1979). *Comp. Bioch. Physiol.* <u>62B</u>, 527–532.

Hildenbrand, K. and Nicolau, C. (1979). *Biochim. biophys. Acta* <u>553</u>, 365–377.

Hill, D.L. (1972). "The Biochemistry and Physiology of Tetrahymena", Academic Press, New York.

Hill, M.W. and Bangham, A.D. (1975). *Adv. Exp. Med. Biol.* <u>59</u>, 1–9.

Hinz, H. and Sturtevant, J.M. (1972a). *J. Biol. Chem.* <u>247</u>, 3697–3701.

Hinz, II. and Sturtevant, J.M. (1972b). *J. Biol. Chem.* <u>247</u>, 6071–6075.

Hiramoto, Y. (1963). *Exp. Cell Res.* <u>32</u>, 59–75.

Hoare, D.G. (1972). *J. Physiol.* 221, 311-329.

Hobbs, P.V. (1974). "Ice Physics", Oxford University Press.

Hochachka, P.W. and Somero, G.N. (1973). "Strategies of Biochemical Adaptation", W.B. Saunders, Philadelphia.

Höchli, M. and Hackenbrock, C.R. (1976). *Proc. Nat. Acad. Sci.* USA. 73, 1636-1640.

Höchli, M. and Hackenbrock, C.R. (1979). *Proc. Nat. Acad. Sci.* USA. 76, 1236-1240.

Hodgkin, A.L. (1958). *Proc. Roy. Soc. B.* 148, 1-37.

Hoffmann, W., Sarzala, G.M. and Chapman, D. (1979). *Proc. Nat. Acad. Sci.* USA. 76, 3860-3864.

Hoffmann, W., Sarzala, G.M., Gomez-Fernandez, J.C., Goni, F.M., Restall, C.J., Chapman, D., Heppeler, G. and Kreutz, W. (1980). *J. Mol. Biol.* 141, 119-132.

Hofmeister, F. (1888). *Arch. Exp. Pathol. Pharmakol.* 24, 247-262.

Holz, G.G. and Conner, R.L. (1973). *In* "Biology of Tetrahymena", (A.M. Elliott, ed.) pp. 99-122. Dowden, Hutchison and Ross, Stroudsburg, Pennsylvania.

Hood, R.D., Foley, C.W. and Martin, T.G. (1970). *J. Anim. Sci.* 30, 91-94.

Horowitz, A.F., Klein, M.P., Michaelson, D.M. and Kohler, S.J. (1973). *Ann. N.Y. Acad. Sci.* 222, 468-488.

Housley, M.D. and Palmer, R.W. (1978). *Biochem. J.* 174, 909-919.

Huang, L., Lorch, S.K., Smith, G.S. and Haug, A. (1973). *FEBS Lett.* 43, 1-5.

Hubbell, W.L. and McConnell, H.M. (1969). *Proc. Nat Acad. Sci.* USA. 64, 20-29.

Hubbell, W.L. and McConnell, H.M. (1971). *J. Amer. Chem. Soc.* 93, 314-326.

Hubbell, W.L., Metcalfe, J.C., Metcalfe, S.M. and McConnell, H.M. (1970). *Biochem. biophys. Acta* 219, 415-427.

Hudson, J.W. (1967). *In* "Mammalian Hibernation III", (K.C. Fisher, A.R. Dawe, C.P. Lyman, E. Schönbaum and F.E. South Jr. eds.), Oliver and Boyd, Toronto.

Hudson, J.W. and Wang, L.C.H. (1979). *Ann. Rev. Physiol.* 41, 287-303.

Huggins, C.E. (1963). *Science* 139, 504-505.

Hughes, R.C. (1975). "Essays in Biochemistry", Vol. XI, pp.1-36. Academic Press, London and New York.

Hui, S.W., Stewart, T.P., Boni, L.T. and Yeagle, P.L. (1981). *Science* 212, 921-923.

Hunter, K. and Rose, A.H. (1972). *Biochim. biophys. Acta* 260, 639-653.

Hunter, M.I.S., Olawoye, T.L. and Saynor, D.A. (1981). *Antonie van Leeuwenhoek* 47, 25-40.

Hymans, J. and Davies, D.R. (1972). *Mutation Research* 14, 381–389.

Iida, H., Maeda, T., Ohki, K., Nozawa, Y. and Ohnishi, S. (1978). *Biochim. biophys. Acta* 508, 55–64.

Inesi, G., Millman, M. and Eletr, S. (1973). *J. Mol. Biol.* 81, 483–504.

Ingraham, J.L. (1973). *In* "Temperature and Life", (H. Precht, J. Christophersen, H. Hensel and W. Larcher, eds.) pp.60–85, Springer, Berlin.

Ingraham, J.L. and Stokes, J.L. (1959). *Bact. Rev.* 23, 97–108.

Ingram, M. and Mackey, B.M. (1976). *Soc. Appl. Bact. Symp. Ser.* 5, 111–151.

Inniss, W.E. and Ingraham, J.L. (1978). *In* "Microbial Life in Extreme Environments", (D.J. Kushner, ed.) pp. 73–104. Academic Press, London.

Israelachvili, J., Sjösten, J., Göran Ericsson, L.E., Ehrström, M., Gräslund, A. and Ehrenberg, A. (1975). *Biochim. biophys. Acta* 382, 125–141.

Israelachvili, J.N., Mitchell, D.J. and Ninham, B.W. (1976). *J. Chem. Soc. (Faraday)* 72, 1525–1568.

Israelachvili, J.N., Marćelja, S. and Horn, R.G. (1980). *Quarterly Reviews of Biophysics* 13, 121–200.

Iype, P.T., Abraham, K.A. and Bhargava, P.M. (1963). *J. Reprod. Fert.* 5, 151–158.

James, R. and Branton, D. (1971). *Biochim. biophys. Acta* 233, 504–512.

Jan, L.Y. and Revel, J.P. (1974). *J. Cell Biol.* 62, 257–273.

Janoff, A.S., Haug, A. and McGroarty, E.J. (1979). *Biochim. biophys. Acta* 555, 56–66.

Janoff, A.S., Gupte, S. and McGroarty, E.J. (1980). *Biochim. biophys. Acta* 598, 641–644.

Jensen, M., Heber, U. and Oettmeier, W. (1981). *Cryobiology* 18, 322–335.

Johnson, L.A., Gerrits, R.J. and Young, E.P. (1969). *Biol. Reprod.* 1, 330–334.

Johnson, L.A., Pursel, V.G. and Gerrits, R.J. (1972). *J. Anim. Sci.* 35, 398–403.

Johnson, S.M. and Buttress, N. (1973). *Biochim. biophys. Acta* 510, 666–675.

Joiner, C.H. and Lauf, P.K. (1979). *Biochim. biophys. Acta* 552, 540–545.

Jones, M.N. (1975). "Biological Interfaces", Elsevier Scientific Publishing Co., Amsterdam.

Jones, P.C.T. (1970). *J. Exp. Bot.* 21, 58–63.

Jones, R.C. (1973). *J. Reprod. Fert.* 33, 145–149.

Jones, R. and Mann, T. (1977). *J. Reprod. Fert.* 50, 261–268.

Jones, R.C. and Martin, I.C.A. (1973). *J. Reprod. Fert.* 35, 311-320.

Jost, P., Libertini, L.J., Herbert, V.C. and Griffith, O.H. (1971). *J. Mol. Biol.* 57, 77-78.

Jost, P., Waggoner, A.S. and Griffith, O.H. (1971). *In* "Structure and Function of Biological Membranes", (L.I. Rothfield, ed.),pp. 83-144, Academic Press, New York and London.

Joyce, G.H., Hammond, R.K. and White, D.C. (1970). *J. Bact.* 104, 323-330.

Kahana, L., Rosenblith, W.A. and Galambos, R. (1950). *Am. J. Physiol.* 163, 213-223.

Kaiser, W.M. and Heber, U. (1981). *Planta* (Submitted).

Kaiser, W.M., Stepper, W. and Urbach, W. (1981a). *Planta* (In press).

Kaiser, W.M., Kaiser, G., Prachuab, P.K., Wildman, S.G. and Heber, U. (1981b). *Planta* (Submitted).

Kamienietzky, A. and Nelson, N. (1975). *Plant Physiol.* 55, 282-287.

Kamm, K.E., Zatzman, M.L., Jones, A.W. and South, F.E. (1979a). *Am. J. Physiol.* 237, C17-C22.

Kamm, K.E., Zatzman, M.L., Jones, A.W. and South, F.E. (1979b). *Am. J. Physiol.* 237, C23-C30.

Kampschmidt, R.F., Mayer, D.T. and Herman, H.A. (1953). *J. Dairy Sci.* 36, 733-742.

Kane, O., Marcellin, P. and Mazliak, P. (1978). *Plant Physiol.* 61, 634-638.

Kanno, H., Speedy, R.J. and Angell, C.A. (1975). *Science* 189, 880-881.

Kang, S., Gutowsky, H.S., Hshung, J.C., Jacobs, R., King, T.E., Rice, D. and Oldfield, E. (1979). *Biochemistry* 18, 3257-3267.

Kasai, R., Kitajima, Y., Martin, C.E., Nozawa, Y., Skriver, L. and Thompson, G.A. Jr. (1976). *Biochemistry* 15, 5228-5233.

Kasai, R., Sekiya, T., Okano, Y., Nagao, S., Ohki, K., Ohnishi, S. and Nozawa, Y. (1977). *Membrane* 2, 301-312.

Kates, M. and Baxter, R.M. (1962). *Can. J. Biochem.* 40, 1213-1227.

Kates, M. and Hagen, P.O. (1964). *Can. J. Biochem.* 42, 481-488.

Kates, M. and Pugh, E.L. (1980). *In* "Membrane Fluidity", (M. Kates and A. Kuksis, eds.), pp. 153-170, Humana, Clifton, N. Jersey.

Kawada, J., Taylor, R.E. and Barker, S.B. (1975). *Comp. Biochem. Physiol.* 50A, 297-302.

Kawaguchi, A., Seyama, Y., Sasaki, K., Okuda, S. and Yamakawa,
    T. (1980). *In* "Membrane Fluidity", (M. Kates and A. Kuksis,
    eds.), pp. 203-211, Humana, Clifton, New Jersey.
Kearns, V. and Toole, E.H. (1939). "Temperature and other
    factors affecting the germination of fescue seed". Tech-
    nical Bulletin No 638. United States Department of Agric-
    ulture, Washington, D.C.
Keith, A.D., Aloia, R.C., Lyons, J., Snipes, W. and Pengelley,
    E.T. (1975). *Biochim. biophys. Acta* 394, 204-210.
Keith, A.D., Mastro, A. and Snipes, W. (1979). *In* "Low
    temperature stress in crop plants: The role of the membrane",
    (J.M. Lyons, D. Graham and J.K. Raison, eds.), pp. 437-452,
    Academic Press, New York and London.
Kernan, R.P. (1980). "Cell Potassium", John Wiley and Sons.
Kimelberg, H.K. (1975). *Biochim. biophys. Acta* 413, 143-156
Kimelberg, H.K. and Papahadjopoulus, D. (1972). *Biochim.
    biophys. Acta* 282, 277-292.
Kimelberg, H.K. and Papahadjopoulos, D. (1974). *J. Biol. Chem.*
    249, 1071-1080.
Kimzey, S.L. and Willis, J.S. (1971a). *J. gen. Physiol.* 58,
    620-633.
Kimzey, S.L. and Willis, J.S. (1971b). *J. gen. Physiol.* 58,
    634-649.
King, M.W. and Roberts, E.H. (1979). "The storage of
    recalcitrant seeds - achievements and possible approaches",
    International Board for Plant Genetic Resources, Rome.
Kinosita, A. Jr., Kawato, S. and Ikegami, A. (1977). *Biophys.
    J.* 20, 289-305.
Kinsky, S.C., Haxby, J., Kinsky, C.B., Demel, R.A. and van
    Deenan, L.L.M. (1968). *Biochim. biophys Acta* 152, 174-
    185.
Kiovsky, T.E. and Pincock, R.E. (1966). *J. Amer. Chem. Soc.*
    88, 4704-4712.
Kitajima, Y. and Thompson, E.A. (1977a). *Biochim. biophys.
    Acta* 468, 73-80.
Kitajima, Y. and Thompson, E.A. (1977b). *J. Cell Biol.* 72,
    744-755.
Kleeman, W. and McConnell, H.M. (1976). *Biochim. biophys.
    Acta* 419, 206-222.
Kleeman, W., Grant, C.W.M. and McConnell, H.M. (1974).
    J. Supramol. Struct. 2, 609-616.
Klein, R.A., Moore, M.J. and Smith, M.W. (1971). *Biochim.
    biophys. Acta* 233, 420-433.
Klosson, R.J. and Krause, G.H. (1981a). *Planta* 151, 339-346.
Klosson, R.J. and Krause, G.H. (1981b). *Planta* 151, 347-353.
Knowles, P.F., Watts, A. and Marsh, D. (1979). *Biochemistry*
    18, 4480-4487.

Knox, J.M., Schwartz, G.S. and Diller, K.R. (1980). *J. Biomech. Engr.* 102, 91-97.

Kol, E. (1968). *In* "Kryobiologie, Biologie und Limnologie des Schnees und des Eises. 1. Kryovegetation". Die Binnengewasser Vol. 24, (H.J. Elster.and W. Ohle, eds.), E. Schweizerbartsche, Stuttgart.

Komarek, R.J., Pickett, B.W., Lanz, R.N. and Jensen, R.G. (1964). *J. Dairy Sci.* 47, 531-534.

Kornberg, R.D. and McConnell, H.M. (1971). *Biochemistry* 10, 1111-1120.

Kovacs, M.I.P. and Simpson, G.M. (1976). *Phytochem.* 15, 455-458.

Krause, G.H. (1973). *Biochim. biophys. Acta* 292, 715-728.

Krog, J.O., Zachariassen, K.E., Larsen, B. and Smidsrød. O. (1979). *Nature* 282, 300-301.

Krogmann, D.W. and Jagendorf. A.T. (1959). *Arch. Biochem. Biophys.* 80, 421-430.

Krause, G.H. (1974). *Biochim. biophys. Acta* 333, 301-313.

Krug, R.R., Hunter, W.G. and Grieger, R.A. (1976). *J. Phys. Chem.* 80, 2335-2342.

de Kruijff, B., Verleij, A.J., Van Echteld, C.J.A., Gerritsen, W.J., Nomers, C., Noordam, P.C. and De Gier, J. (1979). *Biochim. biophys. Acta* 555, 200-209.

de Kruyff, B., Van Dijck, P.W.M., Goldbach, R.W., Demel, R.A. and Van Deenan, L.L.M. (1973). *Biochim. biophys. Acta* 330, 269-282.

Kumamoto, J., Raison, J.K. and Lyons, J.M. (1971). *J. theoret. Biol.* 31, 47-51.

Kuntz, I.D. (1971). *J. Amer. Chem. Soc.* 93, 514-516.

Kuntz, I.D. and Kauzmann, W. (1974). *Adv. Protein Chem.* 28, 239-345.

Ladbrooke, B.D. and Chapman, D. (1969). *Chem. Phys. Lipids* 3, 304-367.

Ladbrooke, B.D., Williams, R.M. and Chapman, D. (1968a). *Biochim. biophys. Acta* 150, 333-340.

Ladbrooke, B.D., Jenkinson, T.J., Kamat, V.B. and Chapman, D. (1968b). *Biochim. biophys. Acta* 164, 101-109.

Lagerspetz, K.Y.H. (1974). *Biol. Rev.* 49, 477-514.

Lagerspetz, K.Y.H. and Talo, A. (1967). *J. Exp. Biol.* 47, 471-480.

Lagerspetz, K.Y.H. and Skytta, M. (1979). *Acta Physiol. Scand.* 106, 151-158.

Lagerspetz, K.Y.H., Kohanen, J. and Tirri, R. (1973). *Comp. Biochem. Physiol.* 44B, 823-827.

Lakowicz, J.R., Prendergast, F.G. and Hogan, D. (1979). *Biochemistry* 18, 508-519.

Lands, W.E.M. (1980). *Trans., Biochem. Soc.* 8, 25-27.

Lands, W.E.M. and Hart, P. (1965). *Biochim. biophys. Acta*
    98, 532-538.
Lang, E. and Lüdemann, H.D. (1977). *J. Chem. Phys.* **67**, 718-
    723.
Langworthy, T.A. (1977). *Biochim. biophys. Acta* **487**, 37-50.
Lardy, H.A. and Phillips, P.H. (1939). *Rec. Proc. Am. Soc.
    Anim. Prod.* **32**, 219.
Larsen, J.W. and Magid, L.J. (1974). *J. Am. Chem. Soc.* **96**,
    5774-5782.
Lasley, J.F. and Bogart, R. (1944a). *J. Anim. Sci.* **3**, 360-
    370.
Lasley, J.F. and Bogart, R. (1944b). *Am. J. Physiol.* **141**,
    619-624.
Lasley, J.F. and Mayer, D.T. (1944). *J. Anim. Sci.* **3**, 129-135.
Lasley, J.F., Easley, G.T. and Bogart, R. (1942). *J. Anim.
    Sci.* **1**, 79.
Lasley, J.F., Easley, G.T. and McKenzie, F.F. (1942). *Anat.
    Rec.* **82**, 167-174.
Laugesen, L.P., Nielsen, J.O.D. and Poulsen, J.H. (1974).
    Acta Physiol. Scand. **91**, 52A.
Lee, A.G. (1975). *Prog. Biophys. Mol. Biol.* **29**, 3-56.
Lee, A.G., Birdsall, N.J.M., Metcalfe, J.C., Toon, P.A. and
    Warren, G.B. (1974). *Biochemistry* **13**, 3699-3705.
Lee, P.E. and Gear, A.R.L. (1974). *J. Biol. Chem.* **249**, 7541-
    7549.
Leibo, S.P. (1977). *In* "The freezing of mammalian embryos",
    (K. Elliot and J. Whelan, eds.), Ciba Foundation symposium
    No 52 (New series), pp 69-96, Elsevier, Amsterdam.
Leibo, S.P. (1980). *J. Membr. Biol.* **53**, 179-188.
Lentz, R.B., Barenholz, Y. and Thompson, T.E. (1976). *Bio-
    chemistry* **15**, 4521-4528.
LeNeveu, D.M., Rand, R.P. and Parsegian, V.A. (1976). *Nature*
    259, 601-603.
Lerner, E., Shug, A.L., Elson, C. and Shrago, E. (1972). *J.
    Biol. Chem.* **247**, 1513-1519.
Levin, R.L., Cravalho, E.G. and Huggins, C.E. (1976a). *Cryo-
    biology* **13**, 415-429.
Levin, R.L., Cravalho, E.G. and Huggins, C.E. (1976b). *Bio-
    phys. J.* **16**, 1411-1426.
Levin, R.L., Cravalho, E.G. and Huggins, C.E. (1977). *J.
    Biomech. Engr.* **99**, 65-73.
Levin, R.L., Steponkus, P.L. and Wiest, S.C. (1978). *Agronomy
    Abstracts* p.80.
Levitt, J. (1966). *In* "Cryobiology", (H.T. Meryman, ed.),
    pp. 495-563, Academic Press, New York.
Levitt, J. (1980). "Responses of Plants to Environmental
    stresses", 2nd Ed., Vol I, Academic Press, New York and
    London.

Lew, V.L. and Ferreira, H.G. (1978). *Curr. Top. Memb. Trans.*
  10, 217-279.
Lieberman, M. (1975). *J. Gen. Physiol.* 65, 527-550.
Lieberman, M., Craft, C.C., Audia, W.V. and Wilcox, M.S. (1958).
  Plant Physiol. 33, 307-311.
Liebman, P.A. and Entine, G. (1974). *Science* 185, 457-459.
Linden, C.D., Keith, A.D. and Fox, C.F. (1973). *J. Supramol.*
  Struct. 1, 523-534.
Lineberger, R.D. and Steponkus, P.L. (1980a). *Plant Physiol.*
  65, 298-304.
Lineberger, R.D. and Steponkus, P.L. (1980b). *Cryobiology*
  17, 486-494.
Liston, J., Holman, M. and Matches, J. (1969). *Bact. Proc.*
  35, 40.
Littleton, J.M. and John, G. (1977). *J. Pharm. Pharmacol.*
  29, 579-580.
Liu, C.-C, Frehn, J.L. and LaPorta, A.D. (1969). *J. Appl.*
  Physiol. 27, 83-89.
Ljungdahl, L.G., Yang, S-S., Lin, M-T. and Wiegel, J. (1978).
  In "Biochemistry of thermophily", (S.M. Friedman, ed.),
  pp. 385-400, Academic Press, New York.
Lockhead, A.G. (1926). *Soil Science* 21, 225-231.
London, I.M. and Schwarz, H. (1953). *J. Clin. Invest.* 32,
  1248-1252.
Lovelock, J.E. (1953a). *Biochim. biophys. Acta* 10, 414-426.
Lovelock, J.E. (1953b). *Biochim. biophys. Acta* 11, 28-36.
Lovelock, J.E. (1954a). *Nature, Lond.* 173, 659-661.
Lovelock, J.E. (1954b). *Biochem. J.* 57, 265-270.
Lovelock, J.E. (1955a). *Biochem. J.* 60, 692-696.
Lovelock, J.E. (1955b). *Brit. J. Haematol.* 1, 117-129.
Lovelock, J.E. (1957). *Proc. Roy. Soc. B.* 147, 427-433.
Lozina-Lozinskii, L.K. (1974). "Studies in Cryobiology",
  John Wiley and Sons, New York.
Lunstra, D.D., Clegg, E.D. and Morré, D.J. (1974). *Prep.*
  Biochem. 4, 341-352.
Lyman, G.H., Papahadjopoulos, D. and Preisler, H.D. (1976).
  Biochim. biophys. Acta 448, 460-473.
Lyons, J.M. (1973). *Ann. Rev. Plant Physiol.* 24, 445-466.
Lyons, J.M. and Asmundson, C.M. (1965). *J. Am. Oil Chem. Soc.*
  42, 1056-1058.
Lyons, J.M. and Raison, J.K. (1970). *Plant Physiol.* 60, 470-
  474.
Lyons, J.M., Wheaton, T.A. and Pratt, H.K. (1964). *Plant*
  Physiol. 39, 262-268.
Lyons, J.M., Graham, D. and Raison, J.K. (1979a). "Low
  temperature stress in crop plants: The role of the membrane",
  Academic Press, New York and London.

Lyons, J.M., Raison, J.K. and Steponkus, P.L. (1979b). *In* "Low temperature stress in crop plants: The role of the membrane", (J.M. Lyons, D. Graham and J.K. Raison, eds.), pp. 1-24, Academic Press, New York, London, Sydney, Toronto, and San Francisco.

McElhaney, R.N. (1974). *J. Supramol. Struct.* 2, 617-628.

McElhaney, R.N. (1976). *In* "Extreme environments: Mechanisms of microbial adaptation", (M.R. Heinrich, ed.), pp. 255-281, Academic Press, New York and London.

McElhaney, R.N. and Tourtellotte, M.E. (1969). *Science* 164, 433-434.

McGrath, J.J. (1977). Ph.D. Dissertation, Massachusetts Institute of Technology.

McGrath, J.J. and Fallahi, A. (1981). *Proceedings of the 9th Annual Northeast Bioengineering Conference* (In press).

McGrath, J.J., Cravalho, E.G. and Huggins, C.E. (1975). *Cryobiology* 12, 540-550.

McGrath, J.J., Cravalho, E.G., and Huggins, C.E. (1978). *In* "1978 Advances in Bioengineering", pp. 195-198, American Society of Mechanical Engineers, New York.

McGrath, J.J., Marcus, J. and Ligon, R. (1980). *In* "1980 Advances in Bioengineering", pp. 201-204, American Society of Mechanical Engineers, New York.

McKersie, B.D. and Stinson, R.H. (1980). *Plant Physiol.* 66, 316-320.

McLean, R.A., Sulzbacher, W.L. and Mudd, S. (1951). *J. Bact.* 62, 723-728.

McMahon, D. (1971). *Molec. gen. Genet.* 112, 80-86.

McMenamin, M.M. (1978). Ph.D thesis, Queen's University, Belfast.

McMurchie, E.J., Raison, J.K. and Cairncross, K.D. (1973). *Comp. Biochem. Physiol.* 44B, 1017-1026.

McMurdo, A.C. (1981). Ph.D. thesis, University College of North Wales, Bangor.

McMurdo, A.C. and Wilson, J.M. (1980). *Cryo-Letters* 1, 231-238.

McWilliam, J.R., Manokaran, W. and Kipnis, T. (1979). *In* "Low temperature stress in crop plants: The role of the membrane", (J.M. Lyons, D. Graham and J.K. Raison, eds.), pp. 491-505, Academic Press, New York and London.

Mabrey, S., Mateo, P.L. and Sturtevant, J.M. (1978). *Biochemistry* 17, 2464-2468.

Madeira, V.M.C., Antunes-Madeira, M.C. and Carvalho, A.D. (1974). *Biochem. Biophys. Res. Commun.* 58, 897-904.

Mahadevan, V., Viswanathan, C.V. and Phillips, F. (1967). *J. Lipid Res.* 8, 2-6.

Maizels, M. and Patterson, J.H. (1940). *Lancet* 2, 417-420.

Maki, L.R., Galyan, E.L., Chang-Chein, M. and Caldwell, D.R. (1974). *Appl. Microbiol.* 28, 456-459.

Malan, A. (1978). *In* "Effectors of Thermogenesis", (L. Girardier and J. Seydoux, eds.), pp. 303-314, Birkhauser, Basel.

Mann, T. (1964). "The biochemistry of semen and of the male reproductive tract", Methuen and Co., Ltd. London.

Mann, T. and Lutwak-Mann, C. (1955). *Arch. Sci. biol.* 39, 578-588.

Marchesi, V.T., Furthmayr, H. and Tomita, M. (1976). *Ann. Rev. Biochem.* 45, 667-698.

Marchiafava, P.L. (1970). *Comp. Biochem. Physiol.* 34, 847-852.

Marr, A.G. and Ingraham, J.L. (1962). *J. Bact.* 84, 1260-1267.

Martin, C.E. and Thompson, G.A. (1978). *Biochemistry* 17, 3581-3586.

Martin, C.E., Hiramitsu, K., Kitajima, Y., Nozawa, Y., Skriver, L. and Thompson, G.A. (1976). *Biochemistry* 15, 5218-5227.

Martonosi, M.A. (1974). *FEBS Lett.* 47, 327-329.

Marx, R. and Brinkmann, K. (1979). *Planta* 144, 359-365.

Mason, G. (1976). Ph.D. thesis, University of Sheffield.

Mayer, D.T. (1955). *In* "Reproduction and Fertility", pp. 45-53, Michigan State University, East Lansing, Michigan.

Mayer, D.T. and Lasley, J.F. (1945). *J. Anim. Sci.* 4, 261-269.

Mazur, P. (1963). *J. Gen Physiol.* 47, 347-369.

Mazur, P. (1966). *In* "Cryobiology", (H.T. Meryman, ed.), pp. 213-315, Academic Press, New York.

Mazur, P. (1969). *Ann. Rev. Plant Physiol.* 20, 419-448.

Mazur, P. (1970). *Science* 168, 939-949.

Mazur, P. (1977a). *Cryobiology* 14, 251-272.

Mazur, P. (1977b). *In* "The freezing of of Mammalian Embryo's", (K. Elliott and J. Whelan, eds.) pp. 19-48. Ciba Foundation Symposium 52. Elsevier, Amsterdam.

Mazur, P., Leibo, S.P., Farrant, J., Chu, E.H.Y., Hanna, M.G. Jr. and Smith, L.H. (1970). *In* "The Frozen Cell", (G.E.W. Wolstenholme and M. O'Conner, eds.), pp. 69-88. Ciba Foundation Symposium, London:Churchill.

Mazur, P., Leibo, S.P. and Chu, E.H.Y. (1972). *Exp. Cell Res.* 71, 345-355.

Melchior, D.L. and Stein, J.M. (1976). *Ann. Rev. Biophys. Bioeng.* 5, 205-238.

Merickel, M. and Kater, S.B. (1974). *J. Comp. Physiol.* 94, 195-206.

Meryman, H.T. (1966). *In* "Cryobiology", pp. 1-114, Academic Press, London.

Meryman, H.T. (1967). *In* "Cellular Injury and Resistance in Freezing Organisms", (E. Asashina, ed.), pp. 231-244, Inst. Low Temp. Sci., Sapporo, Japan.

Meryman, H.T. (1968). *Nature* 218, 333-336.
Meryman, H.T. (1971). *Cryobiology* 8, 489-500.
Meryman, H.T. (1974). *Ann. Rev. Biophys.* 3, 341-363.
Meryman, H.T. and Kafig, E. (1955). *Proc. Soc. Exp. Biol.* 90, 587-589.
Meryman, H.T., Williams, R.J. and Douglas, M.St.J. (1977). *Cryobiology* 14, 287-302.
Metcalfe, J.M. and Warren, G.B. (1977). *In* "International Cell Biology 1976-1977", (B.R. Brinkley and K.R. Porter, eds.), pp. 15-23, Rockefeller, University Press, New York.
Michener, H.D. and Elliott, R.P. (1964). *Adv. Fd. Res.* 13, 349-396.
Miller, L.D. and Mayer, D.T. (1960). *Bull. Univ. Miss. Agr. Exp. Station* No. 742.
Miller, L.D., Mayer, D.T. and Merilan, C.P. (1965). *J. Dairy Sci.* 48, 395-398.
Miller, N.G.A., Hill, M.W. and Smith, M.W. (1976). *Biochim. biophys. Acta* 455, 644-654.
Miller, R.G. (1980). *Nature* 287, 166-167.
Miller, R.W., de la Roche, I. and Pomeroy, M.K. (1974). *Plant Physiol.* 53, 426-433.
Milovanov, V.K. (1934). Iskustvennoe osemenenie S.-L. Zivotnyh, (Artificial Insemination of Livestock), Moscow:Seljhozgiz.
Mitchell, C.D. and Hanahan, D.J. (1966). *Biochemistry* 5, 51-57.
Mitchison, J.M. and Swann, M.M. (1954). *J. Exp. Biol.* 31, 443-472.
Moiroud, A. and Gounod, A.M. (1969). *C.r. hebd. Séanc Acad. Sci.* 269, 2150-2152.
Moor, H. (1964). *Z. Zellforsch. Microscop. Anat.* 62, 546-580.
Moore, H.D.M., Hall, G.A. and Hibbitt, K.G. (1976). *J. Reprod. Fert.* 47, 39-45.
Morita, R.Y. (1966). *Oceanography and Marine Biology Annual Reviews* 4 105-121.
Morita, R.Y. (1975). *Bact. Rev.* 39, 146-167.
Morita, R.Y. and Haight, R.D. (1964). *Limnol. Oceanogr.* 9, 103-106.
Morris, G.J. (1975). *Cryobiology* 12, 192-201.
Morris, G.J. (1980). *In* "Low Temperature Preservation in Medicine and Biology", (M.J. Ashwood-Smith and J. Farrant, eds.), pp.253-283, Pitman Medical Press, Tunbridge Wells.
Morris, G.J. (1981). *Cryobiology* (In press a).
Morris, G.J. (1981). *In* "Biophysics of Water", (F. Franks, ed.), John Wiley and Sons, London (In press b).
Morris, G.J. and Clarke, A. (1978). *Arch. Microbiol.* 119, 153-156.
Morris, G.J. and McGrath, J.J. (1981a). *Cryobiology* 18, 390-398.

Morris, G.J. and McGrath, J.J. (1981b). *Cryo-Letters* (In press).

Morris, G.J., Coulson, G.E. and Clarke, A. (1979). *Cryobiology* 16, 401-410.

Morris, G.J., Coulson, G.E. and Clarke, A. (1981). *Cryo-Letters* 2, 111-116.

Morrison, W.W. and Milkman, R. (1978). *Nature* 273, 49-50.

Morse, P.D., Ruhlig, M., Snipes, W. and Keith, A.D. (1975). *Arch. Biochem. Biophys.* 168, 40-56.

Mosser, J.L., Mosser, A.G. and Brock, T.D. (1977). *J. Phycol.* 13, 22-27.

Muhlrädt, G. and Golecki, J.R. (1975). *Eur. J. Biochem.* 51, 343-352.

Murdoch, R.N. and White, I.G. (1968). *Aust. J. biol. Sci.* 21, 483-490.

Murphy, C. (1981). Ph.D. thesis, University College of North Wales, Bangor.

Murphy, C. and Wilson, J.M. (1981). *Plant, Cell and Environment* (In press).

Nägel, W.C. and Wunderlich, F. (1977). *J. Membr. Biol.* 32, 151-164.

Nakae, T. (1976). *Biochem. Biophys. Res. Commun.* 71, 877-884.

Nakanishi, YuhH., Fujikawa, S. and Nei, T. (1979). *Proc. Japan Acad.* 55, 132-134.

Nakatani, H.Y., Barber, J. and Forrester, J.A. (1978). *Biochim. biophys. Acta* 504, 215-225.

Nakayama, H., Mitsui, T., Nishihara, M. and Kito, M. (1980). *Biochim. biophys. Acta* 601, 1-10.

Nandini-Kishore, S.G., Kitajima, Y. and Thompson, G.A. Jr. (1977). *Biochim. biophys. Acta* 471, 157-161.

Nanney, D.L. (1980). "Experimental Ciliatology", John Wiley and Sons, New York.

Nei, T. (1973). *J. Electronmicrosc.* 22, 371-373.

Nei, T. (1976). *Cryobiology* 13, 287-294.

Neill, A.R. and Masters, C.J. (1972). *Biochem. J.* 127, 375-385.

Neill, A.R. and Masters, C.J. (1973). *J. Reprod. Fert.* 24, 279-287.

Nemat-Gorgani, M. and Meisanii, E. (1979). *J. Neurochem.* 32, 1027-1032.

Nermut, M.V. and Ward, B.J. (1974). *J. Microsc.* 102, 29-39.

Neufeld, A.H. and Levy, H.M. (1970). *J. Biol. Chem.* 245, 4962-4971.

Noggle, G.R. and Fites, R.C. (1974). *In* "Mechanisms of regulation of plant growth", (R.L. Bieleski, A.R. Ferguson and M.M. Cresswell, eds.), pp. 525-531, Royal Society of New Zealand, Wellington.

Nolan, W.G. and Simllie, R.M. (1976). *Biochim. biophys. Acta* <u>440</u>, 461–475.

Nomura, M. (1970). *Bact. Rev.* <u>34</u>, 228–277.

Norris, C.H. (1939). *J. Cell Comp. Physiol.* <u>14</u>, 117–133.

Nozawa, Y. and Kasai, R. (1978). *Biochim. biophys. Acta* <u>529</u>, 54–66.

Nozawa, Y. and Thompson, G.A. (1971). *J. Cell Biol.* <u>49</u>, 712–721.

Nozawa, Y. and Thompson, G.A. (1979). *In* "Biochemistry and Physiology of Protozoa", Vol. 2. (M. Levandowsky and S.H. Hutner, eds.) pp. 275–338. Academic Press, London.

Nozawa, Y., Iida, H., Fukushima, H. and Ohnishi, S. (1974). *Biochim. biophys. Acta* <u>367</u>, 134–147.

Oldfield, E. and Chapman, D. (1972). *FEBS Lett.* <u>23</u>, 285–297.

Oldfield, E., Gilmore, R., Glaser, M., Gutowsky, H.S., Hshung, J.C., Kang, S.J., King, E., Meadows, M. and Rice, D. (1978). *Proc. Nat. Acad. Sci.* USA <u>75</u>, 4657–4660.

Onuma, H. (1963). *Bull. Nat. Inst. Anim. Ind.* <u>3</u>, 105–119.

Opella, S.J., Yesinowski, J.P. and Waugh, J.S. (1976). *Proc. Nat. Acad. Sci.* USA <u>73</u>, 3812–3815.

O'Shea, T. and Wales, R.G. (1966a). *Aust. J. biol. Sci.* <u>19</u>, 871–882.

O'Shea, T. and Wales, R.G. (1966b). *Aust. J. biol. Sci.* <u>20</u>, 447–460.

Osborn, M.J., Gander, J.E. and Paris, E. (1972). *J. Biol. Chem.* <u>267</u>, 3973–3989.

Osipov, Yu. A., Zhelezhnyi, B.V. and Bondarenko, N.F. (1977). *Russ. J. Phys. Chem.* <u>51</u>, 1264–1265.

Overpath, P., Schairer, H.V. and Stoffel, W. (1970). *Proc. Nat. Acad. Sci.* USA <u>67</u>, 606–612.

Pace, M.M. and Graham, E.F. (1974). *J. Anim. Sci.* <u>39</u>, 1144–1149.

Packer, K.J. (1977). *Phil. Trans. Roy. Soc.* <u>B278</u>, 59–86.

Pain, R.H. (1979). *In* "Characterization of Protein Conformation and Function", (F.Franks, ed.), pp.19–36, Symposium Press, London.

Palta, J.P., Levitt, J. and Stadelmann, E.J. (1977). *Plant Physiol.* <u>60</u>, 393–397.

Pandey, G.N., Dorus, E., Davis, J.M. and Tosteson, D.C. (1979). *Arch. Gen. Phychiatry* <u>36</u>, 902–909.

Papahadjopoulos, D., Jacobson, K., Nir, S. and Isac, T. (1973). *Biochim. biophys. Acta* <u>311</u>, 300–348.

Parish, G.R. (1975). *J. Microscopy* <u>104</u>, 245–256.

Park, Y.S. and Hong, S.K. (1976). *Amer. J. Physiol.* <u>231</u>, 1356–1363.

Partridge, L.D. and Connor, J.A. (1978). *Amer. J. Physiol.* <u>234</u>, C155–161.

Patterson, B.D., Murata, T. and Graham, D. (1976). *Aust. J. Plant Physiol.* 3, 435-442.

Patterson, B.D., Graham, D. and Paull, R. (1979). *In* "Low temperature stress in crop plants: The role of the membrane", (J.M. Lyons, D. Graham and J.K. Raison, eds.), pp.25-36, Academic Press, New York and London.

Paul, K. and Morita, R.Y. (1971). *J. Bact.* 108, 835-843.

Penefsky, H.S. and Warner, R.C. (1965). *J. Biol. Chem.* 240, 4694-4702.

Pesce, A.J., Rosen, C-G. and Pasby, T.L. (1971). "Fluorescence Spectroscopy - An Introduction for Biology and Medicine", Marcel Dekker Inc., New York.

Peters, R., Peters, J., Tews, K.H. and Bahr, W. (1974). *Biochim. biophys. Acta* 367, 282-294.

Phillips, M.C. and Finer, E.G. (1974). *Biochim. biophys. Acta* 356, 199-206.

Phillips, P.H. (1939). *J. Biol. Chem.* 130, 415.

Pickett, B.W. and Komarek, R.J. (1967). *J. Dairy Sci.* 50, 753-757.

Pickett, B.W., Komarek, R.J., Gebauer, M.R., Benson, R.W. and Gibson, E.W. (1967). *J. Anim. Sci.* 26, 792-798.

Pinto da Silva, P. and Nicolson, G.L. (1974). *Biochim. biophys. Acta* 363, 311-319.

Platner, W.S., Steven, D.G., Tempel, G. and Musacchia, X.J. (1976). *Comp. Bioch. Physiol.* 53A, 279-283.

Plesner, I.W., Plesner, L., Nørby, J.G. and Klodos, I. (1981). *Biochim. biophys. Acta* 643, 483-494.

Polge, C. (1956). *Vet. Rec.* 68, 62-78.

Pollack, J.D. and Tourtellotte, M.E. (1967). *J. Bact.* 93, 636-641.

Pomeroy, M.K. and Andrews, C.J. (1975). *Plant Physiol.* 56, 703-706.

Pomeroy, M.K. and Siminovitch, D. (1971). *Can. J. Bot.* 49, 787-795.

Ponder, E. (1955). *Protoplasmatologia* 10, 1-123.

Poo, M.M. and Cone, R.A. (1974). *Nature* 247, 438-441.

Porter, V.S., Denning, N.P., Wright, R.C. and Scott, E.M. (1953). *J. Biol. Chem.* 205, 883-891.

Post, R.L. and Jolly, P.C. (1956). *Biochim. biophys. Acta* 25, 118-128.

Potter, J.F. and Ross, G.J.S. (1979). *In* "Low temperature stress in crop plants: The role of the membrane", (J.M. Lyons, D. Graham and J.K. Raison, eds.), pp. 535-542, Academic Press, New York and London.

Poulos, A. and White, I.G. (1973). *J. Reprod. Fert.* 35, 265-272.

Poulos, A., Darin, A.C. and White, I.G. (1972). *Proc. Aust. Biochem. Soc.* 5, 28.

Poulos, A., Darin-Bennett, A. and White, I.G. (1973a). *Comp. Physiol.* 46B, 541-549.

Poulos, A., Voglmayer, J.K. and White, I.G. (1973b). *Biochim. biophys Acta* 306, 194-202.

Poulos, A., Brown-Woodman, P.D.C., White, I.G. and Cox, R.I. (1975). *Biochim. biophys. Acta* 388, 12-18.

Precht, H., Christophersen, J., Hensel, H. and Larcher, W. (1973). "Temperature and Life", Springer Verlag, Berlin.

Prosser, C.L. (1973). "Comparative Animal Physiology", 3rd Edition, Saunders, Philadelphia.

Pursel, V.G. (1979). *Biol. Reprod.* 21, 319-324.

Pursel, V.G. and Graham, E.F. (1967). *J. Reprod. Fert.* 14, 203-211.

Pursel, V.G., Johnson, L.A. and Gerrits, R.J. (1968). *J. Anim. Sci.* 27, 1788.

Pursel, V.G., Johnson, L.A. and Gerrits, R.J. (1969). *J. Anim. Sci.* 29, 196.

Pursel, V.G., Johnson, L.A. and Gerrits, R.J. (1970). *Cryobiology* 7, 141-144.

Pursel, V.G., Johnson, L.A. and Rampacek, G.B. (1972a). *J. Anim. Sci.* 34, 278-283.

Pursel, V.G., Johnson, L.A. and Schulman, L.L. (1972b). *J. Anim. Sci.* 35, 580-584.

Pursel, V.G., Johnson, L.A. and Schulman, L.L. (1973). *J. Anim. Sci.* 37, 528-531.

Quinn, P.J. (1976). "Molecular Biology of Cell Membranes", McMillan, London.

Quinn, P.J. and Chapman, D. (1980). *Crit. Rev. Biochem.* 8, 1-117.

Quinn, P.J. and White, I.G. (1966). *J. Reprod. Fert.* 12, 263-270.

Quinn, P.J. and White, I.G. (1967). *Aust. J. biol. Sci.* 20, 1205-1215.

Quinn, P.J. and White, I.G. (1968). *Exp. Cell Res.* 49, 31-39.

Quinn, P.J., Salamon, S. and White, I.G. (1968a). *Aust. J. agric. Sci.* 19, 119-128.

Quinn, P.J., Salamon, S. and White, I.G. (1968b). *Aust. J. biol. Sci.* 21, 133-140.

Quinn, P.J., White, I.G. and Cleland, K.W. (1969). *J. Reprod. Fert.* 18, 209-220.

Quinn, P.J., Chow, P.Y.W. and White, I.G. (1980). *J. Reprod. Fert.* 60, 403-407.

Raison, J.K. and Chapman, E.A. (1976). *Aust. J. Plant Physiol.* 3, 291-299.

Raison, J.K. and Lyons, J.M. (1971). *Proc. Nat. Acad. Sci. USA* 68, 2092-2094.

Raison, J.K. and McMurchie, E.J. (1974). *Biochim. biophys. Acta* 363, 135-140.

Raison, J.K., Lyons, J.M., Melhorn, R.J. and Keith, A.D.
(1971a). *J. Biol. Chem.* 246, 4036-4040.

Raison, J.K., Lyons, J.M. and Thompson, W.W. (1971b). *Arch.
Biochem. Biophys.* 142, 83-90.

Raison, J.K., Wright, L.C., Chapman, E. and Hannan, G.N. (1978).
Plant Physiol. 61 (supplement) 32.

Raison, J.K., Chapman, E.A., Wright, L.C. and Jacobs, S.W.L.
(1979). *In* "Low temperature stress in crop plants: The
role of the membrane", (J.M. Lyons, D. Graham and J.K.
Raison, eds.), pp. 177-186, Academic Press, New York and
London.

Rall, W.F., Mazur, P. and Souzo, H. (1978). *Biophys. J.* 23,
101-120.

Rand, R.P. and Burton, A.C. (1964). *Biophys. J.* 4, 115-135.

Randle, P.J., Foden, S. and Kanagasuntheram, P. (1979). *In*
"Secretory Mechanisms". Symp. Soc. Exp. Biol. No 33.
(C.R. Hopkins and C.J. Duncan, eds.), pp. 199-223, Camb-
ridge University Press.

Rapatz, G., Sullivan, J. and Luyet, B. (1968). *Cryobiology*
5, 18-25.

Rasmussen, H., Clayberger, C. and Gustin, M.C. (1979). *In*
"Secretory Mechanisms", Symp. Soc. Exp. Biol. No. 33,
(C.R. Hopkins and C.J. Duncan, eds.), pp. 161-197, Camb-
ridge University Press.

Razi-Naqvi, K., Gonzalez-Rodriguez, J., Cherry, R.J. and
Chapman, D. (1973). *Nature* 245, 249-251.

Reeves, R.B. (1977). *Ann. Rev. Physiol.* 39, 559-586.

Reis, H.E. (1979). *Nature* 281, 287-289.

Riordan, J.R. (1980). *In* "Membrane Fluidity", (M. Kates and
A. Kuksis, eds.), pp. 119-129, Humana, Clifton, N. Jersey.

Robards, A.W., Newman, T.M. and Clarkson, D.T. (1980). *In*
"Plant Membrane Transport: Current Conceptual Issues,
(R.M. Spanswick, W.J. Lucas and J. Dainty, eds.), pp. 395-
396, Elsevier/North-Holland, Amsterdam.

Roberts, J.C., Arine, R.M., Rochelle, R.H. and Chaffee, R.R.J.
(1972). *Comp. Biochem. Physiol.* 41B, 127-135.

Roberts, J.C. and Chaffee, R.R.J. (1973). *Comp. Biochem.
Physiol.* 44B, 137-144.

Robinson, J.D. and Flashner, M.S. (1979). *Biochim. biophys.
Acta* 549, 145-176.

Robinson, R.A. and Stokes, R.H. (1959). "Electrolyte Solutions", Second Edition, Butterworth, London.

de la Roche, I.A., Pomeroy, M.K. and Andrew, C.J. (1975).
Cryobiology 12, 506-512.

Rogers, A. (1974). *In* "Spatial and Environmental Systems
Analysis", (R.J. Chorley and D.W. Harvey, eds.), Mono-
graph No 6, Pion Press, London.

Roomans, G.M. (1975). *Exp. Cell Res.* 96, 23-30.

Rose, A.H. and Evison, L.M. (1965). *J. gen. Microbiol.* 38, 131-141.

Rothman, J.E. and Lenard, J. (1977). *Science* 195, 743-753.

Rottem, S., Markowitz, O. and Razin, S. (1978). *Eur. J. Biochem.* 85, 445-450.

Russell, N.J. (1971). *Biochim. biophys. Acta* 231, 254-256.

Ryser, J.J-P. and Hancock, R. (1965). *Science* 150, 501-503.

Sachs, J.R. (1977). *J. Physiol.* 273, 489-514.

Sandercock, S.P. and Russell, N.J. (1980). *Bioch. J.* 188, 585-592.

Sager, R. and Granick, S. (1954). *J. gen. Physiol.* 37, 729-742.

Santarius, K.A. (1969). *Planta* 89, 23-46.

Santarius, K.A. (1971). *Plant Physiol.* 48, 156-162.

Santarius, K.A. (1973a). *Biochim. biophys. Acta* 291, 38-50.

Santarius, K.A. (1973b). *Planta* 113, 105-114.

Santarius, K.A. (1978). *Acta Horticulturae* 81, 9-21.

Santarius, K.A. and Ernst, R. (1967). *Planta* 73, 91-108.

Santarius, K.A. and Heber, U. (1970). *Cryobiology* 7, 71-78.

Santarius, K.A. and Heber, U. (1972). *In* "Proc. Colloquim on the Winter Hardiness of Cereals", pp. 7-29, *Agric. Res.* Inst., Hung. Acad. Sci., Martonvásár.

Santarius, K.A., Heber, U. and Krause, G.H. (1979). *Ber. Deutsch. Bot. Ges.* 92, 209-223.

Scarth, G.W., Levitt, J. and Siminovitch, D. (1940). *Cold Spg. Harbor Sym.* 8, 102-109.

Schatte, C., Rose, C., Durrenburger, J., O'Deem, L. and Swan, H. (1977). *Cryobiology* 14, 443-450.

Schick, B.P., Harpul, L.G. and Conner, R.L. (1979). *Biochim. biophys. Acta* 575, 475-478.

Schneider, A.S., Middaugh, C.R., and Oldewurtel, M.C. (1979). *J. Supramol. Struct.* 10, 265-275.

Schneider, M.J.T. and Schneider, A.S. (1972). *J. Memb. Biol.* 9, 127-140.

Scholander, P.F., Hagg, W., Hack, R.J. and Irving, H. (1953). *Cell Comp. Physiol.* 42, supplement 1.

Schreier-Muccillo, S., Marsh, D., Dugas, H., Schneider, H. and Smith, I.C.P. (1973). *Chem. Phys. Lipids* 10, 11-17.

Schrier, S., Polnaszek, C.F. and Smith, I.C.P. (1978). *Biochim. biophys. Acta* 515, 375-436.

Schroeder, F. (1978). *Biochim. biophys. Acta* 511, 356-376.

Schultz, S.G. (1977). *Amer. J. Physiol.* 233, E249-254.

Scott, T.W. and Dawson, R.M.C. (1968). *Biochem. J.* 108, 457-463.

Scott, T.W., Voglmayr, J.K. and Setchell, B.P. (1967). *Biochem. J.* 102, 456-461.

Seelig, A. and Seelig, J. (1974). *Biochemistry* 13, 4835-4839.
Seelig, J. and Seelig, A. (1970). *Quart. Rev. Biophys.* 13, 19-61.
Seelig, J. and Niederberger, W. (1974). *Biochemistry* 13, 1585-1588.
Selivonchick, D.P., Schmid, P.C., Natarjan, V. and Schmid, H.H.O. (1980). *Biochim. biophys. Acta* 618, 242-254.
Sen, A.K. and Widdas, W.F. (1962). *J. Physiol.* 160, 392-403.
Senser, M. and Beck, E. (1977). *Planta* 137, 195-201.
Shaw, M. and Ingraham, J.L. (1965). *J. Bact.* 90, 141-146.
Shehata, T.E. and Collins, E.B. (1971). *Appl. Microbiol.* 21, 466-469.
Shimonaka, H. and Nozawa, Y. (1977). *Cell Struct. Funct.* 2, 81-89.
Shimonaka, H., Fukushima, H., Kawai, K., Nagao, S., Okano, Y. and Nozawa, Y. (1978). *Experientia* 34, 586-587.
Shimshick, E.J. and McConnell, H.M. (1973a). *Biochemistry* 12, 2351-2360.
Shimshick, E.J. and McConnell, H.M. (1973b). *Biochem. Biophys. Res. Commun.* 53, 446-448.
Shinitzky, M. and Barenholz, Y. (1978). *Biochim. biophys. Acta* 515, 367-394.
Shinitzky, M. and Henkart, P. (1979). *Int. Rev. Cytol.* 60, 121-147.
Shinitzky, M. and Inbar, M. (1976). *Biochim. biophys. Acta* 433, 133-149.
Shinitzky, M., Dianoux, A.C., Gitler, C. and Weber, G. (1971). *Biochemistry* 10, 2106-2113.
Shipiro, H., Prescott, D. and Rabinowitz, J.L. (1978). *Comp. Biochem. Physiol.* 61B, 513-520.
Sidell, B.D. (1977). *J. Exp. Zool.* 199, 233-250.
Sidel, V.W. and Solomon, A.K. (1957). *J. Gen. Phys.* 41, 243-257.
Sieburth, J.McN. (1967). *J. Bact.* 87, 562-565.
Silvares, O.M., Cravalho, E.G., Toscano, W.M. and Huggins, C.E. (1974). *A.S.M.E. paper* No 74-WA/BIO-2.
Silvius, J.R. and McElhaney, R.N. (1980). *Proc. Natl. Acad. Sci.* USA 77, 1255-1259.
Silvius, J.R. and McElhaney, R.N. (1981). *J. Therm. Biol.* 88, 135-152.
Silvius, J.R., Read, B.D. and McElhaney, R.N. (1978). *Science* 199, 902-904.
Silvius, J.R., Mak, N. and McElhaney, R.N. (1980). *In* "Membrane Fluidity", (M. Kates and A. Kuksis, eds.), pp. 213-222, Humana, Clifton, New Jersey.
Silvestroni, L., Sartori, C., Modesti, A. and Frajese, G. (1980). *Archs Androl.* 4, 221-230.

Siminovitch, D. (1979). *Plant Physiol.* 63, 722-725.
Siminovitch, D. and Chapman, D. (1971). *FEBS Lett.* 16, 207-212.
Siminovitch, D. and Chapman, D. (1974). *Cryobiology* 11, 552-553.
Siminovitch, D. and Levitt, J. (1941). *Can. J. Res.* 19, 9-20.
Siminovitch, D., Rheaume, B. and Sacher, R. (1967a). *In* "Molecular Mechanisms of Temperature Adaptation", (C.L. Prosser, ed.), pp. 3-40. Amer. Assoc. Adv. Sci. Washington.
Siminovitch, D., Gfeller, F. and Rheaume, B. (1967b). *In* "Cellular Injury and Resistance in Freezing Organisms", (W. Asashina, Ed.) pp. 93-117. Inst. Low Temp. Sci., Sapporo, Japan.
Siminovitch, D., Singh, J. and de la Roche, I.A. (1975). *Cryobiology* 12, 144-153.
Siminovitch, D., Singh, J., Keller, W.A. and de la Roche, I. A. (1976). *Cryobiology* 13, 670.
Simon, E.W. (1974). *New Phytol.* 73, 377-420.
Simon, E.W. (1978). *In* "Dry biological systems", (J.H. Crowe and J.S. Clegg, eds.) pp. 205-224, Academic Press, New York.
Simon, E.W. (1979). *In* "Low temperature stress in crop plants: The role of the membrane", (J.M. Lyons, D. Graham and J. K. Raison, eds.) pp. 37-45. Academic Press, New York and London.
Simon, E.W., Minchin, A., McMenamin, M.M. and Smith, J.M. (1976). *New Phytol.* 77, 301-311.
Simons, T.J.B. (1974). *J. Physiol.* 237, 123-155.
Sinensky, M. (1971). *J. Bact.* 106, 449-455.
Sinensky, M. (1974). *Proc. Natl. Acad. Sci.* USA 71, 522-525.
Sinensky, M., Pinkerton, F., Sutherland, E. and Simon, F.R. (1979). *Proc. Natl. Acad. Sci.* USA 76, 4893-4897.
Singer, S.J. (1974). *Ann. Rev. Biochem.* 43, 805-834.
Singer, S.J. and Nicholson, G.L. (1972). *Science* 175, 720-731.
Singh, J. (1979a). *Plant Science Letters* 15, 195-201.
Singh, J. (1979b). *Protoplasma* 98, 329-341.
Singh, J. and Miller, R.W. (1980). *Plant Physiol.* 66, 349-352.
Singh, J., de la Roch, I.A. and Siminovitch, D. (1975). *Nature* 257, 669-670.
Simpson, G.M. (1965). *Can. J. Bot.* 43, 793-816.
Sklar, L.A., Miljanich, G.P. and Dratz, E.A. (1979). *Biochemistry* 18, 1707-1716.
Skriver, L. and Thompson, G.A. (1976). *Biochim. biophys. Acta* 431, 180-188.
Sleytr, U.B. and Umrath, W. (1976). *Proc. VIth Europ. Congr. Electron Microscopy*, Jerusalem. Vol. II. 50.

de Smet, M.J., Kingma, J. and Witholt, B. (1978). *Biochim. biophys. Acta* <u>506</u>, 64-80.

Smillie, R.M. (1979). *In* "Low temperature stress in crop plants: The role of the membrane", (J.M. Lyons, D. Graham and J.K. Raison, eds.) pp. 187-202. Academic Press, New York and London.

Smith, A.U. (1950). *Lancet* <u>2</u>, 910-911.

Smith, B.A. and McConnell, H.M. (1978). *Proc. Nat. Acad. Sci* USA <u>75</u>, 2759-2763.

Smith, M.W. (1976). *Biochem. Soc. Symp.* <u>41</u>, 43-60.

Smith, M.W. and Kemp, P. (1969). *Biochem. J.* <u>114</u>, 659-661.

Snyder, F. (1972). "Ether Lipids", Academic Press New York.

Snyder, F., Lee, T-C., Blank, M.L. and Moore, C. (1980). *In* "Membrane Fluidity", (M. Kates and A Kuksis, eds.) pp. 307-323, Humana Press, Clifton, New Jersey.

Soderstrom, N. (1944). *Acta Physiol. Scand.* <u>7</u>, 56-68.

Solomonson, L.P., Liepkalns, V.A. and Spector, A.A. (1976). *Biochemistry* <u>15</u>, 892-897.

South, F.E. (1960). *Am. J. Physiol.* <u>198</u>, 463-466.

Sperelakis, N. and Lee, E.C. (1971). *Biochim. biophys. Acta* <u>233</u>, 562-579.

Speth, V. and Wunderlich, W. (1973). *Biochim. biophys. Acta* <u>291</u>, 621-628.

Stanley, S.O. and Rose, A.H. (1967). *Proc. Roy. Soc.* <u>B252</u>, 199-207.

Stein, J.M., Tourtellotte, M.E., Reinert, J.C., McElhaney, R.N. and Rader, R.L. (1969). *Proc. Nat. Acad. Sci.* USA <u>63</u>, 104-109.

Stekhoven, F.S. and Bonting, S.L. (1981). *Physiol. Rev.* <u>61</u>, 1-76.

Steponkus, P.L. (1973). *Hort. Science* <u>9</u>, 282.

Steponkus, P.L. (1978). *Adv. in Agron.* <u>30</u>, 51-98.

Steponkus, P.L. (1981). *In* "Genetic Engineering of osmoregulation", (D.W. Rains, R.C. Valentine and A. Hollander, eds.) pp. 235-255, Plenum, New York.

Steponkus, P.L. and Dowgert, M.F. (1981). *Cryo-Letters* <u>2</u>, 42-47.

Steponkus, P.L. and Wiest, S.C. (1973). *Cryobiology* <u>10</u>, 532.

Steponkus, P.L. and Wiest, S.C. (1978). *In* "Plant Cold Hardiness and Freezing Stress - Mechanisms and Crop Implications", (P.H. Li and A. Sakai, eds.) pp. 75-91, Academic Press, New York.

Steponkus, P.L. and Wiest, S.C. (1979). *In* "Low temperature stress in crop plants: The role of the membrane", (J.M. Lyons, D. Graham and J.K. Raison, eds.) pp. 231-254, Academic Press.

Steponkus, P.L., Garber, M.P., Myers, S.P. and Lineberger, R.D. (1977). *Cryobiology* <u>14</u>, 303-321.

Steponkus, P.L., Dowgert, M.F. and Roberts, S.R. (1979). *Cryobiology* 16, 594.

Steponkus, P.L., Ferguson, J.R., Levin, R.L. and Dowgert, M.F. (1981). *Plant Physiol.* (In press).

Stewart, G.W., Ellory, J.C. and Klein, R.A. (1980). *Nature* 286, 403–404.

Stewart, J. McD. and Guinn, G. (1969). *Plant Physiol.* 44, 605–608.

Stiles, W. and Cocking, E.C. (1969). "An Introduction to the Principles of Plant Physiology". Methuen, London.

Strauss, G. and Ingenito, E.P. (1980). *Cryobiology* 17, 508–515.

Stokes, P. (1965). *In* "Encyclopedia of Plant Physiology". (W. Ruhland, ed.), Vol. XV/2, pp. 746–803, Springer, Berlin.

Stokes, J.L. and Redmond, M.L. (1966). *Appl. Microbiol.* 14, 74–78.

Swann, A.C. and Albers, R.W. (1979). *J. Biol. Chem.* 254, 4540–4544.

Swann, A.C. and Albers, R.W. (1981). *Biochim. biophys. Acta* 644, 36–40.

Swartz, H.M., Bolton, J.R. and Borg, D.C. (1972). "Biological Applications of Electron Spin Resonance". Wiley - Interscience, New York and London.

Szoka, F. and Papahadjopoulos, D. (1980). *Ann. Rev. Biophys. Bioeng.* 9, 467–508.

Tait, M.J. and Franks, F. (1971). *Nature* 230, 91–94.

Tajima, K., Daiku, K., Ezura, Y., Kimura, J. and Sakai, M. (1974). *In* "Effect of the Ocean Environment on Microbial Activities", (R.R. Colwell and R.Y. Morita, eds.) University Park Press, Baltimore.

Talbot, P., Summers, R.G., Hylander, B.L., Keough, E.M. and Franklin, L.E. (1976). *J. exp. Zool.* 198, 383–392.

Tanaka, R. (1974). *Review of Neuroscience* 1, 181–230.

Tanford, C. (1978). *Science* 200, 1012–1018.

Tanford, C. (1980). "The hydrophobic effect:Formation of micelles and biological membranes", 2nd Edition, Wiley-Interscience, New York.

Tanner, A.C. and Herbert, R.A. (1981). *Kieler Meeresforschungen* (In press).

Tao, K-L. and Khan, A.A. (1974). *Biochim. biophys. Res. Comm.* 59, 764–770.

Tao, K-L. and Khan, A.A. (1976). *Plant Physiol.* 57, 1–4.

Thebud, R. and Santarius, K.A. (1981). *Planta* 152, 242–248.

Thévenot, C. and Come, D. (1973). *Physiol. Vég.* 11, 151–160.

Thilo, L., Trauble, H. and Overath, P. (1977). *Biochemistry* 16, 1283–1290.

Thomas, D.D., Dalton, L.R. and Hyde, J.S. (1976). *J. Chem. Phys.* 65, 3006–3024.

Thomas, E.M. and Syrett, P.J. (1976). *New Phytologist* 76, 409-413.

Thomas, R.C. (1972). *Physiol. Rev.* 52, 563-594.

Thompson, G.A. Jr. and Nozawa, Y. (1977). *Biochim. biophys. Acta* 472, 55-92.

Thompson, G.A., Bambery, R.J. and Nozawa, Y. (1971). *Biochemistry* 10, 4441-4447.

Thompson, G.A., Bambery, R.J. and Nozawa, Y. (1972). *Biochim. biophys. Acta* 260, 630-638.

Thompson, K., Grime, J.P. and Mason, G. (1977). *Nature* 267, 147-149.

Thompson, P.A. (1970a). *Nature* 225, 827-831.

Thompson, P.A. (1970b). *Ann. Bot.* 34, 427-429.

Thompson, P.A. (1973). *In* "Seed Ecology", (W. Heydecker, ed.) pp. 31-58, Butterworths, London.

Tinker, D.O. and Pinteric, L. (1971). *Biochemistry* 10, 860-865.

Toivio-Kinnucan, M.A. and Stushnoff, C. (1981). *Cryobiology* 18, 72-78.

Tornava, S.R. (1939). *Protoplasma* 32, 329-341.

Toscano, W.M., Cravalho, E.G., Silvares, O.M. and Huggins, C.E. (1975). *J. Heat Trans.* 326-332.

Träuble, H. and Eibl, H. (1974). *Proc. Nat. Acad. Sci.* USA 71, 214-219.

Träuble, H. and Haynes, D.H. (1971). *Chem. Phys. Lipids* 7, 324-334.

Tsukagoshi, N. and Fox, C.F. (1973). *Biochemistry* 12, 2816-2822.

Tyankova, L. (1972). *Biochim. biophys. Acta* 274, 75-82.

Tyrell, D.A., Heath, T.D., Colley, E.M. and Ryman, B.E. (1976). *Biochim. biophys. Acta* 457, 259-302.

Uribe, E.G. and Jagendorf, A.T. (1968). *Arch. Biochem. Biophys.* 128, 351-359.

Vegis, A. (1964). *Ann. Rev. Plant Physiol.* 15, 185-224.

Verma, S.P. and Wallach, D.F.H. (1976). *Biochim. biophys. Acta* 436, 307-318.

Verkleij, A.J. and Ververgaert, P.H.J.Th. (1978). *Biochim. biophys. Acta* 515, 303-327.

Verkleij, A.J., Ververgaert, P.H.J., Van Deenan, L.L.M. and Elbers, P.F. (1973). *Biochim. biophys. Acta* 288, 326-332.

Verkleij, A.J., Nauta, I.L.D., Werre, J.M., Mandersloot, J.G., Reinders, B., Ververgaert, P.H.J. Th. and De Grier, J. (1976). *Biochim. biophys. Acta* 436, 366-376.

Verkleij, A.J., Van Echteld, C.J.A., Gerritsen, W.J., Cullis, P.R. and de Kruijff, P. (1980). *Biochim. biophys. Acta* (In press).

Viebke, S.M., Bernson, P.L. and Pettersson, B. (1978). *J. Therm. Biol.* 3, 97.

Volger, H.G. and Heber, U. (1975). *Biochim. biophys. Acta* 412, 335-349.

Volger, H.G. and Santarius, K.A. (1981). *Physiol. Plant* 51, 195-200.

Volger, H.G., Heber, U. and Berzborn, R.J. (1978). *Biochim. biophys. Acta* 511, 455-469.

Vonnegut, B. and Chessin, H. (1971). *Science* 174, 945-946.

Vysochina, T.K. (1977). *Zh. Evol. Biokhim. Fiziol.* 13, 506-507.

Wade. N.L. (1979). *In* "Low temperature stress in crop plants: The role of the membrane", (J.M. Lyons, D. Graham and J.K. Raison, eds.) pp. 81-96, Academic Press, New York and London.

Waldron, C. and Roberts, C.F. (1974). *Molec. gen. Genet.* 134, 115-132.

Wales, R.G. and White, I.G. (1959). *J. Endocr.* 19, 211-220.

Wang, L.C.H. (1979). *Can. J. Zool.* 57, 149-155.

Wang, L.C.H. (1982). *In* "Hibernation and Torpor in Mammals and Birds", (C.P. Lyman, J.S. Willis, A. Malan and L.C.H. Wang, eds.) (In press).

Warren, G.B., Houslay, M.D., Metcalfe, J.C. and Birdsall, N.J.M. (1975). *Nature* 255, 684-687.

Wassink, E.C. (1972). *Meded.* Landbouwlogeschool Wageningen Nederland, 72, 25-37.

Watson, P.F. (1975). *Vet. Rec.* 97, 12-15.

Watson, P.F. (1976). *J. Thermal Biol.* 1, 137-141.

Watson, P.F. (1979). *In* "Oxford Reviews of Reproductive Biology", (C.A. Finn, ed.) Vol. I. pp. 283-350. Oxford University Press.

Watson, P.F. (1981). *J. Reprod. Fert.* 62, 483-492.

Watson, P.F. and Martin, I.C.A. (1976). *Theriogenology* 6, 553-558.

Waugh, R. and Evans, E.A. (1979). *Biophys. J.* 26, 115-132.

Weast, R.C. (1974). "Handbook of Chemistry and Physics", 55th Ed. CRC Press, Cleveland.

Weiser, C.J. (1978). *In* "Plant Cold Hardiness and Freezing Stress: Mechanisms and Crop Implications", (P.H. Li and A. Sakai, eds.) pp. 391-394, Academic Press, New York.

Went, F.W. (1944). *Amer. J. Bot.* 31, 135-150.

Wheaston, T.A. and Morris, L.L. (1967). *Proc. Am. Soc. Hort. Sci.* 91, 529-533.

White, I.G. and Darin-Bennett, A. (1976). *VIIIth Int. Congr. Anim. Reprod. Artif. Insem.* (Cracow) 4, 951-954.

White, I.G. and Wales, R.G. (1961). *J. Reprod. Fert.* 2, 225-237.

Wiest, S.C. (1979). Ph.D. Thesis. Cornell University, Ithaca, New York.
Wiest, S.C. and Steponkus, P.L. (1975). *Cryobiology* 12, 555.
Wiest, S.C. and Steponkus, P.L. (1976). *Cryobiology* 13, 670.
Wiest, S.C. and Steponkus, P.L. (1977). *J. Amer. Soc. Hort. Sci.* 102, 119-123.
Wiest, S.C. and Steponkus, P.L. (1978a). *Plant Physiol.* 62, 599-605.
Wiest, S.C. and Steponkus, P.L. (1978b). *Plant Physiol.* 61, (supplement) 32.
Wiest, S.C. and Steponkus, P.L. (1979). *Cryobiology* 16, 101-104.
Wieth, J.O. (1970). *J. Physiol.* 207, 563-580.
Wiley, J.S. and Cooper, R.A. (1974). *J. Clin. Invest.* 53, 745-755.
Wilkins, M.H.F., Blaurock, A.E. and Engelman, D.M. (1971). *Nature* 230, 72-76.
Wilkins, P.O. (1973). *Can. J. Microbiol.* 19, 909-915.
Wilkinson, D.A., Morowitz, H.J. and Prestegard, J.H. (1977). *Biophysical J.* 20, 169-179.
Wilkinson, H.L. (1976). Ph.D. Thesis, University of Illinois, Urbana.
Willcox, M.E. and Patterson, B.D. (1979). *In* "Low temperature stress in crop plants: The role of the membrane", (J.M. Lyons, D. Graham and J.K. Raison, eds.) pp. 523-526, Academic Press, New York and London.
Williams, R.J. (1981). *In* "Analysis and Improvement of Plant Cold Hardiness", (C.R. Olien and M.N. Smith, eds.) CRC Press, Boca Raton. (In press).
Williams, R.J. and Harris, P. (1977). *Cryobiology* 14, 670-680.
Williams, R.J. and Hope, H.J. (1981). *Cryobiology* 18, 133-145.
Williams, R.J. and Meryman, H.T. (1970). *Plant Physiol.* 45, 752-755.
Williams, R.J. and Shaw, S.K. (1980). *Cryobiology* 17, 530-539.
Williams, R.J. and Takahashi, T. (1978). *Cryobiology* 15, 688.
Williams, R.M. and Chapman, D. (1970). *Prog. Chem. Fats Other Lipids* 11, 1-79.
Willis, J.S. (1966). *J. Gen. Physiol.* 49, 1221-1239.
Willis, J.S. (1967). *In* "Mammalian Hibernation III", (K.C. Fisher, A.R. Dawe, C.P. Lyman, E. Schonbaum and F.E. South, Jr,, eds.) pp 356-381, Oliver and Boyd, Edinburgh.
Willis, J.S. (1968). *Am. J. Physiol.* 214, 923-928.
Willis, J.S. (1972). *Cryobiology* 9, 351-366.
Willis, J.S. (1978). *In* "Strategies in Cold: Natural Torpidity and Thermogenesis", (L.C.H. Wang and J.W. Hudson, eds.) pp 275-286, Academic Press, New York.

Willis, J.S. (1979). *Ann. Rev. Physiol.* 41, 275-286.

Willis, J.S. and Ellory, J.C. (1982). *In* "Current Topics in Membrane Transport", (J.F. Hoffman and B. Forbush III, eds.) Academic Press (In press).

Willis, J.S. and Li, N.M. (1969). *Am. J. Physiol.* 217, 321-326.

Willis, J.S., Goldman, S.S. and Foster, R.F. (1971). *Comp. Biochem. Physiol.* 39A, 437-445.

Willis, J.S., Fang, L.S.T. and Foster, R.F. (1972). *In* "Hibernation and Hypothermia, Perspectives and Challenges", (F.E. South, J.P. Hannon, J.S. Willis, E.T. Pengelley and N.R. Alpert, eds.) pp. 149-166, Elsevier, Amsterdam.

Willis, J.S., Ellory, J.C. and Becker, J.H. (1978). *Am. J. Physiol.* 235, C159-C167.

Willis, J.S., Ellory, J.C. and Wolowyk, M.W. (1980). *J. Comp. Physiol.* 138, 43-47.

Wilson, J.M. (1976). *New Phytol.* 76, 257-270.

Wilson, J.M. (1978). *New Phytol.* 80, 325-334.

Wilson, J.M. and Crawford, R.M.M. (1974a). *J. Exp. Bot.* 25, 121-131.

Wilson, J.M. and Crawford, R.M.M. (1974b). *New Phytol.* 73, 805-820.

Wintrobe, M.M. (1976). *In* "Clinical Hematology", pp. 80-134, Lea and Febiger, Philadelphia.

Wisnieski, B.J., Parkes, J.G., Huang, Y.O. and Fox, C.F. (1974). *Proc. Nat. Acad. Sci.* USA. 4381-4387.

Witter, L.D. (1961). *J. Dairy Sci.* 44, 983-1015.

Wodtke, E. (1976). *J. Comp. Physiol.* 110, 145-157.

Wolf, F.A. and Wolf, F.T. (1947). "The Fungi", Vol. II, Wiley, New York.

Wolfe, J. (1978). *Plant, Cell and Environment* 1, 241-247.

Wolfe, J. and Bagnall, D.J. (1979). *In* "Low temperature stress in crop plants: The role of the membrane", (J.M. Lyons, D. Graham and J.K. Raison, eds.) pp 527-534, Academic Press, New York and London.

Wolfe, J. and Steponkus, P.L. (1981a). *Biochim. biophys. Acta* 643, 663-668.

Wolfe, J. and Steponkus, P.L. (1981b). *Plant Physiol.* (In press).

Wright, M. and Simon, E.W. (1973). *J. Exp. Bot.* 24, 400-411.

Wu, S.H. and McConnell, H.M. (1973). *Biochem. biophys. Res. Commun.* 55, 484-491.

Wu, S.H. and McConnell, H.M. (1975). *Biochemistry* 14, 847-854.

Wunderlich, F. and Ronai, A. (1975). *FEBS Letters* 55, 237-241.

Wunderlich, F., Speth, V., Batz, W. and Kleinig, H. (1973). *Biochim. biophys. Acta* 298, 39–49.

Wunderlich, F., Batz, W., Speth, V. and Wallach, D.F.H. (1974a). *J. Cell Biol.* 61, 633–640.

Wunderlich, F., Wallach, D.F.H., Speth, V. and Fischer, H. (1974b). *Biochim. biophys. Acta* 373, 34–43.

Wunderlich, F., Ronai, A., Speth, V, Seelig, J. and Blume, A. (1975). *Biochemistry.* 14, 3730–3735.

Yamaki, S. and Uritani, I. (1974). *Plant Cell Physiol.* 15, 385–388.

Yanagimachi, R. and Usui, N. (1974). *Exp. Cell Res.* 89, 161–174.

Yankelevitch, B.B. and Nikolaeva, M.G. (1975). *Soviet Plant Physiol.* 22, 535–537.

Yoshida, S. (1976). *Plant Physiol.* 57, 710–715.

Yoshida, S. (1978). *In* "Plant Cold Hardiness and Freezing Stress: Mechanisms and Crop Implications", (P.H. Li and A. Sakai, eds.) pp. 117–135, Academic Press, New York, San Francisco and London.

Yoshida, S. (1979). *Plant Physiol.* 64, 252–256.

Young, J.D., Jones, S.E.M. and Ellory, J.C. (1980). *Proc. Roy. Soc. B.* 209, 355–375.

Yu, J. and Branton, D. (1976). *Proc. Nat. Acad. Sci.* USA 73, 3891–3895.

Zade-Oppen, A.M.M. (1968). *Acta Physiol. Scand.* 73, 341–364.

Zahler, W.L. and Doak, G.A. (1975). *Biochim. biophys. Acta* 406, 479–488.

Zander, J.M., Caspi, E., Pandey, G.N. and Mitra, C.R. (1969). *Phytochemistry* 8, 1597–1563.

Zecevic, D. and Levitan, H. (1980). *Am. J. Physiol.* 239, C47–C57.

Zeidler, R.B. and Willis, J.S. (1976). *Biochim. biophys. Acta* 436, 628–651.

Ziegler, P. and Kandler, O. (1980). *Z. Pflanzenphysiol.* 99, 393–410.

Zimmerman, A.N.E. and Hulsmann, W.C. (1966). *Nature* 211, 646–647.

Zwaal, R.F.A., Roelofsen, B. and Colley, C.M. (1973). *Biochim. biophys. Acta* 300, 159–182.

SPECIES INDEX

SUBJECT INDEX